D1091643

# THERMODYNAMIC PROPERTIES OF STEAM

INCLUDING

## DATA FOR THE LIQUID AND SOLID PHASES

BY

JOSEPH H. KEENAN

Professor of Mechanical Engineering
Massachusetts Institute of Technology

AND

FREDERICK G. KEYES

Professor of Physical Chemistry, Emeritus;
Lecturer in Chemistry
Massachusetts Institute of Technology

**FIRST EDITION**

*Thirty-Ninth Printing*

JOHN WILEY & SONS, INC.

NEW YORK · LONDON · SYDNEY

THIRTY-NINTH PRINTING, JANUARY, 1967

PRINTED IN THE UNITED STATES OF AMERICA

# PREFACE

More than six years have elapsed since the appearance of the Keenan Steam Tables and Mollier Diagram.[1]* The basis of the tables was a considerable body of new data provided in course by the investigators carrying out the steam research program sponsored by the American Society of Mechanical Engineers' Special Committee on the Thermal Properties of Steam.

The available data on water are now considerably increased due to the practical completion of the various series of investigations in progress, not only in this country but abroad. Moreover, in September 1934 a third international conference on steam properties was held during which it became evident that a remarkably close agreement existed in the results of the various investigators. This happy condition is in evidence through the greatly reduced tolerances assigned to the enlarged skeleton table produced by the Special Committee of experts assigned to discuss and compare the data submitted to the conference. The conclusion is therefore justified that our knowledge of the properties of water, after more than a decade of intensive research in the United States, England, Germany and Czechoslovakia, has reached a stage where further delay in preparing revised tables is unnecessary. There is indeed satisfactory evidence that a revised set of tables may be expected to possess a high degree of permanence.

The present tables are computed from entirely new formulations of the properties of water. For the most part the formulations are analytical but there are certain regions where recourse to graphical methods was necessary in default of a knowledge of analytical forms suitable for our requirements. The region where graphical correlation was employed is for volumes less than ten cubic centimeters per gram. The results appear satisfactory, at least as far as can be judged by applying the usual criteria of thermodynamic consistency.

The units employed, the quantities tabulated, the intervals and range of the tables, together with other details, were decided upon largely through correspondence with a considerable number of those who have constructed earlier tables, as well as those who use them for various purposes. We acknowledge with hearty thanks the many valuable suggestions received, albeit the difficulty or even impossibility of carrying out every suggestion will be readily appreciated. Within the limitations of the size of the volume, we have however included tables giving the viscosity, the heat conductivity, compressed liquid properties, data on ice and its vapor, isentropic expansion exponents, and certain thermometric calibration data, while at the same time retaining the temperature-scale conversion table, logarithm tables, and conversion factor tables of the 1930 volume. Certain charts of the specific heat hitherto not available are included. They show the extraordinarily rapid changes and the large values of the quantity corresponding to volumes at and near the critical state. The scales of the charts are selected to permit reasonably accurate reading of the variables. The range of temperature (1600° F) and of pressure (5500 lb./in.$^2$) of the tables is also greater than the 1930 tables, and is sufficient, we believe, for a long period. The inclusion of data for the liquid phase is

* Superior numbers refer to items in the Bibliography on page 25.

an innovation made possible by the recent exact measurements of heat quantities and specific volumes. The sources of all the data employed are briefly discussed with references to the devices of formulation employed and the analytical expressions found to be serviceable.

We take satisfaction in expressing our deep sense of obligation to those who have labored many years to obtain exact measurements on the properties of water and without which the present compilation of properties would evidently be impossible. Their work has without doubt led to a knowledge of the properties of water far exceeding that for any other substance.

The actual labor of computation and the control of the organization set up for the work has been in charge of Doctor Carol Anger Rieke. The accuracy realized in computation and the rapid progress of the work are due to her patience and skill.

JOSEPH H. KEENAN
FREDERICK G. KEYES

CAMBRIDGE, *September*, 1936

# CONTENTS

**BOOKS BY KEENAN (J. H.) AND KEYES (F. G.)**

MOLLIER DIAGRAM—39 inches x 29½ inches
THERMODYNAMIC PROPERTIES OF STEAM

**BY JOSEPH H. KEENAN**

THERMODYNAMICS

**BY KEENAN (J. H.) AND KAYE (J.)**

GAS TABLES

# FIGURES

# INTRODUCTION

## 1. The Temperature Scale and Some General Equations

The concept of a scale of temperature independent of the properties of substances is one of the accompaniments of the development of the second empirical principle of thermodynamics. Sadi Carnot first showed (1824) that for a reversibly operating machine withdrawing heat in amount $Q_1$ at fixed temperature $t_1$ and rejecting heat $Q_2$ at a lower fixed temperature $t_2$, the ratio $Q_1/Q_2$ is the same for all working substances. It is a universal function, $\phi(t_1, t_2)$, of the fixed temperatures.

The accepted solution of the function $\phi(t_1, t_2)$ is $Q_1/Q_2 = T_1/T_2$, and $T_1$ and $T_2$, called thermodynamic temperatures, must be determined on the basis of the exact validity of Carnot's proposition. Further progress requires a knowledge of the physical properties of substances, and the accumulated knowledge of gas properties has led to the concept of the ideal gas whose pressure-volume properties are given in relation to temperature by the simple equation $pv = RT$ where $R$ is a universal constant when the mole is the unit of mass. The energy of the ideal gas is moreover dependent on temperature alone. With these specifications it is possible to identify the $T$ used to describe the properties of the ideal gas with the thermodynamic temperature.

The hypothetical ideal gas represents a state of matter into which an actual gas may be brought at zero pressure, as far as the pressure may be given by the relation $pv = RT$. This statement implies that the actual gas will show approximately ideal properties at finite pressures. Nitrogen, for example, at the ice point and one atmosphere pressure is smaller in volume compared to the ideal gas by a part in 1450. At 100°, however, it is larger in volume by a part in 7300. Some gases show much larger deviations at the same temperatures but the example suffices to indicate the possibility of using suitably chosen gases as thermometric fluids. The expansion of a gas at constant pressure or the pressure increase at constant volume offer alternate modes of use although the constant volume type is the more frequently used. It is important to observe, in accordance with the ideal gas concept, that if the indications of temperature with the gas thermometer are carried out for any fixed temperature at a series of successively lower pressures the indicated temperatures will converge toward a fixed value. In the limit of zero pressure, arrived at by extrapolation of the indicated temperatures, the number obtained will be the thermodynamic temperature. The test of the entire scheme depends evidently on whether *different* thermometric gases lead to the same limiting temperature.* The increasingly precise measurements made over more than a century have only served to deepen the conviction that the entire structure of ideas and concepts is sound.

* In practice a "correction" is applied to the gas thermometer reading to obtain the thermodynamic centigrade temperature reading. The correction may be and usually is obtained analytically from a formulation of the high pressure measurements of $p$, $v$ properties as a function of the gas scale temperature. The formulation of whatever kind must yield the relation $pv = RT$ for zero pressure to keep within the circle of ideas outlined above.

Of course any of the thermodynamic equations involving the second principle will contain $T$ and may be used to determine the temperature, but convenience and ultimate precision have led to the use of gas thermometers where the direct indications are very close to the thermodynamic temperatures.*

The primary thermometry fixed points are the equilibrium temperature of ice (zero centigrade) in contact with water saturated with air at one atmosphere pressure, and the boiling point of water at one international† atmosphere (100° centigrade). Using these fixed points and the gas thermometer, the boiling point of sulphur has been measured by many observers in the past fifty years. The values found vary, showing a maximum difference of about 0.3° C for the constant volume thermometer. In 1927 by international agreement 444.60° C was selected to serve as the normal or thermodynamic centigrade temperature for boiling sulphur. The international scale[2] is reproduced by a platinum resistance thermometer of specified characteristics calibrated at the ice point and normal boiling points of water and sulphur and interpolated by the Callendar difference formula.

The precise degree of approximation of the international scale to the thermodynamic scale is not yet definitely known. It now seems likely, however, that some years will elapse before sufficient new findings have accumulated to bring about improved specifications for an international temperature scale.‡

The international temperature scale has been used in all the steam investigations of recent years and is the scale of the present tables. A required fundamental constant is the thermodynamic temperature of the ice point. We have used 273.16 (491.69° F) for this temperature and our thermodynamic temperature $T$ is given by the equation $T = (273.16 + t° \text{ C})$ where $t$ is the centigrade temperature on the international scale. On the Fahrenheit thermodynamic scale we have $T = (459.69 + t° \text{ F})$. The uncertainty in $T_0$ or the number 273.16 may be as much as 0.02°.

The principal thermodynamic function employed in the tables is the chi ($\chi$) function of Gibbs for which we use the symbol $h$, and the name enthalpy. The same quantity is often called total heat and heat content. The following equations relate the quantity to the energy $u$, the entropy $s$, the pressure $p$, the volume $v$, the reciprocal thermodynamic temperature tau ($\tau$) and the usual specific heats $c_p$, $c_v$, and $c_{sat}$.

$$h = u + pv \quad \text{or} \quad h_1 - h_2 = u_1 - u_2 + (pv)_1 - (pv)_2 \tag{1}$$

$$\left.\begin{aligned} dh &= T\,ds + v\,dp \\ dh &= c_p\,dT - \left[T\left(\frac{\partial v}{\partial T}\right)_p - v\right]dp \\ &= c_p\,dT + \left(\frac{\partial(v\tau)}{\partial \tau}\right)_p dp \\ dh &= \left[c_v + v\left(\frac{\partial p}{\partial T}\right)_v\right]dT + \left(\frac{\partial(v\tau)}{\partial \tau}\right)_p\left(\frac{\partial p}{\partial v}\right)_T dv \end{aligned}\right\} \tag{2}$$

* At 500° C the constant volume nitrogen thermometer filled to one atmosphere pressure at the ice point will read, at most, 0.14° C lower than the thermodynamic scale.

† The international atmosphere is the pressure produced by a column of mercury 760 mm. high at zero centigrade where the gravitational acceleration is 980.665 cm./sec.$^2$

‡ Investigations directed to a solution of this problem for temperatures above zero centigrade were started at the Reichsanstalt with refinements in apparatus following the war. In 1920 similar work was begun at the Research Laboratory of Physical Chemistry at M. I. T. Professor James A. Beattie, who is carrying on the investigations, informs us that his data bearing on the relation between the international scale and the thermodynamic scale will be published shortly.

$$
\left.
\begin{aligned}
c_p &= \left(\frac{\partial u}{\partial T}\right)_p + p\left(\frac{\partial v}{\partial T}\right)_p; \; c_v = \left(\frac{\partial u}{\partial T}\right)_v \\
c_{\text{sat}} &= c_p - T\left(\frac{\partial v}{\partial T}\right)_p \frac{dp}{dt} = c_v + T\left(\frac{\partial p}{\partial T}\right)_v \frac{dv}{dT}
\end{aligned}
\right\} \tag{3}
$$

A few derived quantities are also useful and may be easily formed. They are as follows:

$$
(h \text{ constant}); \; \left(\frac{\partial T}{\partial p}\right)_h = \mu = -\frac{\left(\frac{\partial (v\tau)}{\partial \tau}\right)_p}{c_p} \tag{4}
$$

$$
c_p = \left(\frac{\partial h}{\partial T}\right)_p; \quad -\mu c_p = \left(\frac{\partial h}{\partial p}\right)_T = \left(\frac{\partial (v\tau)}{\partial \tau}\right)_p \tag{5}
$$

$$
c_p = c_{p_r} - \int T\left(\frac{\partial^2 v}{\partial T^2}\right)_p dp = c_{p_r} - \frac{\partial}{\partial T}\int\left(\frac{\partial (v\tau)}{\partial \tau}\right)_p dp \tag{6}
$$

$$
\begin{gathered}
s - s_r = \int \frac{dh}{T} - \int v \, dp \\
\left(\frac{\partial s}{\partial p}\right)_T = -\left(\frac{\partial v}{\partial T}\right)_p; \; \left(\frac{\partial s}{\partial T}\right)_p = \frac{c_p}{T}
\end{gathered} \tag{7}
$$

$$
v = \left[v_r \frac{T}{T_r} - T\int \mu c_p \, d\tau\right]_p \tag{8}
$$

The subscript $r$ refers to a reference state at the constant pressure $p$. The relation between similar quantities corresponding to two phases in equilibrium is frequently required. The following equations are of interest wherein subscript $g$ refers to the saturated vapor and $f$ to the saturated liquid. The letter $L$ designates enthalpy change of evaporation. Subscript small $s$ refers to the saturation condition.

$$
\begin{gathered}
c_{pg} - c_{pf} = \frac{dL}{dT} - \left[\left(\frac{\partial (v\tau)}{\partial \tau}\right)_{pg} - \left(\frac{\partial (v\tau)}{\partial \tau}\right)_{pf}\right]\frac{dp}{dT} \\
L = T\frac{dp}{dT}(v_g - v_f) \\
c_{sg} - c_{sf} = \frac{dL}{dT} - \frac{L}{T} = -\tau\left(\frac{d(L\tau)}{d\tau}\right)
\end{gathered} \tag{9}
$$

The first of equations (9) may be written in terms of the Joule-Thomson coefficient, using the second of equations (5), as follows:

$$
c_{pg} - c_{pf} = \frac{dL}{dT} + \left[\,(\mu c_p)_g - (\mu c_p)_f\,\right]\frac{dp}{dT}. \tag{10}
$$

The sections to follow contain a brief discussion of the data upon which the tables are based and the methods used in deriving properties from them. The first exact measurements on steam and water properties were made by Regnault about ninety years ago. During the two decades following 1890, Rowland, Dieterici, Smith, Griffiths, Henning, Joly, Knoblauch, Holborn, Scheel, Heuse, Jacob and others contributed data which after thorough correlation by L. S. Marks and H. N. Davis supplied the basis for the Marks and Davis tables of 1908. Since a comprehensive review of the data available at that time is contained in the Marks and Davis volume it is unnecessary to refer here to the work on steam properties which preceded its publication.

## 2. The Vapor Pressure of Liquid Water and of Ice

The measurements of Henning,[3] Holborn and Henning,[4] Holborn and Baumann,[5] Scheel and Heuse,[6] Egerton and Callendar,[7] Osborn, Stimson, Fioch and Ginnings,[8] and Smith, Keyes and Gerry[9] on the vapor pressure of water are in remarkably close agreement to the critical temperature (374.11° M. I. T.; 374.20° Reichsanstalt). Below 100° the results of Henning, and Scheel and Heuse as compiled in the Wärmetabellen[10] may be accepted as satisfactory, though the results of measurements obtained prior to 1928 should of course be changed to conform to the international scale of temperature.

The data above 100° C have been discussed in detail by the recent investigators and the great bulk of the measurements are included in a range of pressure variation corresponding to ±0.02° C. This degree of accord in the work done in different laboratories has never before been attained and is convincing evidence of the satisfactory state of the pressure measuring technique employed and of the reproducible qualities of the modern platinum resistance thermometer. The smoothed results of any of the investigations would be satisfactory. We have used the formulas (1) and (2) taken from the paper by Smith, Keyes and Gerry[9] which reproduce pressures in satisfactory agreement with the pressures given in the Wärmetabellen down to 10° C. Below 10° we have taken the values directly from the Wärmetabellen.

$$\log_{10}\frac{p_c}{p} = \frac{x}{T}\left[\frac{a + bx + cx^3 + ex^4}{1 + dx}\right] \quad \text{Formula (1) S, K, and G. } 50° \text{ C to } 374.11° \text{ C} \tag{11}$$

where $p$ is the vapor pressure in int. atm.

$p_c$ is the critical pressure = 218.167 int. atm.

$T$ is the temperature in degrees Kelvin = $t$(°C) + 273.16

$x$ is $(T_c - T)$ where $T_c$ is the critical temperature in deg. K. (647.27 deg. K.)

$a = 3.3463130$, $b = 4.14113 \times 10^{-2}$, $c = 7.515484 \times 10^{-9}$

$d = 1.3794481 \times 10^{-2}$, $e = 6.56444 \times 10^{-11}$

$$\log_{10}\frac{p_c}{p} = \frac{x}{T}\left[\frac{a' + b'x + c'x^3}{1 + d'x}\right] \quad \text{Formula (2) S, K, and G. } 10° \text{ C to } 150° \text{ C} \tag{12}$$

where $p$ is the vapor pressure in int. atm.

$p_c$ = 218.167 int. atm.

$T = t°$ C + 273.16

$x = (T_c - T)$ $\qquad T_c = 647.27$

$a' = 3.2437814$, $\qquad b' = 5.86826 \times 10^{-3}$

$c' = 1.1702379 \times 10^{-8}$, $\qquad d' = 2.1878462 \times 10^{-3}$

Derivatives of the pressure with respect to temperature were required for the thermodynamic intercomparison of data and these were obtained from the differentiated form of equations (11) and (12).[11,12]

E. W. Washburn[13] has correlated the vapor-pressure data for sub-cooled water and for ice. The primary data are not as consistent as would be desirable for many purposes but supply every need of the tables excluding possibly applications in low temperature humidity control.

## 3. Properties of Water Vapor

Prior to the recent American steam research program the 1905 measurements of Knoblauch, Linde and Klebe were the only specific volume measurements available. They extend from about 100° C to 190° C and to 6 atm. The recent measurements by Keyes, Smith and Gerry[12] cover the range 195° C to 460° C and extend to pressures of more than 350 atm.

The published formulation of the data expresses the volume as a function of pressure and temperature valid to 10 cc. per gram, which corresponds to about 344° C on the vapor saturation side of the steam dome. Formula 9 of the Keyes, Smith and Gerry paper is as follows:

$$v = \frac{4.55504\, T}{p} + B \qquad \begin{array}{l} p \text{ in int. atm.} \\ v \text{ in cm.}^3/\text{g.} \\ T = 273.16 + t^\circ \text{ C} = \tau^{-1} \\ \text{mol. wt. } H_2O \text{ 18.0154} \end{array}$$

$$B = B_0 + B_0^2 g_1(\tau)\,\tau p + B_0^4 g_2(\tau)\,\tau^3 p^3 - B_0^{13} g_3(\tau)\,\tau^{12} p^{12} \tag{13}$$

$$B_0 = 1.89 - 2641.62\,\tau\,10^{80870\tau^2}$$

$$g_1(\tau) = 82.546\tau - 1.6246 \cdot 10^5 \tau^2$$

$$g_2(\tau) = 0.21828 - 1.2697 \cdot 10^5 \tau^2$$

$$g_3(\tau) = 3.635 \cdot 10^{-4} - 6.768 \cdot 10^{64} \tau^{24}$$

The average deviation of the observed volume data from this formula above 10 cm.$^3$/g is on the whole less than one part in a thousand. The formula is used to compute all volumes in the table in that region.

The enthalpy of the vapor in this same region is obtained through equation (13) by integrating the relationship,

$$\left(\frac{\partial h}{\partial p}\right)_t = \left(\frac{\partial(v\tau)}{\partial \tau}\right)_p.$$

It is evident from equation (13) that

$$\left(\frac{\partial(v\tau)}{\partial \tau}\right)_p = \left(\frac{\partial(B\tau)}{\partial \tau}\right)_p,$$

and that $B$ may be expressed as

$$B = B_0 + \phi_1 p + \phi_3 p^3 + \phi_{12} p^{12}, \tag{13A}$$

where $B_0$, $\phi_1$, $\phi_3$ and $\phi_{12}$ are pure temperature functions. Then the constant temperature coefficient becomes

$$\left(\frac{\partial h}{\partial p}\right)_t = F_0 + F_1 p + F_3 p^3 + F_{12} p^{12},$$

where $F_1$, $F_3$ and $F_{12}$ are each equal to $\partial(\phi\tau)/\partial\tau$ in terms of the corresponding $\phi$, and $F_0$ is $\partial(B_0\tau)/\partial\tau$. Integrating, we get

$$h = F_0 p + \frac{F_1}{2} p^2 + \frac{F_3}{4} p^4 + \frac{F_{12}}{13} p^{13} + F', \tag{14}$$

where $F'$ is a pure temperature function and is identified upon substituting $p = 0$ in the enthalpy equation (14). Therefore, we may write

$$F' = h_{p=0} = \int_{T=273.16}^{T} c_{p_0}\, dT + h',$$

where $c_{p_0}$ is the specific heat at zero pressure and $h'$ is an arbitrary constant.

The direct measurement of $c_{p_0}$, a pure temperature function, is of course an impossibility, but the quantity may be obtained by using measurements of pressure, volume and temperature to compute values of $\int T \left(\frac{\partial^2 v}{\partial T^2}\right)_p dp$ needed to obtain it from $c_p$ measurements by use of the Clausius relation

$$c_{p_0} = c_p + \int T \left(\frac{\partial^2 v}{\partial T^2}\right)_p dp.$$

The $c_p$ measurements may be conveniently reduced otherwise provided Joule-Thomson data are available.[14, 15] Another method of obtaining $c_{p_0}$ values is through the use of enthalpy data and the expression for enthalpy (14) obtained from the equation of state. Thus,

$$c_{p_0} = \frac{\partial}{\partial t}\left[h - \left(F_0 p + \frac{F_1}{2}p^2 + \frac{F_3}{4}p^4 + \frac{F_{12}}{13}p^{13}\right)\right]_{p\to 0}.$$

All three methods were applied to the available data recently[16] in an attempt to obtain a reliable expression for $c_{p_0}$. However, in the present development all of these methods were discarded in favor of a new, more fundamental method of determining $c_{p_0}$.

The development of quantum mechanics and of the science of interpreting band spectra makes possible the calculation of the energy of molecules. From the band spectra investigations on water vapor carried out to date, particularly recent important ones by Mecke,[17] Baumann and Mecke,[18] and Freudenberg and Mecke,[19] we possess sufficient data from which to compute the energy. The temperature derivative of the energy plus a constant gives of course the desired $c_{p_0}$ values. A. R. Gordon[20] has recently made the computations and we believe the results represent the most satisfactory information on $c_{p_0}$ now available. The present tables make use of the Gordon $c_{p_0}$ values which are represented by the following formula:*

$$c_{p_0} = 1.4720 + 7.5566 \cdot 10^{-4}\, T + 47.8365\, \tau \text{ int. joules/g. °C} \qquad (15)$$

Completion of the enthalpy equation for the vapor depends only on the selection of a value for $h'$. It is customary to make the enthalpy zero for the saturated liquid state at 32° F,† and it would be convenient in determining $h'$ to have a value for the enthalpy of the saturated vapor corresponding to this temperature. The magnitude in this case would equal numerically the heat of evaporation for which unfortunately no new direct measurements exist. Any accurate value of the enthalpy of the vapor will however suffice for determining $h'$, provided of course that it is based on the convention of zero enthalpy for the liquid at 32° in accordance with the steam tables usage. Since the 212° F saturation value has been verified by more experimenters than any other one state it was chosen for the purpose. Osborne's most recently published value[21] of the enthalpy, 2675.35 int. joules/g., was adopted.

* Doctor E. Bright Wilson, Jr. has called to the attention of one of us the fact that the "coupling term" between the rotational and vibrational states of the water molecule may contribute sufficient energy to raise Gordon's values slightly.

† It is realized that the saturated liquid state at 32° F is below the triple point by .02° F, and hence is not a state of true equilibrium. Nevertheless, because extrapolation of properties to that state is easy, no real difficulty arises in using it as a reference condition.

The corresponding enthalpy values for that state obtained from latent heat measurements from three other sources by adding the Osborne enthalpy of the saturated liquid are as follows:

| | |
|---|---|
| Richards and Matthews[22] | 2670.1 |
| Henning[23, 24] | 2674.82 |
| A. W. Smith[25] | 2680.31 |

The average of these three figures, 2675.08, is in extraordinarily good agreement with the value chosen. Upon substitution in the enthalpy equation it is found that $h'$ is 2502.36 int. joules/g. The enthalpy equation with this value of $h'$ is

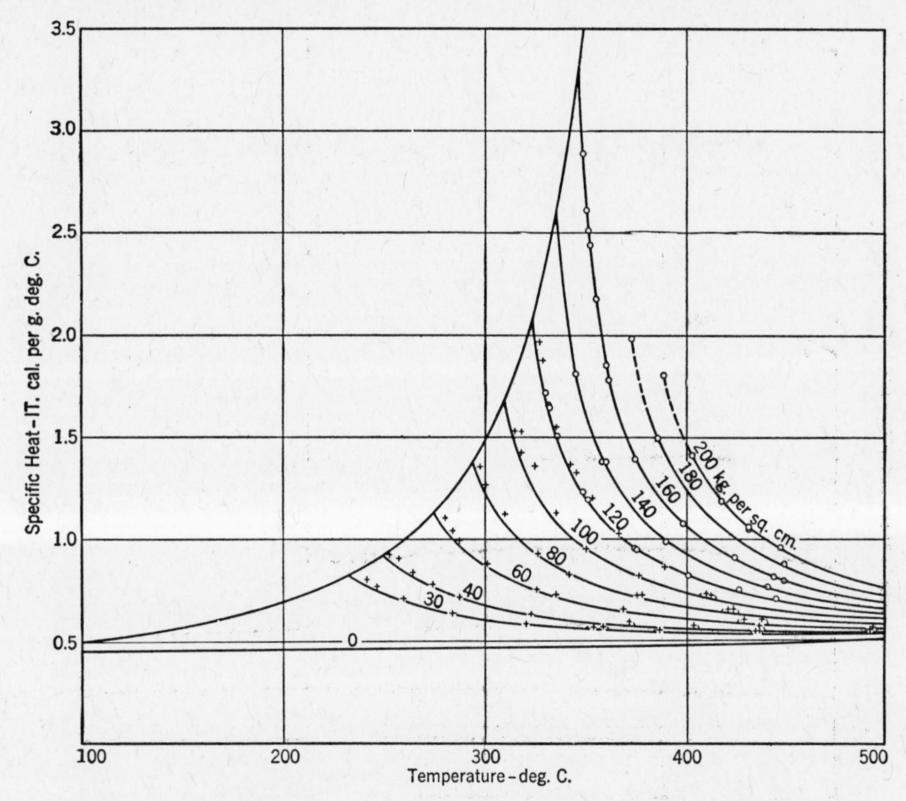


FIGURE 1. Comparison of the heat capacities computed from the equation of state with those measured by Knoblauch and Koch, and by Koch.

Lines represent the equation of state values.
Circles represent Koch measurements, crosses represent Knoblauch and Koch measurements.

the source of all enthalpies tabulated except for states having specific volumes less than 10 $cm.^3/g$.

Since the enthalpy values are derived quantities* it is fortunate that there is available a wealth of experimental material by which they may be tested. The agreement between the enthalpies computed by equation (14) and Osborne's

* It is of interest to note that a recent German steam table (A. Knoblauch, E. Raisch, H. Hausen and W. Koch, Tabellen and Diagramme für Wasserdampf, R. Oldenbourg, Munich and Berlin, 1932) reverses this process and derives volumes from calorimetric data. See also H. Hausen, Forsch. a.d. Geb. d. Ing. **2**, 319 (1931).

directly measured enthalpies of the saturated vapor is better than one part in a thousand throughout and is usually better than one part in two thousand.* The excellent latent heat measurements by Henning,[23] Jakob,[26] Jakob and Fritz,[27] and Koch[28] all serve to verify the work of Osborne and, in turn, equation (14).

Havliček and Miškovský[29] have published directly measured values of enthalpy that are admirable in quality and extent. The values from equation (14) differ from these measurements less than 1 cal./g. or about 1 part in 1000, with few exceptions.

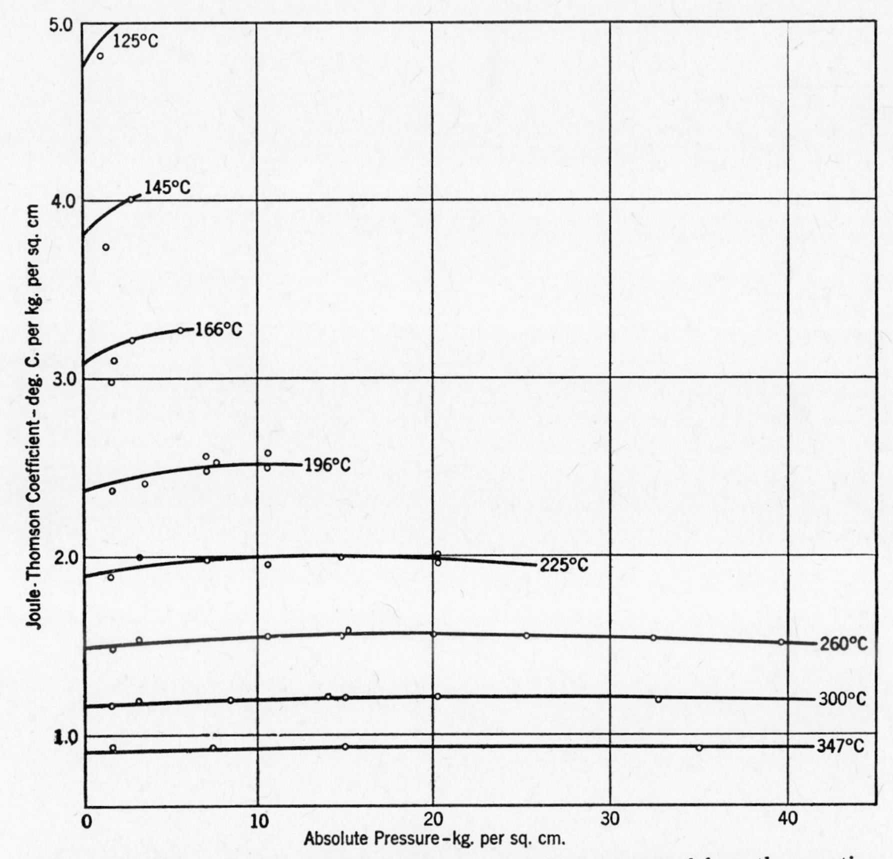


FIGURE 2. Comparison of the Joule-Thomson coefficients computed from the equation of state with those measured by Davis and Kleinschmidt.

Lines represent the equation of state. Circles represent the measurements.

The Koch[30] and Knoblauch and Koch[31] measurements† of specific heat have been compared with the corresponding derivative of equation (14) and the agreement, particularly with the latest measurements, is all that could be desired. The measurements and calculated values are represented in Fig. 1 which makes clear

* Numerical comparisons are given in Proc. Am. Acad. Arts Sci., **70**, 345 (1936).

† References to earlier work on specific heat and critical discussions of the data will be found in the papers of the yearly steam conferences of the A. S. M. E. since 1922. The following papers are of special interest: H. N. Davis, Mech. Eng., **46**, 85 (1924), J. H. Keenan, Mech. Eng., **48**, **144** (1926), H. N. Davis and J. H. Keenan, Mech. Eng., **51**, 291 (1929).

the superior accord of the latest observations with the computed values. Another very sensitive test is provided at moderate temperatures and pressures by the Joule-Thomson coefficients measured at Harvard University by Davis and Kleinschmidt.[32, 33, 34, 35] In Fig. 2 the full lines represent the coefficients obtained by differentiating equation (14) at constant enthalpy and the circles represent the actual measurements. The experimental values lying below the full lines at 125°, 145° and 166° are, according to the investigators, difficult to obtain and are of lower precision than those at higher pressures. The agreement is entirely satisfactory.

The entropy of the vapor can be found from the equation of state through the relationship,

$$\left(\frac{\partial s}{\partial p}\right)_T = -\left(\frac{\partial v}{\partial T}\right)_p.$$

Thus the entropy becomes

$$\begin{aligned} s &= -\int\left(\frac{\partial v}{\partial T}\right)_p dp + M' \\ &= -R\ln p + M_0 p + \frac{M_1}{2}p^2 + \frac{M_3}{4}p^4 + \frac{M_{12}}{13}p^{13} + M', \end{aligned} \tag{16}$$

where $M_1$, $M_3$, and $M_{12}$ are each equal to $(\phi - F)_T$ in terms of the corresponding $\phi$ and $F$ of equations (13A) and (14), $M_0$ is $(B_0 - F_0)_T$, and $M'$ is a pure temperature function. Upon differentiating $s$ with respect to $T$ at constant $p$ it becomes evident that

$$\frac{dM'}{dT} = \frac{c_{p0}}{T};$$

whence

$$s = -R\ln p + M_0 p + \frac{M_1}{2}p^2 + \frac{M_3}{4}p^4 + \frac{M_{12}}{13}p^{13} + \int\frac{c_{p0}}{T}\,dT + s' \tag{16A}$$

in which $s'$ is an arbitrary constant.

As in the case of the enthalpy, it is customary to make the entropy zero for the saturated liquid state at 32° F. Consequently $s'$ must be so chosen that the vapor entropies represent the difference between the entropies of the vapor states and the entropy of the saturated liquid at 32° F. The change in entropy between the reference state and the saturated vapor state at 212° F was calculated from Osborne's enthalpy measurements* by integrating the relationship

$$ds = \frac{dh}{T} - \frac{v}{T}dp$$

along the saturated liquid line from 32° to 212° F and then at constant pressure and temperature from liquid state to vapor state. The resulting entropy of the saturated vapor at 212° F when substituted into the entropy equation (16A) determines $s'$. Its value is $-1.48847$ int. joules/g. °C.

An alternative and simpler method of determining $s'$ consists of substituting in the entropy equation $h_g/T$ for $s_g$ at 32° F and the corresponding vapor pressure. The result is an $s'$ which differs by 3 in the last figure given in the tables or about one part in 7000 of the entropy of the vapor at 32° F. Though this difference is

* Specific volumes also enter into this calculation; but since the volume term is small, precise values of the liquid volume are not necessary.

negligible for all uses of the tables, it signifies certain defects in our knowledge of water vapor at low temperatures.

With the determination of $s'$ the entropy expression is complete. It serves like the equation of state and the enthalpy equation for all states in which the specific volume exceeds 10 cm.$^3$/g.

## 4. The Liquid

The liquid volume data of Amagat[36] and of Smith and Keyes[37] were used by Keenan[38] in the development of a table of properties. Other recent experimental data by Trautz and Steyer[39] are in agreement with the other sources except at high temperatures where the measurements of Trautz and Steyer are probably of less precision.*

The method used by Keenan was a combination of algebraic and graphic devices. The specific volumes were formulated by means of the equation

$$v = 3.086 - 0.899017\, z^{0.147166} - 0.4\, y^{-1.6}\, (p - 218.5) + \delta \tag{17}$$

where $v$ is specific volume in cm.$^3$/g.,
$z$ is $374.1 - t$,
$y$ is $385 - t$,
$t$ is temperature in °C,
$p$ is pressure in int. atm., and
$\delta$ is a function of $p$ and $t$ to be treated graphically.

Except for the immediate neighborhood of the critical point the graphic term was less than 6% of the volume. Since the publication of Keenan's analysis of the liquid phase Koch[40] has used his apparatus for measuring the heat capacity and Havliček[29] has measured enthalpy in that region. The agreement between the specific heats found from Keenan's integrations together with Osborne's saturation data and the heat capacities measured by Koch is extremely gratifying. The agreement with Havliček's work is equally good. All specific volumes in the table of properties of the compressed liquid at temperatures up to 680° F were obtained from equation (17) and faired charts of $\delta$.

The change in enthalpy at constant temperature from saturation was found from the relation,

$$h - h_f = \int_{p_f}^{p} \left(\frac{\partial(v\tau)}{\partial \tau}\right)_p dp = \int_{p_f}^{p} \left(\frac{\partial(f\tau)}{\partial \tau}\right)_p dp + \int_{p_f}^{p} \left(\frac{\partial(\delta\tau)}{\partial \tau}\right)_p dp,$$

where $f$ is the algebraic portion of the formulation or the first three terms in the expression for $v$ (equation 17). The integral involving the graphic term is evaluated graphically.

The change in entropy at constant temperature from saturation was found similarly from the relation

$$s - s_f = -\int_{p_f}^{p} \left(\frac{\partial v}{\partial t}\right)_p dp = -\int_{p_f}^{p} \left(\frac{\partial f}{\partial t}\right)_p dp - \int_{p_f}^{p} \left(\frac{\partial \delta}{\partial t}\right)_p dp.$$

The compressed liquid table is taken from the Keenan development except for small modifications near the critical temperature.

* For a critical discussion of the work of Trautz and Steyer, see M. Jakob, Engineering (London) **132**, 143 (1931).

The specific volume of the saturated liquid up to 680° F was obtained from the formulation by Smith and Keyes[37] which represents the values obtained by extrapolation of their measured values for the compressed liquid and is in excellent accord at low temperatures with the measurements by Amagat.[36] According to this formula the saturated liquid volume is given in cm.³/g. by

$$v_s = \frac{v_c + a(t_c - t)^{\frac{1}{3}} + b(t_c - t) + c(t_c - t)^4}{1 + d(t_c - t)^{\frac{1}{3}} + e(t_c - t)}, \tag{18}$$

where

$$\begin{aligned} v_c &= 3.1975 \frac{\text{cm.}^3}{\text{g.}}, \\ t_c &= \text{critical temperature} = 374.11^\circ \text{ C}, \\ t &= \text{temperature in degrees C}, \\ a &= -0.3151548, \\ b &= -1.203374 \times 10^{-3}, \\ c &= +7.48908 \times 10^{-13}, \\ d &= +0.1342489, \\ e &= -3.946263 \times 10^{-3}. \end{aligned}$$

Though the Smith-Keyes formula was used to 680° F without change in any of its constants, it should be noted that the value of $v_c$ given there is not the critical specific volume chosen for the present table. The selection of saturation values for temperatures above 680° F is discussed in section 5 below.

The enthalpy of the saturated liquid from 32° F to the critical point is taken from the work of Osborne[21] and his colleagues. The values correspond to their most recent publication except near the critical temperature where revised figures privately communicated have been substituted.

The Clapeyron equation offers an alternative method of determining the enthalpy of the saturated liquid through use of the specific volumes previously chosen for saturated liquid and vapor and the vapor pressure relationship. The agreement between the results of the two methods is enough to make it unnecessary to choose between them.*

The entropy of the saturated liquid is found above 212° F by subtracting $h_{fg}/T$ from the entropy of the saturated vapor. A second computation of the entropy change along the saturated liquid line using the relation

$$s = \int_{32}^{t} \frac{dh}{T} - \int_{32}^{t} \frac{v}{T}\, dp$$

verified the results obtained in the first calculation within one in the last figure in the table. Below 212° F the second method is used in order not to concentrate the low temperature inconsistency discussed in section 3 in these small numbers.

## 5. Values Near the Critical Point

Properties change so rapidly near the critical point that it is difficult to maintain precision in experiment or analysis. In the present state of our knowledge there remain some uncertainties in the data given for saturation states within a few degrees of the critical point.

* This agreement is evident from comparisons on p. 345 of Proc. Am. Acad. Arts Sci. **70,** (1936).

The sources of information drawn upon for data near the critical point were

1. the Keyes vapor pressure equation (1);
2. the Third International Steam Table Conference[41] values for the specific volume of saturated liquid and vapor;
3. saturated liquid enthalpies by Osborne[21] and his colleagues, and the Third International Steam Table Conference[41] values;
4. saturated vapor enthalpies by Osborne,[21] and Third International Steam Table Conference[41] values.

The revised Osborne liquid enthalpies and the Keyes vapor pressures were adhered to. All other values were chosen to give agreement with the most reliable sources, to give consistency through the Clapeyron equation, and to give mean density and mean enthalpy curves with normal characteristics to the critical point.

The region encompassing the critical point, extending from a volume of 10 cm.$^3$/g. on the vapor side to a temperature of 680° F on the liquid side, includes such rapidly changing derivatives that it has resisted formulation. Between volumes of 10 and 5 cm.$^3$/g. it is possible to apply a graphic correction to the equation of state and calculate the properties by combined algebraic and graphic means as in the case of the compressed liquid.

The Keyes, Smith and Gerry[12] volume data occur at such wide intervals over the remaining area that they fail to guide a graphic development in sufficient detail. Havlíček and Miškovský,[29] on the other hand, provide measured values of enthalpy at gratifyingly frequent intervals through this most difficult region. Consequently a semi-graphic formulation was based on enthalpy measurements. The formula used is,

$$h = m - \frac{n}{p} + \frac{1000\,t}{p} + \delta, \tag{19}$$

where $h$ is the enthalpy in IT. cal./g.,
$m$ is 265,
$n$ is 318,000,
$t$ is the temperature in °C,
$p$ is the pressure in kg./cm.$^2$ and
$\delta$ is a graphically formulated function of $p$ and $t$.

Because it is found that the Havlíček-Miškovský values fall slightly below both vapor and liquid enthalpies of the present table at more moderate conditions, the formulation is made consistently higher than their measurements by 1 cal./g. The agreement between enthalpies and specific heat values taken from this formulation and Koch's measurements above the critical point is entirely satisfactory.

Values of the constant temperature coefficient corresponding to formulation (19) were computed and integrated between the observed isometrics of the Keyes-Smith measurements, using the following relation,

$$\Delta(v\tau) = \int \left(\frac{\partial h}{\partial p}\right)_t d\tau.$$

Slight adjustments in the graphic formulation of $\delta$ brought agreement with the volume measurements to about one part in 1000 throughout the region formulated.

The same type of integration provided volumes for the table while the enthalpy formulation provided enthalpies. Entropies were computed either by constant pressure integration, using the equation,

$$\Delta s = \int_{p=c} \frac{dh}{T},$$

between a state at a temperature below 680° F and the required state, or by a constant temperature integration, by the following equation,

$$\Delta s = \frac{1}{T}\left[\Delta h - \int v\,dp\right],$$

between a state in the region covered by the equation of state and the required condition. It is evident that these computations offer two tests of the consistency of the work: first, the constant pressure integration may be carried through the formulated region to a temperature so high that the equation of state entropy is valid; second, the entropy of a given state in the formulated region can be computed either from a constant pressure integration starting from a compressed liquid state or from a constant temperature integration starting from an equation of state value of the entropy. In no case did the two methods disagree by more than 2 in the last figure tabulated.

## 6. The Solid and Its Vapor

In addition to the vapor pressure of ice, values for the specific heat[42] of the solid phase, its coefficient of expansion,[43] and its latent heat of fusion[44] are required. The properties of the vapor in equilibrium with the solid are found by substituting the vapor pressure into the equation of state at various temperatures. The enthalpy of the solid is obtained by integrating specific heats from 32° F, where the enthalpy is the negative of the latent heat of fusion. Specific volumes of the solid follow from the chosen coefficient of expansion with temperature.

## 7. The Viscosity and Heat Conductivity of Water Vapor

One of the striking successes of the classical kinetic theory of gases was the deduction that the viscosity of a gas is independent of pressure and a function of temperature only. The proposition was early submitted to test at low pressure and its truth established. Later measurements on carbon dioxide at high pressures showed, however, that a pressure effect existed, and as a general proposition the viscosity must be assumed to be a function of both temperature and pressure. The viscosity function, $\eta$, may for moderate pressures be assumed to consist of a temperature function and a pressure-temperature function as follows:

$$\eta = \eta_0 + f_1(p,\tau) + f_2(p,T). \qquad (20)$$

The function $\eta_0$ in the case of steam may safely be assumed to conform to the deduction from theory using a van der Waals molecular model.[45] The data[46, 47, 48, 49, 53, 54] for the viscosity of steam based on the recent results[55] of Hawkins, Solberg, and Potter lead to the following equation for $\eta_0$:

$$\eta_0\,(\text{poise}) = \frac{1.501 \cdot 10^{-5}\,(T)^{\frac{1}{2}}}{1 + 446.8\tau} \qquad (21)$$

The value of $\eta_0$ was subtracted from the measured viscosity and the differences assumed to be $f_1(p,\tau) + f_2(p,T)$. The viscosity may be represented by the following equation, which includes the deduced forms of the functions $f_1$ and $f_2$:

$$\eta = \eta_0 + 10^{-4}\,[\tau\,(a - b\,10^{m\tau})\,p + (d\,10^{-nT})\,p^2], \tag{22}$$

where $p$ denotes the pressure in kg. cm.$^{-2}$ The numbers in Table 6 were derived using equation (21) in (22), the constants of which for g. cm.$^{-1}$ sec.$^{-1}$ units are as follows: $a = 6.36$, $b = 2.31 \cdot 10^{-3}$, $m = 1340$, $d = 3.89 \cdot 10^{-2}$, $n = 5.476 \cdot 10^{-3}$.

The only available earlier work on the heat conductivity of steam is that of Moser[50] and of Milverton[51] and of Timroth and Vargaftik.[52] Additional measurements[62] have been made to 150 kg. per cm.$^2$ and a correlation obtained using the following form of equation from which the numbers given in Table 7 were computed:

$$\lambda = \lambda_0 + c\,(e^{\alpha p \tau^4} - 1) \tag{23}$$

where $\lambda_0 = 1.546 \cdot 10^{-5}\,(T)^{\frac{1}{2}}/(1 + 1737.3\tau/10^{12\tau})$ and $p$ denotes the pressure in kg. cm.$^{-2}$. The constants in equation (23) which give $\lambda$ in IT cal. cm.$^{-1}$ sec.$^{-1}$ °C$^{-1}$ are as follows: $\alpha = 2.082 \cdot 10^9$, $c = 1.096 \cdot 10^{-5}$.

## Comments, 1950

Since the appearance of these tables in 1936 a number of publications have appeared containing important contributions to our knowledge of the thermodynamic properties of steam.[56, 57, 58, 59, 60, 61] Also new measurements on the transport properties, viscosity[53, 54] and heat conductivity,[62] have been reported. The new transport information has been correlated (section 7) and used to prepare revised Tables 6 and 7.

A comprehensive test of the thermodynamic consistency of the very accurate and extensive data on steam properties was postponed until the intercomparison of the international scale of temperature of 1927 with the gas scale, referred to in the footnote on page 12, had been completed by Professor James A. Beattie, whose publications are listed in the latest of two recent papers.[63] These two papers report the results of a test of the consistency of the available steam data using the new information on the temperature scale. It is shown therein that the present tables list the most acceptable values that can be compiled on the basis of the existing information. A slight thermodynamic inconsistency, which is of insignificant practical importance, does exist in the tables. It will be removed when international agreement is reached relative to a new temperature scale. There develops also from the consistency study definite evidence of a small but systematic difference between the measured enthalpy of steam extrapolated to the ideal gas state and the same property deduced from equation (15) amended to include the term calculated by E. Bright Wilson, Jr.,[59] and confirmed by Stephenson and McMahon.[60] This difference is not serious to 1100° F. At 1600° F, however, the difference caused in the low pressure enthalpy amounts by extrapolation to 1 part in 300.

# BIBLIOGRAPHY


1. J. H. Keenan. — Steam Tables and Mollier Diagram, Am. Soc. Mech. Eng., (1930).
2. G. K. Burgess. — U. S. Bur. of Stand., Jour. of Res., **1**, 635 (1928).
3. F. Henning. — Ann. d. Phys. (4), **22**, 609 (1907).
4. L. Holborn and F. Henning. — Ann. d. Phys. (4), **26**, 833 (1908).
5. L. Holborn and A. Baumann. — Ann. d. Phys. (4), **31**, 945 (1910).
6. K. Scheel and W. Heuse. — Ann. d. Phys. (4), **29**, 723 (1909); **31**, 715 (1910).
7. A. Egerton and G. S. Callendar. — Phil. Trans. Roy. Soc. London, **231**, 147 (1932).
8. N. S. Osborne, H. F. Stimson, E. F. Fioch and D. C. Ginnings. — U. S. Bur. of Stand., Jour. of Res., **10**, 155 (1933).
9. L. B. Smith, F. G. Keyes and H. T. Gerry. — Proc. Am. Acad. Arts and Sci., **69**, 137 (1934).
10. Wärmetabellen, Friedr. Vieweg U. Sohn, Braunschweig, 1919.
11. N. S. Osborne and C. H. Meyers. — U. S. Bur. of Stand., Jour. of Res., **13**, 1 (1934).
12. F. G. Keyes, L. B. Smith and H. T. Gerry. — Proc. Am. Acad. Arts and Sci., **70**, 319 (1936).
13. E. W. Washburn. — Monthly Weather Rev., **52**, 488 (1924).
14. H. N. Davis. — Mech. Eng., **46**, 85 (1924).
15. J. H. Keenan. — Mech. Eng., **48**, 144 (1926).
16. F. G. Keyes, L. B. Smith and H. T. Gerry. — Mech. Eng., **56**, 87 (1934).
17. R. Mecke. — Zeit. f. Phys., **81**, 313 (1933).
18. W. Baumann and R. Mecke. — Zeit. f. Phys., **81**, 445 (1933).
19. K. Freudenberg and R. Mecke. — Zeit. f. Phys., **81**, 465 (1933).
20. A. R. Gordon. — J. Chem. Phys., **2**, 65 (1934).
21. N. S. Osborne, H. F. Stimson and D. C. Ginnings. — Mech. Eng., **57**, 162 (1935); see also N. S. Osborne, H. F. Stimson and E. F. Fioch. — U. S. Bur. of Stand., Jour. of Res., **5**, 411 (1930); Trans. Am. Soc. Mech. Eng., F. S. P., **52**, 191 (1930); N. S. Osborne, *Ibid.*, 221; E. F. Fioch, *Ibid.*, 231.
22. T. W. Richards and J. H. Matthews. — J. Am. Chem. Soc., **33**, 863 (1911).
23. F. Henning. — Ann. d. Phys., (4), **21**, 849 (1906).
24. F. Henning. — Ann. d. Phys., (4), **29**, 441 (1909).
25. A. W. Smith. — Phys. Rev., **25**, 145 (1907).
26. M. Jakob. — Forsch. a. d. Geb. d. Ing., Heft 310 (1928).
27. M. Jakob and W. Fritz. — Zeit. Ver. Deutsch. Ing., **73**, 629 (1929), and Phys. Zeit., **36**, 651 (1935).
28. W. Koch. — Forsch. a. d. Geb. d. Ing., **5**, 257 (1934).
29. J. Havlíček and L. Miškovský. — Hel. Phys. Acta., **9**, 161 (1936).
30. W. Koch. — Forsch. a. d. Geb. d. Ing., **3**, 1 (1932); **3**, 189 (1932).
31. O. Knoblauch and W. Koch. — Zeit. Ver. Deutsch. Ing., **72**, 1733 (1928).
32. R. V. Kleinschmidt. — Mech. Eng., **45**, 165 (1923).
33. H. N. Davis. — Mech. Eng., **47**, 107 (1925).
34. R. V. Kleinschmidt. — Mech. Eng., **48**, 155 (1926).
35. H. N. Davis and J. H. Keenan. — World Eng. Conf. Report No. 455, Tokyo (1929).
36. M. E. H. Amagat. — Ann. d. Chim. Phys., **29**, 68, 505 (1893).
37. L. B. Smith and F. G. Keyes. — Proc. Am. Acad. Arts and Sci., **69**, 285 (1934).
38. J. H. Keenan. — Mech. Eng., **53**, 127 (1931).
39. M. Trautz and H. Steyer. — Forsch. a. d. Geb. d. Ing., **2**, 45 (1931).
40. W. Koch. — Forsch. a. d. Geb. d. Ing., **5**, 138 (1934).
41. Third International Conference on Steam Tables, Mech. Eng., **57**, 710 (1935).

42. W. F. Giauque and J. W. Stout. — J. Am. Chem. Soc., **58**, 1144 (1936).
43. Int. Crit. Tables, National Research Council, McGraw-Hill, 1928.
44. H. C. Dickinson, D. R. Harper and N. S. Osborne. — Bull. U. S. Bur. of Stand., **10**, 235 (1914).
45. F. G. Keyes. — Zeit. f. Phys. Chim., Cohen Festschrift, 709 (1927).
46. R. Plank. — Forsch. a. d. Geb. d. Ing., **4**, 1 (1933).
47. K. Sigwart. — Forsch. a. d. Geb. d. Ing., **7**, 125 (1936).
48. D. L. Timroth. — J. Phys. (U.S.S.R.), **2**, 419 (1940).
49. G. A. Hawkins, H. L. Solberg and A. A. Potter. — Trans. Am. Soc. Mech. Eng., **57**, 395 (1935).
50. Moser. — Dissertation, Berlin (1913).
51. S. W. Milverton. — Phil. Mag., **17**, 397 (1934); Roy. Soc. Proc. A **150**, 287 (1935).
52. D. L. Timroth and N. B. Vargaftik. — J. Phys. (U.S.S.R.), **2**, 101 (1940).
53. G. A. Hawkins, H. L. Solberg, and A. A. Potter. — Trans. A.S.M.E., **62**, 677 (1940).
54. G. A. Hawkins, W. L. Sibbitt and H. L. Solberg. — Trans. A.S.M.E., **70**, 19 (1948).
55. E. F. Leib. — Combustion, **12**, 45 (1940).
56. N. S. Osborne, H. F. Stimson, and D. C. Ginnings. — U. S. Bur. of Stand. Jour. of Res., **23**, 197, 261 (1939).
57. N. S. Osborne, H. F. Stimson, and D. C. Ginnings. — U. S. Bur. of Stand. Jour. of Res., **18**, 389 (1937).
58. Jan Jůza. — Engineering, **146**, 1, 3 (1938).
59. E. Bright Wilson, Jr. — J. Chem. Phys., **4**, 526 (1936).
60. C. C. Stephenson and H. O. McMahon. — J. Chem. Phys., **7**, 614 (1939).
61. S. C. Collins and F. G. Keyes. — Proc. Am. Acad. Arts and Sci., **72**, 283 (1938).
62. F. G. Keyes and D. J. Sandell, Jr. — Trans. Amer. Soc. Mech. Eng., **72**, 767–778 (1950).
63. F. G. Keyes. — J. Chem. Phys., **15**, 602 (1947); **17**, 923 (1949).

# SYMBOLS USED IN TABLES

$t$ — temperature, deg. fahr.

$p$ — absolute pressure, lb. per sq. in. or in. Hg.

$v$ — specific volume, cu. ft. per lb.

$h$ — enthalpy, Btu.* per lb.

$s$ — entropy, Btu.* per deg. fahr. per lb.

$u$ — internal energy, Btu.* per lb.

$f$ (subscript) — refers to a property of the saturated liquid.

$g$ (subscript) — refers to a property of the saturated vapor.

$i$ (subscript) — refers to a property of the saturated solid.

$fg$ (subscript) — refers to a change by evaporation.

$ig$ (subscript) — refers to a change by sublimation.

* 1 Btu. = 778.26 ft. lb., 1 kw. hr. = 3412.75 Btu.
The Btu. is derived from the IT. calorie through the relation 1 IT. cal./g. = 1.8 Btu./lb.
The IT. calorie is defined in the report of the First International Steam Tables Conference (Mechanical Engineering, v. 52 (1930) pp. 120–122) as 1/860 international watt-hour.

# Table 1. Saturation: Temperatures

| Temp. Fahr. | Abs. Pressure | | Specific Volume | | | Enthalpy | | | Entropy | | | Temp. Fahr. |
|---|---|---|---|---|---|---|---|---|---|---|---|---|
| | Lb./Sq. In. | In. Hg. | Sat. Liquid | Evap. | Sat. Vapor | Sat. Liquid | Evap. | Sat. Vapor | Sat. Liquid | Evap. | Sat. Vapor | |
| t | p | | $v_f$ | $v_{fg}$ | $v_g$ | $h_f$ | $h_{fg}$ | $h_g$ | $s_f$ | $s_{fg}$ | $s_g$ | t |
| 32° | 0.08854 | 0.1803 | 0.01602 | 3306 | 3306 | 0.00 | 1075.8 | 1075.8 | 0.0000 | 2.1877 | 2.1877 | 32° |
| 33 | 0.09223 | 0.1878 | 0.01602 | 3180 | 3180 | 1.01 | 1075.2 | 1076.2 | 0.0020 | 2.1821 | 2.1841 | 33 |
| 34 | 0.09603 | 0.1955 | 0.01602 | 3061 | 3061 | 2.02 | 1074.7 | 1076.7 | 0.0041 | 2.1764 | 2.1805 | 34 |
| 35° | 0.09995 | 0.2035 | 0.01602 | 2947 | 2947 | 3.02 | 1074.1 | 1077.1 | 0.0061 | 2.1709 | 2.1770 | 35° |
| 36 | 0.10401 | 0.2118 | 0.01602 | 2837 | 2837 | 4.03 | 1073.6 | 1077.6 | 0.0081 | 2.1654 | 2.1735 | 36 |
| 37 | 0.10821 | 0.2203 | 0.01602 | 2732 | 2732 | 5.04 | 1073.0 | 1078.0 | 0.0102 | 2.1598 | 2.1700 | 37 |
| 38 | 0.11256 | 0.2292 | 0.01602 | 2632 | 2632 | 6.04 | 1072.4 | 1078.4 | 0.0122 | 2.1544 | 2.1666 | 38 |
| 39 | 0.11705 | 0.2383 | 0.01602 | 2536 | 2536 | 7.04 | 1071.9 | 1078.9 | 0.0142 | 2.1489 | 2.1631 | 39 |
| 40° | 0.12170 | 0.2478 | 0.01602 | 2444 | 2444 | 8.05 | 1071.3 | 1079.3 | 0.0162 | 2.1435 | 2.1597 | 40° |
| 41 | 0.12652 | 0.2576 | 0.01602 | 2356 | 2356 | 9.05 | 1070.7 | 1079.7 | 0.0182 | 2.1381 | 2.1563 | 41 |
| 42 | 0.13150 | 0.2677 | 0.01602 | 2271 | 2271 | 10.05 | 1070.1 | 1080.2 | 0.0202 | 2.1327 | 2.1529 | 42 |
| 43 | 0.13665 | 0.2782 | 0.01602 | 2190 | 2190 | 11.06 | 1069.5 | 1080.6 | 0.0222 | 2.1274 | 2.1496 | 43 |
| 44 | 0.14199 | 0.2891 | 0.01602 | 2112 | 2112 | 12.06 | 1068.9 | 1081.0 | 0.0242 | 2.1220 | 2.1462 | 44 |
| 45° | 0.14752 | 0.3004 | 0.01602 | 2036.4 | 2036.4 | 13.06 | 1068.4 | 1081.5 | 0.0262 | 2.1167 | 2.1429 | 45° |
| 46 | 0.15323 | 0.3120 | 0.01602 | 1964.3 | 1964.3 | 14.06 | 1067.8 | 1081.9 | 0.0282 | 2.1113 | 2.1395 | 46 |
| 47 | 0.15914 | 0.3240 | 0.01603 | 1895.1 | 1895.1 | 15.07 | 1067.3 | 1082.4 | 0.0302 | 2.1060 | 2.1362 | 47 |
| 48 | 0.16525 | 0.3364 | 0.01603 | 1828.6 | 1828.6 | 16.07 | 1066.7 | 1082.8 | 0.0321 | 2.1008 | 2.1329 | 48 |
| 49 | 0.17157 | 0.3493 | 0.01603 | 1764.7 | 1764.7 | 17.07 | 1066.1 | 1083.2 | 0.0341 | 2.0956 | 2.1297 | 49 |
| 50° | 0.17811 | 0.3626 | 0.01603 | 1703.2 | 1703.2 | 18.07 | 1065.6 | 1083.7 | 0.0361 | 2.0903 | 2.1264 | 50° |
| 51 | 0.18486 | 0.3764 | 0.01603 | 1644.2 | 1644.2 | 19.07 | 1065.0 | 1084.1 | 0.0380 | 2.0852 | 2.1232 | 51 |
| 52 | 0.19182 | 0.3906 | 0.01603 | 1587.6 | 1587.6 | 20.07 | 1064.4 | 1084.5 | 0.0400 | 2.0799 | 2.1199 | 52 |
| 53 | 0.19900 | 0.4052 | 0.01603 | 1533.3 | 1533.3 | 21.07 | 1063.9 | 1085.0 | 0.0420 | 2.0747 | 2.1167 | 53 |
| 54 | 0.20642 | 0.4203 | 0.01603 | 1481.0 | 1481.0 | 22.07 | 1063.3 | 1085.4 | 0.0439 | 2.0697 | 2.1136 | 54 |
| 55° | 0.2141 | 0.4359 | 0.01603 | 1430.7 | 1430.7 | 23.07 | 1062.7 | 1085.8 | 0.0459 | 2.0645 | 2.1104 | 55° |
| 56 | 0.2220 | 0.4520 | 0.01603 | 1382.4 | 1382.4 | 24.06 | 1062.2 | 1086.3 | 0.0478 | 2.0594 | 2.1072 | 56 |
| 57 | 0.2302 | 0.4686 | 0.01603 | 1335.9 | 1335.9 | 25.06 | 1061.6 | 1086.7 | 0.0497 | 2.0544 | 2.1041 | 57 |
| 58 | 0.2386 | 0.4858 | 0.01604 | 1291.1 | 1291.1 | 26.06 | 1061.0 | 1087.1 | 0.0517 | 2.0493 | 2.1010 | 58 |
| 59 | 0.2473 | 0.5035 | 0.01604 | 1248.1 | 1248.1 | 27.06 | 1060.5 | 1087.6 | 0.0536 | 2.0443 | 2.0979 | 59 |
| 60° | 0.2563 | 0.5218 | 0.01604 | 1206.6 | 1206.7 | 28.06 | 1059.9 | 1088.0 | 0.0555 | 2.0393 | 2.0948 | 60° |
| 61 | 0.2655 | 0.5407 | 0.01604 | 1166.8 | 1166.8 | 29.06 | 1059.3 | 1088.4 | 0.0574 | 2.0343 | 2.0917 | 61 |
| 62 | 0.2751 | 0.5601 | 0.01604 | 1128.4 | 1128.4 | 30.05 | 1058.8 | 1088.9 | 0.0593 | 2.0293 | 2.0886 | 62 |
| 63 | 0.2850 | 0.5802 | 0.01604 | 1091.4 | 1091.4 | 31.05 | 1058.2 | 1089.3 | 0.0613 | 2.0243 | 2.0856 | 63 |
| 64 | 0.2951 | 0.6009 | 0.01605 | 1055.7 | 1055.7 | 32.05 | 1057.6 | 1089.7 | 0.0632 | 2.0194 | 2.0826 | 64 |
| 65° | 0.3056 | 0.6222 | 0.01605 | 1021.4 | 1021.4 | 33.05 | 1057.1 | 1090.2 | 0.0651 | 2.0145 | 2.0796 | 65° |
| 66 | 0.3164 | 0.6442 | 0.01605 | 988.4 | 988.4 | 34.05 | 1056.5 | 1090.6 | 0.0670 | 2.0096 | 2.0766 | 66 |
| 67 | 0.3276 | 0.6669 | 0.01605 | 956.6 | 956.6 | 35.05 | 1056.0 | 1091.0 | 0.0689 | 2.0047 | 2.0736 | 67 |
| 68 | 0.3390 | 0.6903 | 0.01605 | 925.9 | 925.9 | 36.04 | 1055.5 | 1091.5 | 0.0708 | 1.9998 | 2.0706 | 68 |
| 69 | 0.3509 | 0.7144 | 0.01605 | 896.3 | 896.3 | 37.04 | 1054.9 | 1091.9 | 0.0726 | 1.9950 | 2.0676 | 69 |
| 70° | 0.3631 | 0.7392 | 0.01606 | 867.8 | 867.9 | 38.04 | 1054.3 | 1092.3 | 0.0745 | 1.9902 | 2.0647 | 70° |
| 71 | 0.3756 | 0.7648 | 0.01606 | 840.4 | 840.4 | 39.04 | 1053.8 | 1092.8 | 0.0764 | 1.9854 | 2.0618 | 71 |
| 72 | 0.3886 | 0.7912 | 0.01606 | 813.9 | 813.9 | 40.04 | 1053.2 | 1093.2 | 0.0783 | 1.9805 | 2.0588 | 72 |
| 73 | 0.4019 | 0.8183 | 0.01606 | 788.3 | 788.4 | 41.03 | 1052.6 | 1093.6 | 0.0802 | 1.9757 | 2.0559 | 73 |
| 74 | 0.4156 | 0.8462 | 0.01606 | 763.7 | 763.8 | 42.03 | 1052.1 | 1094.1 | 0.0820 | 1.9710 | 2.0530 | 74 |
| 75° | 0.4298 | 0.8750 | 0.01607 | 740.0 | 740.0 | 43.03 | 1051.5 | 1094.5 | 0.0839 | 1.9663 | 2.0502 | 75° |
| 76 | 0.4443 | 0.9046 | 0.01607 | 717.1 | 717.1 | 44.03 | 1050.9 | 1094.9 | 0.0858 | 1.9615 | 2.0473 | 76 |
| 77 | 0.4593 | 0.9352 | 0.01607 | 694.9 | 694.9 | 45.02 | 1050.4 | 1095.4 | 0.0876 | 1.9569 | 2.0445 | 77 |
| 78 | 0.4747 | 0.9666 | 0.01607 | 673.6 | 673.6 | 46.02 | 1049.8 | 1095.8 | 0.0895 | 1.9521 | 2.0416 | 78 |
| 79 | 0.4906 | 0.9989 | 0.01608 | 653.0 | 653.0 | 47.02 | 1049.2 | 1096.2 | 0.0913 | 1.9475 | 2.0388 | 79 |
| 80° | 0.5069 | 1.0321 | 0.01608 | 633.1 | 633.1 | 48.02 | 1048.6 | 1096.6 | 0.0932 | 1.9428 | 2.0360 | 80° |
| 81 | 0.5237 | 1.0664 | 0.01608 | 613.9 | 613.9 | 49.02 | 1048.1 | 1097.1 | 0.0950 | 1.9382 | 2.0332 | 81 |
| 82 | 0.5410 | 1.1016 | 0.01608 | 595.3 | 595.3 | 50.01 | 1047.5 | 1097.5 | 0.0969 | 1.9335 | 2.0304 | 82 |
| 83 | 0.5588 | 1.1378 | 0.01609 | 577.4 | 577.4 | 51.01 | 1046.9 | 1097.9 | 0.0987 | 1.9290 | 2.0277 | 83 |
| 84 | 0.5771 | 1.1750 | 0.01609 | 560.1 | 560.2 | 52.01 | 1046.4 | 1098.4 | 0.1005 | 1.9244 | 2.0249 | 84 |
| 85° | 0.5959 | 1.2133 | 0.01609 | 543.4 | 543.5 | 53.00 | 1045.8 | 1098.8 | 0.1024 | 1.9198 | 2.0222 | 85° |
| 86 | 0.6152 | 1.2527 | 0.01609 | 527.3 | 527.3 | 54.00 | 1045.2 | 1099.2 | 0.1042 | 1.9153 | 2.0195 | 86 |
| 87 | 0.6351 | 1.2931 | 0.01610 | 511.7 | 511.7 | 55.00 | 1044.7 | 1099.7 | 0.1060 | 1.9108 | 2.0168 | 87 |
| 88 | 0.6556 | 1.3347 | 0.01610 | 496.6 | 496.7 | 56.00 | 1044.1 | 1100.1 | 0.1079 | 1.9062 | 2.0141 | 88 |
| 89 | 0.6766 | 1.3775 | 0.01610 | 482.1 | 482.1 | 56.99 | 1043.5 | 1100.5 | 0.1097 | 1.9017 | 2.0114 | 89 |

# Table 1. Saturation: Temperatures

| Temp. Fahr. | Abs. Pressure Lb./Sq. In. | Abs. Pressure In. Hg. | Specific Volume Sat. Liquid | Specific Volume Evap. | Specific Volume Sat. Vapor | Enthalpy Sat. Liquid | Enthalpy Evap. | Enthalpy Sat. Vapor | Entropy Sat. Liquid | Entropy Evap. | Entropy Sat. Vapor | Temp. Fahr. |
|---|---|---|---|---|---|---|---|---|---|---|---|---|
| t | p | p | $v_f$ | $v_{fg}$ | $v_g$ | $h_f$ | $h_{fg}$ | $h_g$ | $s_f$ | $s_{fg}$ | $s_g$ | t |
| **90°** | 0.6982 | 1.4215 | 0.01610 | 468.0 | 468.0 | 57.99 | 1042.9 | 1100.9 | 0.1115 | 1.8972 | 2.0087 | **90°** |
| **91** | 0.7204 | 1.4667 | 0.01611 | 454.4 | 454.4 | 58.99 | 1042.4 | 1101.4 | 0.1133 | 1.8927 | 2.0060 | **91** |
| **92** | 0.7432 | 1.5131 | 0.01611 | 441.2 | 441.3 | 59.99 | 1041.8 | 1101.8 | 0.1151 | 1.8883 | 2.0034 | **92** |
| **93** | 0.7666 | 1.5608 | 0.01611 | 428.5 | 428.5 | 60.98 | 1041.2 | 1102.2 | 0.1169 | 1.8838 | 2.0007 | **93** |
| **94** | 0.7906 | 1.6097 | 0.01612 | 416.2 | 416.2 | 61.98 | 1040.7 | 1102.6 | 0.1187 | 1.8794 | 1.9981 | **94** |
| **95°** | 0.8153 | 1.6600 | 0.01612 | 404.3 | 404.3 | 62.98 | 1040.1 | 1103.1 | 0.1205 | 1.8750 | 1.9955 | **95°** |
| **96** | 0.8407 | 1.7117 | 0.01612 | 392.8 | 392.8 | 63.98 | 1039.5 | 1103.5 | 0.1223 | 1.8706 | 1.9929 | **96** |
| **97** | 0.8668 | 1.7647 | 0.01612 | 381.7 | 381.7 | 64.97 | 1038.9 | 1103.9 | 0.1241 | 1.8662 | 1.9903 | **97** |
| **98** | 0.8935 | 1.8192 | 0.01613 | 370.9 | 370.9 | 65.97 | 1038.4 | 1104.4 | 0.1259 | 1.8618 | 1.9877 | **98** |
| **99** | 0.9210 | 1.8751 | 0.01613 | 360.4 | 360.5 | 66.97 | 1037.8 | 1104.8 | 0.1277 | 1.8575 | 1.9852 | **99** |
| **100°** | 0.9492 | 1.9325 | 0.01613 | 350.3 | 350.4 | 67.97 | 1037.2 | 1105.2 | 0.1295 | 1.8531 | 1.9826 | **100°** |
| **101** | 0.9781 | 1.9915 | 0.01614 | 340.6 | 340.6 | 68.96 | 1036.6 | 1105.6 | 0.1313 | 1.8488 | 1.9801 | **101** |
| **102** | 1.0078 | 2.0519 | 0.01614 | 331.1 | 331.1 | 69.96 | 1036.1 | 1106.1 | 0.1330 | 1.8445 | 1.9775 | **102** |
| **103** | 1.0382 | 2.1138 | 0.01614 | 321.9 | 321.9 | 70.96 | 1035.5 | 1106.5 | 0.1348 | 1.8402 | 1.9750 | **103** |
| **104** | 1.0695 | 2.1775 | 0.01615 | 313.1 | 313.1 | 71.96 | 1034.9 | 1106.9 | 0.1366 | 1.8359 | 1.9725 | **104** |
| **105°** | 1.1016 | 2.2429 | 0.01615 | 304.5 | 304.5 | 72.95 | 1034.3 | 1107.3 | 0.1383 | 1.8317 | 1.9700 | **105°** |
| **106** | 1.1345 | 2.3099 | 0.01615 | 296.1 | 296.2 | 73.95 | 1033.8 | 1107.8 | 0.1401 | 1.8274 | 1.9675 | **106** |
| **107** | 1.1683 | 2.3786 | 0.01616 | 288.1 | 288.1 | 74.95 | 1033.3 | 1108.2 | 0.1419 | 1.8232 | 1.9651 | **107** |
| **108** | 1.2029 | 2.4491 | 0.01616 | 280.3 | 280.3 | 75.95 | 1032.7 | 1108.6 | 0.1436 | 1.8190 | 1.9626 | **108** |
| **109** | 1.2384 | 2.5214 | 0.01616 | 272.7 | 272.7 | 76.94 | 1032.1 | 1109.0 | 0.1454 | 1.8147 | 1.9601 | **109** |
| **110°** | 1.2748 | 2.5955 | 0.01617 | 265.3 | 265.4 | 77.94 | 1031.6 | 1109.5 | 0.1471 | 1.8106 | 1.9577 | **110°** |
| **111** | 1.3121 | 2.6715 | 0.01617 | 258.2 | 258.3 | 78.94 | 1031.0 | 1109.9 | 0.1489 | 1.8064 | 1.9553 | **111** |
| **112** | 1.3504 | 2.7494 | 0.01617 | 251.3 | 251.4 | 79.94 | 1030.4 | 1110.3 | 0.1506 | 1.8023 | 1.9529 | **112** |
| **113** | 1.3896 | 2.8293 | 0.01618 | 244.6 | 244.7 | 80.94 | 1029.8 | 1110.7 | 0.1524 | 1.7981 | 1.9505 | **113** |
| **114** | 1.4298 | 2.9111 | 0.01618 | 238.2 | 238.2 | 81.93 | 1029.2 | 1111.1 | 0.1541 | 1.7940 | 1.9481 | **114** |
| **115°** | 1.4709 | 2.9948 | 0.01618 | 231.9 | 231.9 | 82.93 | 1028.7 | 1111.6 | 0.1559 | 1.7898 | 1.9457 | **115°** |
| **116** | 1.5130 | 3.0806 | 0.01619 | 225.8 | 225.8 | 83.93 | 1028.1 | 1112.0 | 0.1576 | 1.7857 | 1.9433 | **116** |
| **117** | 1.5563 | 3.1687 | 0.01619 | 219.9 | 219.9 | 84.93 | 1027.5 | 1112.4 | 0.1593 | 1.7816 | 1.9409 | **117** |
| **118** | 1.6006 | 3.2589 | 0.01620 | 214.2 | 214.2 | 85.92 | 1026.9 | 1112.8 | 0.1610 | 1.7776 | 1.9386 | **118** |
| **119** | 1.6459 | 3.3512 | 0.01620 | 208.6 | 208.7 | 86.92 | 1026.3 | 1113.2 | 0.1628 | 1.7735 | 1.9363 | **119** |
| **120°** | 1.6924 | 3.4458 | 0.01620 | 203.25 | 203.27 | 87.92 | 1025.8 | 1113.7 | 0.1645 | 1.7694 | 1.9339 | **120°** |
| **121** | 1.7400 | 3.5427 | 0.01621 | 198.02 | 198.03 | 88.92 | 1025.2 | 1114.1 | 0.1662 | 1.7654 | 1.9316 | **121** |
| **122** | 1.7888 | 3.6420 | 0.01621 | 192.93 | 192.95 | 89.92 | 1024.6 | 1114.5 | 0.1679 | 1.7614 | 1.9293 | **122** |
| **123** | 1.8387 | 3.7436 | 0.01622 | 188.01 | 188.02 | 90.91 | 1024.0 | 1114.9 | 0.1696 | 1.7574 | 1.9270 | **123** |
| **124** | 1.8897 | 3.8475 | 0.01622 | 183.23 | 183.25 | 91.91 | 1023.4 | 1115.3 | 0.1714 | 1.7533 | 1.9247 | **124** |
| **125°** | 1.9420 | 3.9539 | 0.01622 | 178.59 | 178.61 | 92.91 | 1022.9 | 1115.8 | 0.1731 | 1.7493 | 1.9224 | **125°** |
| **126** | 1.9955 | 4.0629 | 0.01623 | 174.09 | 174.10 | 93.91 | 1022.3 | 1116.2 | 0.1748 | 1.7454 | 1.9202 | **126** |
| **127** | 2.0503 | 4.1745 | 0.01623 | 169.71 | 169.72 | 94.91 | 1021.7 | 1116.6 | 0.1765 | 1.7414 | 1.9179 | **127** |
| **128** | 2.1064 | 4.2887 | 0.01624 | 165.46 | 165.47 | 95.91 | 1021.1 | 1117.0 | 0.1782 | 1.7374 | 1.9156 | **128** |
| **129** | 2.1638 | 4.4055 | 0.01624 | 161.33 | 161.35 | 96.90 | 1020.5 | 1117.4 | 0.1799 | 1.7335 | 1.9134 | **129** |
| **130°** | 2.2225 | 4.5251 | 0.01625 | 157.32 | 157.34 | 97.90 | 1020.0 | 1117.9 | 0.1816 | 1.7296 | 1.9112 | **130°** |
| **131** | 2.2826 | 4.6474 | 0.01625 | 153.43 | 153.44 | 98.90 | 1019.4 | 1118.3 | 0.1833 | 1.7257 | 1.9090 | **131** |
| **132** | 2.3440 | 4.7725 | 0.01626 | 149.65 | 149.66 | 99.90 | 1018.8 | 1118.7 | 0.1849 | 1.7218 | 1.9067 | **132** |
| **133** | 2.4069 | 4.9005 | 0.01626 | 145.97 | 145.99 | 100.90 | 1018.2 | 1119.1 | 0.1866 | 1.7179 | 1.9045 | **133** |
| **134** | 2.4712 | 5.0314 | 0.01626 | 142.40 | 142.42 | 101.90 | 1017.6 | 1119.5 | 0.1883 | 1.7141 | 1.9023 | **134** |
| **135°** | 2.5370 | 5.1653 | 0.01627 | 138.93 | 138.95 | 102.90 | 1017.0 | 1119.9 | 0.1900 | 1.7102 | 1.9002 | **135°** |
| **136** | 2.6042 | 5.3022 | 0.01627 | 135.56 | 135.58 | 103.90 | 1016.4 | 1120.3 | 0.1917 | 1.7063 | 1.8980 | **136** |
| **137** | 2.6729 | 5.4421 | 0.01628 | 132.29 | 132.30 | 104.89 | 1015.9 | 1120.8 | 0.1934 | 1.7024 | 1.8958 | **137** |
| **138** | 2.7432 | 5.5852 | 0.01628 | 129.10 | 129.12 | 105.89 | 1015.3 | 1121.2 | 0.1950 | 1.6987 | 1.8937 | **138** |
| **139** | 2.8151 | 5.7316 | 0.01629 | 126.00 | 126.02 | 106.89 | 1014.7 | 1121.6 | 0.1967 | 1.6948 | 1.8915 | **139** |
| **140°** | 2.8886 | 5.8812 | 0.01629 | 122.99 | 123.01 | 107.89 | 1014.1 | 1122.0 | 0.1984 | 1.6910 | 1.8894 | **140°** |
| **141** | 2.9637 | 6.0341 | 0.01630 | 120.06 | 120.08 | 108.89 | 1013.5 | 1122.4 | 0.2000 | 1.6873 | 1.8873 | **141** |
| **142** | 3.0404 | 6.1903 | 0.01630 | 117.22 | 117.23 | 109.89 | 1012.9 | 1122.8 | 0.2016 | 1.6835 | 1.8851 | **142** |
| **143** | 3.1188 | 6.3500 | 0.01631 | 114.45 | 114.46 | 110.89 | 1012.3 | 1123.2 | 0.2033 | 1.6797 | 1.8830 | **143** |
| **144** | 3.1990 | 6.5132 | 0.01631 | 111.75 | 111.77 | 111.89 | 1011.7 | 1123.6 | 0.2049 | 1.6760 | 1.8809 | **144** |

## Table 1. Saturation: Temperatures

| Temp. Fahr. | Abs. Pressure | | Specific Volume | | | Enthalpy | | | Entropy | | | Temp. Fahr. |
|---|---|---|---|---|---|---|---|---|---|---|---|---|
| | Lb./Sq. In. | In. Hg. | Sat. Liquid | Evap. | Sat. Vapor | Sat. Liquid | Evap. | Sat. Vapor | Sat. Liquid | Evap. | Sat. Vapor | |
| t | p | | $v_f$ | $v_{fg}$ | $v_g$ | $h_f$ | $h_{fg}$ | $h_g$ | $s_f$ | $s_{fg}$ | $s_g$ | t |
| **145°** | 3.281 | 6.680 | 0.01632 | 109.13 | 109.15 | 112.89 | 1011.2 | 1124.1 | 0.2066 | 1.6722 | 1.8788 | **145°** |
| **146** | 3.365 | 6.850 | 0.01632 | 106.58 | 106.60 | 113.89 | 1010.6 | 1124.5 | 0.2083 | 1.6685 | 1.8768 | **146** |
| **147** | 3.450 | 7.024 | 0.01633 | 104.10 | 104.12 | 114.89 | 1010.0 | 1124.9 | 0.2099 | 1.6648 | 1.8747 | **147** |
| **148** | 3.537 | 7.202 | 0.01633 | 101.69 | 101.71 | 115.89 | 1009.4 | 1125.3 | 0.2116 | 1.6610 | 1.8726 | **148** |
| **149** | 3.627 | 7.384 | 0.01634 | 99.34 | 99.36 | 116.89 | 1008.8 | 1125.7 | 0.2133 | 1.6573 | 1.8706 | **149** |
| **150°** | 3.718 | 7.569 | 0.01634 | 97.06 | 97.07 | 117.89 | 1008.2 | 1126.1 | 0.2149 | 1.6537 | 1.8685 | **150°** |
| **151** | 3.811 | 7.759 | 0.01635 | 94.83 | 94.85 | 118.89 | 1007.6 | 1126.5 | 0.2165 | 1.6500 | 1.8665 | **151** |
| **152** | 3.906 | 7.952 | 0.01635 | 92.67 | 92.68 | 119.89 | 1007.0 | 1126.9 | 0.2182 | 1.6463 | 1.8645 | **152** |
| **153** | 4.003 | 8.150 | 0.01636 | 90.56 | 90.57 | 120.89 | 1006.4 | 1127.3 | 0.2198 | 1.6427 | 1.8624 | **153** |
| **154** | 4.102 | 8.351 | 0.01636 | 88.51 | 88.52 | 121.89 | 1005.8 | 1127.7 | 0.2214 | 1.6390 | 1.8604 | **154** |
| **155°** | 4.203 | 8.557 | 0.01637 | 86.51 | 86.52 | 122.89 | 1005.2 | 1128.1 | 0.2230 | 1.6354 | 1.8584 | **155°** |
| **156** | 4.306 | 8.767 | 0.01637 | 84.56 | 84.58 | 123.89 | 1004.7 | 1128.6 | 0.2246 | 1.6318 | 1.8564 | **156** |
| **157** | 4.411 | 8.981 | 0.01638 | 82.67 | 82.69 | 124.89 | 1004.1 | 1129.0 | 0.2263 | 1.6282 | 1.8545 | **157** |
| **158** | 4.519 | 9.200 | 0.01638 | 80.82 | 80.84 | 125.89 | 1003.5 | 1129.4 | 0.2279 | 1.6246 | 1.8525 | **158** |
| **159** | 4.629 | 9.424 | 0.01639 | 79.03 | 79.04 | 126.89 | 1002.9 | 1129.8 | 0.2295 | 1.6210 | 1.8505 | **159** |
| **160°** | 4.741 | 9.652 | 0.01639 | 77.27 | 77.29 | 127.89 | 1002.3 | 1130.2 | 0.2311 | 1.6174 | 1.8485 | **160°** |
| **161** | 4.855 | 9.885 | 0.01640 | 75.57 | 75.58 | 128.89 | 1001.7 | 1130.6 | 0.2327 | 1.6138 | 1.8466 | **161** |
| **162** | 4.971 | 10.122 | 0.01640 | 73.91 | 73.92 | 129.89 | 1001.1 | 1131.0 | 0.2343 | 1.6103 | 1.8446 | **162** |
| **163** | 5.090 | 10.364 | 0.01641 | 72.29 | 72.30 | 130.89 | 1000.5 | 1131.4 | 0.2360 | 1.6067 | 1.8427 | **163** |
| **164** | 5.212 | 10.611 | 0.01641 | 70.71 | 70.73 | 131.89 | 999.9 | 1131.8 | 0.2376 | 1.6032 | 1.8408 | **164** |
| **165°** | 5.335 | 10.863 | 0.01642 | 69.17 | 69.19 | 132.89 | 999.3 | 1132.2 | 0.2392 | 1.5997 | 1.8388 | **165°** |
| **166** | 5.461 | 11.120 | 0.01643 | 67.67 | 67.69 | 133.89 | 998.7 | 1132.6 | 0.2408 | 1.5961 | 1.8369 | **166** |
| **167** | 5.590 | 11.382 | 0.01643 | 66.21 | 66.23 | 134.89 | 998.1 | 1133.0 | 0.2424 | 1.5926 | 1.8350 | **167** |
| **168** | 5.721 | 11.649 | 0.01644 | 64.79 | 64.80 | 135.90 | 997.5 | 1133.4 | 0.2440 | 1.5891 | 1.8331 | **168** |
| **169** | 5.855 | 11.921 | 0.01644 | 63.40 | 63.41 | 136.90 | 996.9 | 1133.8 | 0.2455 | 1.5857 | 1.8312 | **169** |
| **170°** | 5.992 | 12.199 | 0.01645 | 62.04 | 62.06 | 137.90 | 996.3 | 1134.2 | 0.2472 | 1.5822 | 1.8293 | **170°** |
| **171** | 6.131 | 12.483 | 0.01645 | 60.72 | 60.74 | 138.90 | 995.7 | 1134.6 | 0.2488 | 1.5787 | 1.8275 | **171** |
| **172** | 6.273 | 12.772 | 0.01646 | 59.43 | 59.45 | 139.90 | 995.1 | 1135.0 | 0.2503 | 1.5753 | 1.8256 | **172** |
| **173** | 6.417 | 13.066 | 0.01647 | 58.18 | 58.20 | 140.90 | 994.5 | 1135.4 | 0.2519 | 1.5718 | 1.8237 | **173** |
| **174** | 6.565 | 13.366 | 0.01647 | 56.96 | 56.97 | 141.90 | 993.9 | 1135.8 | 0.2535 | 1.5684 | 1.8219 | **174** |
| **175°** | 6.715 | 13.671 | 0.01648 | 55.76 | 55.78 | 142.91 | 993.3 | 1136.2 | 0.2551 | 1.5649 | 1.8200 | **175°** |
| **176** | 6.868 | 13.983 | 0.01648 | 54.60 | 54.61 | 143.91 | 992.7 | 1136.6 | 0.2567 | 1.5615 | 1.8182 | **176** |
| **177** | 7.024 | 14.301 | 0.01649 | 53.46 | 53.48 | 144.91 | 992.1 | 1137.0 | 0.2583 | 1.5581 | 1.8164 | **177** |
| **178** | 7.183 | 14.625 | 0.01650 | 52.35 | 52.37 | 145.91 | 991.5 | 1137.4 | 0.2599 | 1.5547 | 1.8146 | **178** |
| **179** | 7.345 | 14.955 | 0.01650 | 51.27 | 51.29 | 146.92 | 990.8 | 1137.7 | 0.2614 | 1.5513 | 1.8127 | **179** |
| **180°** | 7.510 | 15.291 | 0.01651 | 50.21 | 50.23 | 147.92 | 990.2 | 1138.1 | 0.2630 | 1.5480 | 1.8109 | **180°** |
| **181** | 7.678 | 15.633 | 0.01651 | 49.18 | 49.20 | 148.92 | 989.6 | 1138.5 | 0.2645 | 1.5446 | 1.8091 | **181** |
| **182** | 7.850 | 15.982 | 0.01652 | 48.18 | 48.19 | 149.92 | 989.0 | 1138.9 | 0.2661 | 1.5412 | 1.8073 | **182** |
| **183** | 8.024 | 16.337 | 0.01653 | 47.19 | 47.21 | 150.93 | 988.4 | 1139.3 | 0.2676 | 1.5379 | 1.8055 | **183** |
| **184** | 8.202 | 16.699 | 0.01653 | 46.24 | 46.25 | 151.93 | 987.8 | 1139.7 | 0.2692 | 1.5346 | 1.8038 | **184** |
| **185°** | 8.383 | 17.068 | 0.01654 | 45.29 | 45.31 | 152.93 | 987.2 | 1140.1 | 0.2708 | 1.5312 | 1.8020 | **185°** |
| **186** | 8.567 | 17.443 | 0.01654 | 44.39 | 44.40 | 153.94 | 986.6 | 1140.5 | 0.2723 | 1.5279 | 1.8002 | **186** |
| **187** | 8.755 | 17.825 | 0.01655 | 43.50 | 43.51 | 154.94 | 986.0 | 1140.9 | 0.2739 | 1.5246 | 1.7985 | **187** |
| **188** | 8.946 | 18.214 | 0.01656 | 42.62 | 42.64 | 155.94 | 985.4 | 1141.3 | 0.2754 | 1.5213 | 1.7967 | **188** |
| **189** | 9.141 | 18.611 | 0.01656 | 41.77 | 41.79 | 156.95 | 984.8 | 1141.7 | 0.2770 | 1.5180 | 1.7950 | **189** |
| **190°** | 9.339 | 19.014 | 0.01657 | 40.94 | 40.96 | 157.95 | 984.1 | 1142.0 | 0.2785 | 1.5147 | 1.7932 | **190°** |
| **191** | 9.541 | 19.425 | 0.01658 | 40.13 | 40.15 | 158.95 | 983.4 | 1142.4 | 0.2801 | 1.5114 | 1.7915 | **191** |
| **192** | 9.746 | 19.843 | 0.01658 | 39.34 | 39.36 | 159.96 | 982.8 | 1142.8 | 0.2816 | 1.5082 | 1.7898 | **192** |
| **193** | 9.955 | 20.269 | 0.01659 | 38.57 | 38.58 | 160.96 | 982.2 | 1143.2 | 0.2831 | 1.5049 | 1.7880 | **193** |
| **194** | 10.168 | 20.703 | 0.01659 | 37.81 | 37.83 | 161.97 | 981.6 | 1143.6 | 0.2846 | 1.5017 | 1.7863 | **194** |
| **195°** | 10.385 | 21.144 | 0.01660 | 37.07 | 37.09 | 162.97 | 981.0 | 1144.0 | 0.2862 | 1.4984 | 1.7846 | **195°** |
| **196** | 10.605 | 21.593 | 0.01661 | 36.35 | 36.37 | 163.97 | 980.4 | 1144.4 | 0.2877 | 1.4952 | 1.7829 | **196** |
| **197** | 10.830 | 22.050 | 0.01661 | 35.64 | 35.66 | 164.98 | 979.7 | 1144.7 | 0.2892 | 1.4920 | 1.7812 | **197** |
| **198** | 11.058 | 22.515 | 0.01662 | 34.95 | 34.97 | 165.98 | 979.1 | 1145.1 | 0.2907 | 1.4888 | 1.7795 | **198** |
| **199** | 11.290 | 22.987 | 0.01663 | 34.28 | 34.30 | 166.99 | 978.5 | 1145.5 | 0.2923 | 1.4856 | 1.7779 | **199** |

# Table 1. Saturation: Temperatures

| | Abs. Pressure | | Specific Volume | | | Enthalpy | | | Entropy | | | |
|---|---|---|---|---|---|---|---|---|---|---|---|---|
| Temp. Fahr. | Lb./Sq. In. | In. Hg. | Sat. Liquid | Evap. | Sat. Vapor | Sat. Liquid | Evap. | Sat. Vapor | Sat. Liquid | Evap. | Sat. Vapor | Temp. Fahr. |
| t | p | | $v_f$ | $v_{fg}$ | $v_g$ | $h_f$ | $h_{fg}$ | $h_g$ | $s_f$ | $s_{fg}$ | $s_g$ | t |
| **200°** | 11.526 | 23.467 | 0.01663 | 33.62 | 33.64 | 167.99 | 977.9 | 1145.9 | 0.2938 | 1.4824 | 1.7762 | **200°** |
| **202** | 12.011 | 24.455 | 0.01665 | 32.35 | 32.37 | 170.00 | 976.6 | 1146.6 | 0.2969 | 1.4760 | 1.7729 | **202** |
| **204** | 12.512 | 25.475 | 0.01666 | 31.14 | 31.15 | 172.02 | 975.4 | 1147.4 | 0.2999 | 1.4697 | 1.7696 | **204** |
| **206** | 13.031 | 26.531 | 0.01667 | 29.97 | 29.99 | 174.03 | 974.2 | 1148.2 | 0.3029 | 1.4634 | 1.7663 | **206** |
| **208** | 13.568 | 27.625 | 0.01669 | 28.86 | 28.88 | 176.04 | 972.9 | 1148.9 | 0.3059 | 1.4571 | 1.7630 | **208** |
| **210°** | 14.123 | 28.755 | 0.01670 | 27.80 | 27.82 | 178.05 | 971.6 | 1149.7 | 0.3090 | 1.4508 | 1.7598 | **210°** |
| **212** | 14.696 | 29.922 | 0.01672 | 26.78 | 26.80 | 180.07 | 970.3 | 1150.4 | 0.3120 | 1.4446 | 1.7566 | **212** |
| **214** | 15.289 | 31.129 | 0.01673 | 25.81 | 25.83 | 182.08 | 969.0 | 1151.1 | 0.3149 | 1.4385 | 1.7534 | **214** |
| **216** | 15.901 | 32.375 | 0.01674 | 24.88 | 24.90 | 184.10 | 967.8 | 1151.9 | 0.3179 | 1.4323 | 1.7502 | **216** |
| **218** | 16.533 | 33.662 | 0.01676 | 23.99 | 24.01 | 186.11 | 966.5 | 1152.6 | 0.3209 | 1.4262 | 1.7471 | **218** |
| **220°** | 17.186 | 34.992 | 0.01677 | 23.13 | 23.15 | 188.13 | 965.2 | 1153.4 | 0.3239 | 1.4201 | 1.7440 | **220°** |
| **222** | 17.861 | 36.365 | 0.01679 | 22.31 | 22.33 | 190.15 | 963.9 | 1154.1 | 0.3268 | 1.4141 | 1.7409 | **222** |
| **224** | 18.557 | 37.782 | 0.01680 | 21.53 | 21.55 | 192.17 | 962.6 | 1154.8 | 0.3298 | 1.4080 | 1.7378 | **224** |
| **226** | 19.275 | 39.244 | 0.01682 | 20.78 | 20.79 | 194.18 | 961.3 | 1155.5 | 0.3328 | 1.4020 | 1.7348 | **226** |
| **228** | 20.016 | 40.753 | 0.01683 | 20.06 | 20.07 | 196.20 | 960.1 | 1156.3 | 0.3357 | 1.3961 | 1.7318 | **228** |
| **230°** | 20.780 | 42.308 | 0.01684 | 19.365 | 19.382 | 198.23 | 958.8 | 1157.0 | 0.3387 | 1.3901 | 1.7288 | **230°** |
| **232** | 21.567 | 43.911 | 0.01686 | 18.703 | 18.720 | 200.25 | 957.4 | 1157.7 | 0.3416 | 1.3842 | 1.7258 | **232** |
| **234** | 22.379 | 45.564 | 0.01688 | 18.067 | 18.084 | 202.27 | 956.1 | 1158.4 | 0.3444 | 1.3784 | 1.7228 | **234** |
| **236** | 23.217 | 47.269 | 0.01689 | 17.456 | 17.473 | 204.29 | 954.8 | 1159.1 | 0.3473 | 1.3725 | 1.7199 | **236** |
| **238** | 24.080 | 49.027 | 0.01691 | 16.869 | 16.886 | 206.32 | 953.5 | 1159.8 | 0.3502 | 1.3667 | 1.7169 | **238** |
| **240°** | 24.969 | 50.837 | 0.01692 | 16.306 | 16.323 | 208.34 | 952.2 | 1160.5 | 0.3531 | 1.3609 | 1.7140 | **240°** |
| **242** | 25.884 | 52.701 | 0.01694 | 15.765 | 15.782 | 210.37 | 950.8 | 1161.2 | 0.3560 | 1.3551 | 1.7111 | **242** |
| **244** | 26.827 | 54.620 | 0.01696 | 15.245 | 15.262 | 212.39 | 949.5 | 1161.9 | 0.3589 | 1.3494 | 1.7083 | **244** |
| **246** | 27.798 | 56.597 | 0.01697 | 14.745 | 14.762 | 214.42 | 948.2 | 1162.6 | 0.3618 | 1.3436 | 1.7054 | **246** |
| **248** | 28.797 | 58.631 | 0.01699 | 14.265 | 14.282 | 216.45 | 946.8 | 1163.3 | 0.3647 | 1.3379 | 1.7026 | **248** |
| **250°** | 29.825 | 60.725 | 0.01700 | 13.804 | 13.821 | 218.48 | 945.5 | 1164.0 | 0.3675 | 1.3323 | 1.6998 | **250°** |
| **252** | 30.884 | 62.880 | 0.01702 | 13.360 | 13.377 | 220.51 | 944.2 | 1164.7 | 0.3704 | 1.3266 | 1.6970 | **252** |
| **254** | 31.973 | 65.098 | 0.01704 | 12.933 | 12.950 | 222.54 | 942.8 | 1165.3 | 0.3732 | 1.3210 | 1.6942 | **254** |
| **256** | 33.093 | 67.378 | 0.01705 | 12.522 | 12.539 | 224.58 | 941.4 | 1166.0 | 0.3761 | 1.3154 | 1.6915 | **256** |
| **258** | 34.245 | 69.723 | 0.01707 | 12.127 | 12.144 | 226.61 | 940.1 | 1166.7 | 0.3789 | 1.3099 | 1.6888 | **258** |
| **260°** | 35.429 | 72.134 | 0.01709 | 11.746 | 11.763 | 228.64 | 938.7 | 1167.3 | 0.3817 | 1.3043 | 1.6860 | **260°** |
| **262** | 36.646 | 74.612 | 0.01710 | 11.379 | 11.396 | 230.68 | 937.3 | 1168.0 | 0.3845 | 1.2988 | 1.6833 | **262** |
| **264** | 37.897 | 77.159 | 0.01712 | 11.026 | 11.043 | 232.72 | 936.0 | 1168.7 | 0.3874 | 1.2933 | 1.6807 | **264** |
| **266** | 39.182 | 79.775 | 0.01714 | 10.687 | 10.704 | 234.76 | 934.5 | 1169.3 | 0.3902 | 1.2878 | 1.6780 | **266** |
| **268** | 40.502 | 82.463 | 0.01715 | 10.359 | 10.376 | 236.80 | 933.2 | 1170.0 | 0.3930 | 1.2824 | 1.6753 | **268** |
| **270°** | 41.858 | 85.225 | 0.01717 | 10.044 | 10.061 | 238.84 | 931.8 | 1170.6 | 0.3958 | 1.2769 | 1.6727 | **270°** |
| **272** | 43.252 | 88.062 | 0.01719 | 9.739 | 9.756 | 240.88 | 930.3 | 1171.2 | 0.3986 | 1.2715 | 1.6701 | **272** |
| **274** | 44.682 | 90.974 | 0.01721 | 9.446 | 9.463 | 242.92 | 929.0 | 1171.9 | 0.4014 | 1.2661 | 1.6675 | **274** |
| **276** | 46.150 | 93.963 | 0.01722 | 9.163 | 9.181 | 244.96 | 927.5 | 1172.5 | 0.4041 | 1.2608 | 1.6649 | **276** |
| **278** | 47.657 | 97.031 | 0.01724 | 8.891 | 8.908 | 247.01 | 926.1 | 1173.1 | 0.4069 | 1.2554 | 1.6623 | **278** |
| **280°** | 49.203 | 100.18 | 0.01726 | 8.628 | 8.645 | 249.06 | 924.7 | 1173.8 | 0.4096 | 1.2501 | 1.6597 | **280°** |
| **282** | 50.790 | 103.41 | 0.01728 | 8.374 | 8.391 | 251.10 | 923.3 | 1174.4 | 0.4124 | 1.2448 | 1.6572 | **282** |
| **284** | 52.418 | 106.72 | 0.01730 | 8.129 | 8.146 | 253.15 | 921.8 | 1175.0 | 0.4152 | 1.2395 | 1.6547 | **284** |
| **286** | 54.088 | 110.12 | 0.01732 | 7.892 | 7.910 | 255.20 | 920.4 | 1175.6 | 0.4179 | 1.2343 | 1.6522 | **286** |
| **288** | 55.800 | 113.61 | 0.01733 | 7.664 | 7.682 | 257.26 | 918.9 | 1176.2 | 0.4207 | 1.2290 | 1.6497 | **288** |
| **290°** | 57.556 | 117.19 | 0.01735 | 7.444 | 7.461 | 259.31 | 917.5 | 1176.8 | 0.4234 | 1.2238 | 1.6472 | **290°** |
| **292** | 59.356 | 120.85 | 0.01737 | 7.231 | 7.248 | 261.36 | 916.0 | 1177.4 | 0.4261 | 1.2186 | 1.6447 | **292** |
| **294** | 61.201 | 124.61 | 0.01739 | 7.025 | 7.043 | 263.42 | 914.6 | 1178.0 | 0.4288 | 1.2134 | 1.6422 | **294** |
| **296** | 63.091 | 128.46 | 0.01741 | 6.827 | 6.844 | 265.48 | 913.1 | 1178.6 | 0.4315 | 1.2083 | 1.6398 | **296** |
| **298** | 65.028 | 132.40 | 0.01743 | 6.635 | 6.652 | 267.53 | 911.6 | 1179.1 | 0.4343 | 1.2031 | 1.6374 | **298** |
| **300°** | 67.013 | 136.44 | 0.01745 | 6.449 | 6.466 | 269.59 | 910.1 | 1179.7 | 0.4369 | 1.1980 | 1.6350 | **300°** |
| **302** | 69.046 | 140.58 | 0.01747 | 6.269 | 6.287 | 271.66 | 908.6 | 1180.3 | 0.4397 | 1.1929 | 1.6326 | **302** |
| **304** | 71.127 | 144.82 | 0.01749 | 6.096 | 6.114 | 273.72 | 907.2 | 1180.9 | 0.4424 | 1.1878 | 1.6302 | **304** |
| **306** | 73.259 | 149.16 | 0.01751 | 5.928 | 5.946 | 275.78 | 905.6 | 1181.4 | 0.4450 | 1.1828 | 1.6278 | **306** |
| **308** | 75.442 | 153.60 | 0.01753 | 5.766 | 5.783 | 277.85 | 904.1 | 1182.0 | 0.4477 | 1.1777 | 1.6254 | **308** |

## Table 1. Saturation: Temperatures

| Temp. Fahr. t | Abs. Pressure Lb./Sq. In. p | Specific Volume Sat. Liquid $v_f$ | Specific Volume Evap. $v_{fg}$ | Specific Volume Sat. Vapor $v_g$ | Enthalpy Sat. Liquid $h_f$ | Enthalpy Evap. $h_{fg}$ | Enthalpy Sat. Vapor $h_g$ | Entropy Sat. Liquid $s_f$ | Entropy Evap. $s_{fg}$ | Entropy Sat. Vapor $s_g$ | Temp. Fahr. t |
|---|---|---|---|---|---|---|---|---|---|---|---|
| 310° | 77.68 | 0.01755 | 5.609 | 5.626 | 279.92 | 902.6 | 1182.5 | 0.4504 | 1.1727 | 1.6231 | 310° |
| 312 | 79.96 | 0.01757 | 5.457 | 5.474 | 281.99 | 901.0 | 1183.1 | 0.4530 | 1.1677 | 1.6207 | 312 |
| 314 | 82.30 | 0.01759 | 5.310 | 5.327 | 284.06 | 899.5 | 1183.6 | 0.4557 | 1.1627 | 1.6184 | 314 |
| 316 | 84.70 | 0.01761 | 5.167 | 5.185 | 286.13 | 898.0 | 1184.1 | 0.4584 | 1.1577 | 1.6161 | 316 |
| 318 | 87.15 | 0.01763 | 5.030 | 5.047 | 288.20 | 896.5 | 1184.7 | 0.4611 | 1.1527 | 1.6138 | 318 |
| 320° | 89.66 | 0.01765 | 4.896 | 4.914 | 290.28 | 894.9 | 1185.2 | 0.4637 | 1.1478 | 1.6115 | 320° |
| 322 | 92.22 | 0.01768 | 4.767 | 4.785 | 292.36 | 893.3 | 1185.7 | 0.4664 | 1.1428 | 1.6092 | 322 |
| 324 | 94.84 | 0.01770 | 4.642 | 4.660 | 294.43 | 891.8 | 1186.2 | 0.4690 | 1.1379 | 1.6069 | 324 |
| 326 | 97.52 | 0.01772 | 4.521 | 4.538 | 296.52 | 890.2 | 1186.7 | 0.4717 | 1.1330 | 1.6047 | 326 |
| 328 | 100.26 | 0.01774 | 4.403 | 4.421 | 298.60 | 888.6 | 1187.2 | 0.4743 | 1.1281 | 1.6024 | 328 |
| 330° | 103.06 | 0.01776 | 4.289 | 4.307 | 300.68 | 887.0 | 1187.7 | 0.4769 | 1.1233 | 1.6002 | 330° |
| 332 | 105.92 | 0.01778 | 4.179 | 4.197 | 302.77 | 885.4 | 1188.2 | 0.4795 | 1.1184 | 1.5979 | 332 |
| 334 | 108.85 | 0.01781 | 4.072 | 4.090 | 304.86 | 883.8 | 1188.7 | 0.4821 | 1.1136 | 1.5957 | 334 |
| 336 | 111.84 | 0.01783 | 3.968 | 3.986 | 306.95 | 882.2 | 1189.2 | 0.4847 | 1.1088 | 1.5935 | 336 |
| 338 | 114.89 | 0.01785 | 3.868 | 3.886 | 309.04 | 880.6 | 1189.6 | 0.4873 | 1.1040 | 1.5913 | 338 |
| 340° | 118.01 | 0.01787 | 3.770 | 3.788 | 311.13 | 879.0 | 1190.1 | 0.4900 | 1.0992 | 1.5891 | 340° |
| 342 | 121.20 | 0.01790 | 3.675 | 3.693 | 313.23 | 877.4 | 1190.6 | 0.4926 | 1.0944 | 1.5870 | 342 |
| 344 | 124.45 | 0.01792 | 3.584 | 3.602 | 315.33 | 875.7 | 1191.0 | 0.4952 | 1.0896 | 1.5848 | 344 |
| 346 | 127.77 | 0.01794 | 3.495 | 3.513 | 317.43 | 874.1 | 1191.5 | 0.4978 | 1.0848 | 1.5826 | 346 |
| 348 | 131.17 | 0.01797 | 3.408 | 3.426 | 319.53 | 872.4 | 1191.9 | 0.5004 | 1.0801 | 1.5805 | 348 |
| 350° | 134.63 | 0.01799 | 3.324 | 3.342 | 321.63 | 870.7 | 1192.3 | 0.5029 | 1.0754 | 1.5783 | 350° |
| 352 | 138.16 | 0.01801 | 3.243 | 3.261 | 323.74 | 869.1 | 1192.8 | 0.5055 | 1.0707 | 1.5762 | 352 |
| 354 | 141.77 | 0.01804 | 3.164 | 3.182 | 325.85 | 867.3 | 1193.2 | 0.5081 | 1.0660 | 1.5741 | 354 |
| 356 | 145.45 | 0.01806 | 3.087 | 3.105 | 327.96 | 865.6 | 1193.6 | 0.5106 | 1.0613 | 1.5719 | 356 |
| 358 | 149.21 | 0.01808 | 3.012 | 3.030 | 330.07 | 863.9 | 1194.0 | 0.5132 | 1.0566 | 1.5698 | 358 |
| 360° | 153.04 | 0.01811 | 2.939 | 2.957 | 332.18 | 862.2 | 1194.4 | 0.5158 | 1.0519 | 1.5677 | 360° |
| 362 | 156.95 | 0.01813 | 2.869 | 2.887 | 334.30 | 860.5 | 1194.8 | 0.5183 | 1.0473 | 1.5656 | 362 |
| 364 | 160.93 | 0.01816 | 2.801 | 2.819 | 336.42 | 858.8 | 1195.2 | 0.5209 | 1.0426 | 1.5635 | 364 |
| 366 | 165.00 | 0.01818 | 2.734 | 2.752 | 338.54 | 857.1 | 1195.6 | 0.5235 | 1.0380 | 1.5615 | 366 |
| 368 | 169.15 | 0.01821 | 2.669 | 2.687 | 340.66 | 855.3 | 1196.0 | 0.5260 | 1.0334 | 1.5594 | 368 |
| 370° | 173.37 | 0.01823 | 2.606 | 2.625 | 342.79 | 853.5 | 1196.3 | 0.5286 | 1.0287 | 1.5573 | 370° |
| 372 | 177.68 | 0.01826 | 2.545 | 2.564 | 344.91 | 851.8 | 1196.7 | 0.5311 | 1.0241 | 1.5553 | 372 |
| 374 | 182.07 | 0.01829 | 2.486 | 2.504 | 347.04 | 850.0 | 1197.0 | 0.5336 | 1.0196 | 1.5532 | 374 |
| 376 | 186.55 | 0.01831 | 2.428 | 2.446 | 349.18 | 848.2 | 1197.4 | 0.5362 | 1.0150 | 1.5512 | 376 |
| 378 | 191.12 | 0.01834 | 2.372 | 2.390 | 351.31 | 846.4 | 1197.7 | 0.5388 | 1.0104 | 1.5492 | 378 |
| 380° | 195.77 | 0.01836 | 2.317 | 2.335 | 353.45 | 844.6 | 1198.1 | 0.5413 | 1.0059 | 1.5471 | 380° |
| 382 | 200.50 | 0.01839 | 2.264 | 2.282 | 355.59 | 842.8 | 1198.4 | 0.5438 | 1.0013 | 1.5451 | 382 |
| 384 | 205.33 | 0.01842 | 2.212 | 2.231 | 357.73 | 841.0 | 1198.7 | 0.5463 | 0.9968 | 1.5431 | 384 |
| 386 | 210.25 | 0.01844 | 2.162 | 2.180 | 359.88 | 839.1 | 1199.0 | 0.5488 | 0.9923 | 1.5411 | 386 |
| 388 | 215.26 | 0.01847 | 2.113 | 2.131 | 362.02 | 837.3 | 1199.3 | 0.5514 | 0.9877 | 1.5391 | 388 |
| 390° | 220.37 | 0.01850 | 2.0651 | 2.0836 | 364.17 | 835.4 | 1199.6 | 0.5539 | 0.9832 | 1.5371 | 390° |
| 392 | 225.56 | 0.01853 | 2.0187 | 2.0372 | 366.33 | 833.6 | 1199.9 | 0.5564 | 0.9787 | 1.5351 | 392 |
| 394 | 230.85 | 0.01855 | 1.9734 | 1.9920 | 368.48 | 831.7 | 1200.2 | 0.5589 | 0.9742 | 1.5331 | 394 |
| 396 | 236.24 | 0.01858 | 1.9293 | 1.9479 | 370.64 | 829.9 | 1200.5 | 0.5614 | 0.9698 | 1.5311 | 396 |
| 398 | 241.73 | 0.01861 | 1.8864 | 1.9050 | 372.80 | 827.9 | 1200.7 | 0.5639 | 0.9653 | 1.5292 | 398 |
| 400° | 247.31 | 0.01864 | 1.8447 | 1.8633 | 374.97 | 826.0 | 1201.0 | 0.5664 | 0.9608 | 1.5272 | 400° |
| 405 | 261.71 | 0.01871 | 1.7448 | 1.7635 | 380.39 | 821.2 | 1201.6 | 0.5726 | 0.9497 | 1.5223 | 405 |
| 410 | 276.75 | 0.01878 | 1.6512 | 1.6700 | 385.83 | 816.3 | 1202.1 | 0.5788 | 0.9386 | 1.5174 | 410 |
| 415 | 292.45 | 0.01886 | 1.5635 | 1.5823 | 391.29 | 811.3 | 1202.6 | 0.5850 | 0.9276 | 1.5126 | 415 |
| 420 | 308.83 | 0.01894 | 1.4811 | 1.5000 | 396.77 | 806.3 | 1203.1 | 0.5912 | 0.9166 | 1.5078 | 420 |
| 425° | 325.92 | 0.01902 | 1.4036 | 1.4226 | 402.27 | 801.2 | 1203.5 | 0.5974 | 0.9056 | 1.5030 | 425° |
| 430 | 343.72 | 0.01910 | 1.3308 | 1.3499 | 407.79 | 796.0 | 1203.8 | 0.6035 | 0.8947 | 1.4982 | 430 |
| 435 | 362.27 | 0.01918 | 1.2623 | 1.2815 | 413.34 | 790.8 | 1204.1 | 0.6097 | 0.8838 | 1.4935 | 435 |
| 440 | 381.59 | 0.01926 | 1.1979 | 1.2171 | 418.90 | 785.4 | 1204.3 | 0.6158 | 0.8730 | 1.4887 | 440 |
| 445 | 401.68 | 0.01935 | 1.1371 | 1.1565 | 424.49 | 780.0 | 1204.5 | 0.6219 | 0.8622 | 1.4840 | 445 |

## Table 1. Saturation: Temperatures

| Temp. Fahr. t | Abs. Pressure Lb./Sq. In. p | Specific Volume Sat. Liquid $v_f$ | Evap. $v_{fg}$ | Sat. Vapor $v_g$ | Enthalpy Sat. Liquid $h_f$ | Evap. $h_{fg}$ | Sat. Vapor $h_g$ | Entropy Sat. Liquid $s_f$ | Evap. $s_{fg}$ | Sat. Vapor $s_g$ | Temp. Fahr. t |
|---|---|---|---|---|---|---|---|---|---|---|---|
| 450° | 422.6 | 0.0194 | 1.0799 | 1.0993 | 430.1 | 774.5 | 1204.6 | 0.6280 | 0.8513 | 1.4793 | 450° |
| 455 | 444.3 | 0.0195 | 1.0258 | 1.0453 | 435.7 | 768.9 | 1204.6 | 0.6341 | 0.8406 | 1.4746 | 455 |
| 460 | 466.9 | 0.0196 | 0.9748 | 0.9944 | 441.4 | 763.2 | 1204.6 | 0.6402 | 0.8298 | 1.4700 | 460 |
| 465 | 490.3 | 0.0197 | 0.9266 | 0.9463 | 447.1 | 757.4 | 1204.5 | 0.6463 | 0.8190 | 1.4653 | 465 |
| 470 | 514.7 | 0.0198 | 0.8811 | 0.9009 | 452.8 | 751.5 | 1204.3 | 0.6523 | 0.8083 | 1.4606 | 470 |
| 475° | 539.9 | 0.0199 | 0.8380 | 0.8579 | 458.6 | 745.4 | 1204.0 | 0.6584 | 0.7976 | 1.4560 | 475° |
| 480 | 566.1 | 0.0200 | 0.7972 | 0.8172 | 464.4 | 739.4 | 1203.7 | 0.6645 | 0.7868 | 1.4513 | 480 |
| 485 | 593.3 | 0.0201 | 0.7586 | 0.7787 | 470.2 | 733.1 | 1203.3 | 0.6705 | 0.7761 | 1.4466 | 485 |
| 490 | 621.4 | 0.0202 | 0.7221 | 0.7423 | 476.0 | 726.8 | 1202.8 | 0.6766 | 0.7653 | 1.4419 | 490 |
| 495 | 650.6 | 0.0203 | 0.6874 | 0.7077 | 481.9 | 720.4 | 1202.3 | 0.6826 | 0.7546 | 1.4372 | 495 |
| 500° | 680.8 | 0.0204 | 0.6545 | 0.6749 | 487.8 | 713.9 | 1201.7 | 0.6887 | 0.7438 | 1.4325 | 500° |
| 505 | 712.0 | 0.0205 | 0.6233 | 0.6438 | 493.8 | 707.1 | 1200.9 | 0.6948 | 0.7331 | 1.4278 | 505 |
| 510 | 744.3 | 0.0207 | 0.5935 | 0.6142 | 499.8 | 700.3 | 1200.1 | 0.7008 | 0.7223 | 1.4231 | 510 |
| 515 | 777.8 | 0.0208 | 0.5653 | 0.5861 | 505.8 | 693.4 | 1199.2 | 0.7069 | 0.7115 | 1.4184 | 515 |
| 520 | 812.4 | 0.0209 | 0.5385 | 0.5594 | 511.9 | 686.4 | 1198.2 | 0.7130 | 0.7006 | 1.4136 | 520 |
| 525° | 848.1 | 0.0210 | 0.5130 | 0.5340 | 518.0 | 679.1 | 1197.1 | 0.7191 | 0.6897 | 1.4088 | 525° |
| 530 | 885.0 | 0.0212 | 0.4886 | 0.5098 | 524.1 | 671.8 | 1195.9 | 0.7252 | 0.6788 | 1.4040 | 530 |
| 535 | 923.2 | 0.0213 | 0.4655 | 0.4868 | 530.3 | 664.3 | 1194.6 | 0.7313 | 0.6679 | 1.3991 | 535 |
| 540 | 962.5 | 0.0215 | 0.4434 | 0.4649 | 536.6 | 656.6 | 1193.2 | 0.7374 | 0.6568 | 1.3942 | 540 |
| 545 | 1003.2 | 0.0216 | 0.4224 | 0.4440 | 542.9 | 648.8 | 1191.7 | 0.7435 | 0.6458 | 1.3893 | 545 |
| 550° | 1045.2 | 0.0218 | 0.4022 | 0.4240 | 549.3 | 640.8 | 1190.0 | 0.7497 | 0.6346 | 1.3843 | 550° |
| 555 | 1088.5 | 0.0219 | 0.3831 | 0.4050 | 555.7 | 632.6 | 1188.3 | 0.7559 | 0.6234 | 1.3793 | 555 |
| 560 | 1133.1 | 0.0221 | 0.3647 | 0.3868 | 562.2 | 624.2 | 1186.4 | 0.7621 | 0.6121 | 1.3742 | 560 |
| 565 | 1179.1 | 0.0222 | 0.3472 | 0.3694 | 568.8 | 615.5 | 1184.3 | 0.7683 | 0.6007 | 1.3690 | 565 |
| 570 | 1226.5 | 0.0224 | 0.3304 | 0.3528 | 575.4 | 606.7 | 1182.1 | 0.7746 | 0.5893 | 1.3638 | 570 |
| 575° | 1275.4 | 0.0226 | 0.3143 | 0.3369 | 582.1 | 597.7 | 1179.8 | 0.7809 | 0.5777 | 1.3585 | 575° |
| 580 | 1325.8 | 0.0228 | 0.2989 | 0.3217 | 588.9 | 588.4 | 1177.3 | 0.7872 | 0.5659 | 1.3532 | 580 |
| 585 | 1377.7 | 0.0230 | 0.2841 | 0.3071 | 595.8 | 578.8 | 1174.6 | 0.7936 | 0.5541 | 1.3477 | 585 |
| 590 | 1431.2 | 0.0232 | 0.2700 | 0.2931 | 602.8 | 569.0 | 1171.8 | 0.8001 | 0.5421 | 1.3422 | 590 |
| 595 | 1486.2 | 0.0234 | 0.2563 | 0.2797 | 609.8 | 558.9 | 1168.7 | 0.8066 | 0.5299 | 1.3365 | 595 |
| 600° | 1542.9 | 0.0236 | 0.2432 | 0.2668 | 617.0 | 548.5 | 1165.5 | 0.8131 | 0.5176 | 1.3307 | 600° |
| 605 | 1601.2 | 0.0239 | 0.2306 | 0.2545 | 624.3 | 537.7 | 1162.0 | 0.8197 | 0.5051 | 1.3248 | 605 |
| 610 | 1661.2 | 0.0241 | 0.2185 | 0.2426 | 631.6 | 526.7 | 1158.4 | 0.8264 | 0.4924 | 1.3188 | 610 |
| 615 | 1723.0 | 0.0244 | 0.2068 | 0.2312 | 639.1 | 515.3 | 1154.4 | 0.8331 | 0.4795 | 1.3126 | 615 |
| 620 | 1786.6 | 0.0247 | 0.1955 | 0.2201 | 646.7 | 503.6 | 1150.3 | 0.8398 | 0.4664 | 1.3062 | 620 |
| 625° | 1852.0 | 0.0250 | 0.1845 | 0.2095 | 654.4 | 491.4 | 1145.8 | 0.8467 | 0.4530 | 1.2997 | 625° |
| 630 | 1919.3 | 0.0253 | 0.1740 | 0.1992 | 662.3 | 478.8 | 1141.1 | 0.8536 | 0.4394 | 1.2930 | 630 |
| 635 | 1988.5 | 0.0256 | 0.1637 | 0.1893 | 670.4 | 465.6 | 1136.0 | 0.8607 | 0.4254 | 1.2861 | 635 |
| 640 | 2059.7 | 0.0260 | 0.1538 | 0.1798 | 678.6 | 452.0 | 1130.5 | 0.8679 | 0.4110 | 1.2789 | 640 |
| 645 | 2132.9 | 0.0264 | 0.1441 | 0.1705 | 687.0 | 437.7 | 1124.7 | 0.8752 | 0.3962 | 1.2715 | 645 |
| 650° | 2208.2 | 0.0268 | 0.1348 | 0.1616 | 695.7 | 422.8 | 1118.5 | 0.8828 | 0.3809 | 1.2637 | 650° |
| 655 | 2285.7 | 0.0273 | 0.1256 | 0.1528 | 704.8 | 406.9 | 1111.7 | 0.8906 | 0.3651 | 1.2557 | 655 |
| 660 | 2365.4 | 0.0278 | 0.1165 | 0.1442 | 714.2 | 390.2 | 1104.4 | 0.8987 | 0.3485 | 1.2472 | 660 |
| 665 | 2447.4 | 0.0283 | 0.1076 | 0.1359 | 724.1 | 372.4 | 1096.4 | 0.9071 | 0.3311 | 1.2382 | 665 |
| 670 | 2531.8 | 0.0290 | 0.0987 | 0.1277 | 734.4 | 353.2 | 1087.7 | 0.9159 | 0.3127 | 1.2285 | 670 |
| 675° | 2618.7 | 0.0297 | 0.0899 | 0.1196 | 745.4 | 332.6 | 1078.0 | 0.9251 | 0.2931 | 1.2183 | 675° |
| 680 | 2708.1 | 0.0305 | 0.0810 | 0.1115 | 757.3 | 309.9 | 1067.2 | 0.9351 | 0.2719 | 1.2071 | 680 |
| 685 | 2800.2 | 0.0315 | 0.0719 | 0.1034 | 770.1 | 284.7 | 1054.8 | 0.9459 | 0.2487 | 1.1946 | 685 |
| 690 | 2895.1 | 0.0328 | 0.0625 | 0.0953 | 784.4 | 256.0 | 1040.4 | 0.9578 | 0.2227 | 1.1805 | 690 |
| 695 | 2992.9 | 0.0344 | 0.0520 | 0.0864 | 801.2 | 220.7 | 1021.9 | 0.9719 | 0.1911 | 1.1630 | 695 |
| 700° | 3093.7 | 0.0369 | 0.0392 | 0.0761 | 823.3 | 172.1 | 995.4 | 0.9905 | 0.1484 | 1.1389 | 700° |
| 702 | 3134.9 | 0.0385 | 0.0325 | 0.0710 | 835.4 | 145.2 | 980.6 | 1.0006 | 0.1249 | 1.1256 | 702 |
| 704 | 3176.7 | 0.0410 | 0.0234 | 0.0645 | 852.7 | 106.0 | 958.7 | 1.0152 | 0.0911 | 1.1063 | 704 |
| 705 | 3197.7 | 0.0438 | 0.0152 | 0.0589 | 869.2 | 69.1 | 938.4 | 1.0293 | 0.0593 | 1.0886 | 705 |
| 705.4 | 3206.2 | 0.0503 | 0 | 0.0503 | 902.7 | 0 | 902.7 | 1.0580 | 0 | 1.0580 | 705.4 |

# Table 2. Saturation: Pressures

| Abs. Press. In. Hg. p | Temp. Fahr. t | Specific Volume Sat. Liquid $v_f$ | Specific Volume Sat. Vapor $v_g$ | Enthalpy Sat. Liquid $h_f$ | Enthalpy Evap. $h_{fg}$ | Enthalpy Sat. Vapor $h_g$ | Entropy Sat. Liquid $s_f$ | Entropy Evap. $s_{fg}$ | Entropy Sat. Vapor $s_g$ | Internal Energy Sat. Liquid $u_f$ | Internal Energy Evap. $u_{fg}$ | Internal Energy Sat. Vapor $u_g$ | Abs. Press. In. Hg. p |
|---|---|---|---|---|---|---|---|---|---|---|---|---|---|
| *0.25* | 40.23 | 0.01602 | 2423.7 | 8.28 | 1071.1 | 1079.4 | 0.0166 | 2.1423 | 2.1589 | 8.28 | 1016.0 | 1024.3 | *0.25* |
| *0.50* | 58.80 | 0.01604 | 1256.4 | 26.86 | 1060.6 | 1087.5 | 0.0532 | 2.0453 | 2.0985 | 26.86 | 1003.5 | 1030.4 | *0.50* |
| *0.75* | 70.43 | 0.01606 | 856.1 | 38.47 | 1054.0 | 1092.5 | 0.0754 | 1.9881 | 2.0635 | 38.47 | 995.7 | 1034.2 | *0.75* |
| *1.00* | 79.03 | 0.01608 | 652.3 | 47.05 | 1049.2 | 1096.3 | 0.0914 | 1.9473 | 2.0387 | 47.05 | 990.0 | 1037.0 | *1.00* |
| *1.5* | 91.72 | 0.01611 | 444.9 | 59.71 | 1042.0 | 1101.7 | 0.1147 | 1.8894 | 2.0041 | 59.71 | 981.4 | 1041.1 | *1.5* |
| *2.0* | 101.14 | 0.01614 | 339.2 | 69.10 | 1036.6 | 1105.7 | 0.1316 | 1.8481 | 1.9797 | 69.10 | 974.9 | 1044.0 | *2.0* |
| *2.5* | 108.71 | 0.01616 | 274.9 | 76.65 | 1032.3 | 1108.9 | 0.1449 | 1.8160 | 1.9609 | 76.65 | 969.8 | 1046.4 | *2.5* |
| *3.0* | 115.06 | 0.01618 | 231.6 | 82.99 | 1028.6 | 1111.6 | 0.1560 | 1.7896 | 1.9456 | 82.99 | 965.5 | 1048.5 | *3.0* |
| *4.0* | 125.43 | 0.01622 | 176.7 | 93.34 | 1022.7 | 1116.0 | 0.1738 | 1.7476 | 1.9214 | 93.33 | 958.5 | 1051.8 | *4.0* |
| *5* | 133.76 | 0.01626 | 143.25 | 101.66 | 1017.7 | 1119.4 | 0.1879 | 1.7150 | 1.9028 | 101.65 | 952.6 | 1054.3 | *5* |
| *6* | 140.78 | 0.01630 | 120.72 | 108.67 | 1013.6 | 1122.3 | 0.1996 | 1.6881 | 1.8877 | 108.66 | 947.8 | 1056.5 | *6* |
| *7* | 146.86 | 0.01633 | 104.46 | 114.75 | 1010.0 | 1124.8 | 0.2097 | 1.6653 | 1.8750 | 114.74 | 943.7 | 1058.4 | *7* |
| *8* | 152.24 | 0.01635 | 92.16 | 120.13 | 1006.9 | 1127.0 | 0.2186 | 1.6454 | 1.8640 | 120.12 | 939.9 | 1060.0 | *8* |
| *9* | 157.09 | 0.01638 | 82.52 | 124.97 | 1004.0 | 1129.0 | 0.2264 | 1.6279 | 1.8543 | 124.96 | 936.5 | 1061.5 | *9* |
| *10* | 161.49 | 0.01640 | 74.76 | 129.38 | 1001.4 | 1130.8 | 0.2335 | 1.6121 | 1.8456 | 129.37 | 933.4 | 1062.8 | *10* |
| *11* | 165.54 | 0.01642 | 68.38 | 133.43 | 999.0 | 1132.4 | 0.2400 | 1.5978 | 1.8378 | 133.41 | 930.6 | 1064.0 | *11* |
| *12* | 169.28 | 0.01644 | 63.03 | 137.18 | 996.7 | 1133.9 | 0.2460 | 1.5847 | 1.8307 | 137.16 | 928.0 | 1065.2 | *12* |
| *13* | 172.78 | 0.01646 | 58.47 | 140.68 | 994.6 | 1135.3 | 0.2516 | 1.5725 | 1.8241 | 140.66 | 925.5 | 1066.2 | *13* |
| *14* | 176.05 | 0.01648 | 54.55 | 143.96 | 992.6 | 1136.6 | 0.2568 | 1.5613 | 1.8181 | 143.94 | 923.3 | 1067.2 | *14* |
| *15* | 179.14 | 0.01650 | 51.14 | 147.06 | 990.7 | 1137.8 | 0.2616 | 1.5508 | 1.8125 | 147.04 | 921.1 | 1068.1 | *15* |
| *16* | 182.05 | 0.01652 | 48.14 | 149.98 | 988.9 | 1138.9 | 0.2662 | 1.5410 | 1.8072 | 149.96 | 918.9 | 1068.9 | *16* |
| *17* | 184.82 | 0.01654 | 45.48 | 152.75 | 987.3 | 1140.0 | 0.2705 | 1.5318 | 1.8023 | 152.72 | 917.0 | 1069.7 | *17* |
| *18* | 187.45 | 0.01655 | 43.11 | 155.39 | 985.7 | 1141.1 | 0.2746 | 1.5231 | 1.7977 | 155.36 | 915.1 | 1070.5 | *18* |
| *19* | 189.96 | 0.01657 | 40.99 | 157.91 | 984.2 | 1142.1 | 0.2784 | 1.5148 | 1.7933 | 157.88 | 913.4 | 1071.3 | *19* |
| *20* | 192.37 | 0.01658 | 39.07 | 160.33 | 982.7 | 1143.0 | 0.2822 | 1.5069 | 1.7891 | 160.30 | 911.7 | 1072.0 | *20* |
| *21* | 194.68 | 0.01660 | 37.32 | 162.65 | 981.2 | 1143.9 | 0.2857 | 1.4994 | 1.7851 | 162.62 | 910.1 | 1072.7 | *21* |
| *22* | 196.90 | 0.01661 | 35.73 | 164.87 | 979.8 | 1144.7 | 0.2891 | 1.4923 | 1.7814 | 164.84 | 908.5 | 1073.3 | *22* |
| *23* | 199.03 | 0.01663 | 34.28 | 167.02 | 978.5 | 1145.5 | 0.2923 | 1.4855 | 1.7779 | 166.99 | 906.9 | 1073.9 | *23* |
| *24* | 201.09 | 0.01664 | 32.94 | 169.09 | 977.2 | 1146.3 | 0.2955 | 1.4789 | 1.7744 | 169.05 | 905.4 | 1074.5 | *24* |
| *25* | 203.08 | 0.01666 | 31.70 | 171.09 | 975.9 | 1147.0 | 0.2985 | 1.4726 | 1.7711 | 171.05 | 904.0 | 1075.0 | *25* |
| *26* | 205.00 | 0.01667 | 30.56 | 173.02 | 974.8 | 1147.8 | 0.3014 | 1.4665 | 1.7679 | 172.98 | 902.6 | 1075.6 | *26* |
| *27* | 206.87 | 0.01668 | 29.50 | 174.90 | 973.6 | 1148.5 | 0.3042 | 1.4607 | 1.7649 | 174.86 | 901.2 | 1076.1 | *27* |
| *28* | 208.67 | 0.01669 | 28.52 | 176.72 | 972.5 | 1149.2 | 0.3069 | 1.4550 | 1.7619 | 176.68 | 899.9 | 1076.6 | *28* |
| *29* | 210.43 | 0.01671 | 27.60 | 178.48 | 971.4 | 1149.9 | 0.3096 | 1.4495 | 1.7591 | 178.44 | 898.7 | 1077.1 | *29* |
| *30* | 212.13 | 0.01672 | 26.74 | 180.19 | 970.3 | 1150.5 | 0.3122 | 1.4442 | 1.7564 | 180.14 | 897.5 | 1077.6 | *30* |

| Lb./Sq. In. | t | $v_f$ | $v_g$ | $h_f$ | $h_{fg}$ | $h_g$ | $s_f$ | $s_{fg}$ | $s_g$ | $u_f$ | $u_{fg}$ | $u_g$ | Lb./Sq. In. |
|---|---|---|---|---|---|---|---|---|---|---|---|---|---|
| **0.20** | 53.14 | 0.01603 | 1526.0 | 21.21 | 1063.8 | 1085.0 | 0.0422 | 2.0741 | 2.1163 | 21.21 | 1007.3 | 1028.5 | **0.20** |
| **0.25** | 59.30 | 0.01604 | 1235.3 | 27.36 | 1060.3 | 1087.7 | 0.0542 | 2.0428 | 2.0970 | 27.36 | 1003.2 | 1030.6 | **0.25** |
| **0.30** | 64.47 | 0.01605 | 1039.5 | 32.52 | 1057.4 | 1090.0 | 0.0641 | 2.0171 | 2.0812 | 32.52 | 999.8 | 1032.3 | **0.30** |
| **0.35** | 68.93 | 0.01605 | 898.5 | 36.97 | 1054.9 | 1091.9 | 0.0725 | 1.9953 | 2.0678 | 36.97 | 996.7 | 1033.7 | **0.35** |
| **0.40** | 72.86 | 0.01606 | 791.9 | 40.89 | 1052.7 | 1093.6 | 0.0799 | 1.9764 | 2.0563 | 40.89 | 994.1 | 1035.0 | **0.40** |
| **0.45** | 76.38 | 0.01607 | 708.5 | 44.41 | 1050.7 | 1095.1 | 0.0865 | 1.9597 | 2.0462 | 44.41 | 991.7 | 1036.1 | **0.45** |
| **0.50** | 79.58 | 0.01608 | 641.4 | 47.60 | 1048.8 | 1096.4 | 0.0924 | 1.9448 | 2.0372 | 47.60 | 989.5 | 1037.1 | **0.50** |
| **0.60** | 85.21 | 0.01609 | 540.0 | 53.21 | 1045.7 | 1098.9 | 0.1028 | 1.9188 | 2.0216 | 53.21 | 985.7 | 1038.9 | **0.60** |
| **0.70** | 90.08 | 0.01610 | 466.9 | 58.07 | 1042.9 | 1101.0 | 0.1117 | 1.8968 | 2.0085 | 58.07 | 982.4 | 1040.5 | **0.70** |
| **0.80** | 94.38 | 0.01612 | 411.7 | 62.36 | 1040.4 | 1102.8 | 0.1194 | 1.8777 | 1.9971 | 62.36 | 979.5 | 1041.9 | **0.80** |
| **0.90** | 98.24 | 0.01613 | 368.4 | 66.21 | 1038.3 | 1104.5 | 0.1263 | 1.8608 | 1.9871 | 66.21 | 976.9 | 1043.1 | **0.90** |
| **1.0** | 101.74 | 0.01614 | 333.6 | 69.70 | 1036.3 | 1106.0 | 0.1326 | 1.8456 | 1.9782 | 69.70 | 974.6 | 1044.3 | **1.0** |
| **1.2** | 107.92 | 0.01616 | 280.9 | 75.87 | 1032.7 | 1108.6 | 0.1435 | 1.8193 | 1.9628 | 75.87 | 970.3 | 1046.2 | **1.2** |
| **1.4** | 113.26 | 0.01618 | 243.0 | 81.20 | 1029.6 | 1110.8 | 0.1528 | 1.7971 | 1.9498 | 81.20 | 966.7 | 1047.9 | **1.4** |
| **1.6** | 117.99 | 0.01620 | 214.3 | 85.91 | 1026.9 | 1112.8 | 0.1610 | 1.7776 | 1.9386 | 85.91 | 963.5 | 1049.4 | **1.6** |
| **1.8** | 122.23 | 0.01621 | 191.8 | 90.14 | 1024.5 | 1114.6 | 0.1683 | 1.7605 | 1.9288 | 90.13 | 960.6 | 1050.7 | **1.8** |
| **2.0** | 126.08 | 0.01623 | 173.73 | 93.99 | 1022.2 | 1116.2 | 0.1749 | 1.7451 | 1.9200 | 93.98 | 957.9 | 1051.9 | **2.0** |
| **2.2** | 129.62 | 0.01624 | 158.85 | 97.52 | 1020.2 | 1117.7 | 0.1809 | 1.7311 | 1.9120 | 97.51 | 955.5 | 1053.0 | **2.2** |
| **2.4** | 132.89 | 0.01626 | 146.38 | 100.79 | 1018.3 | 1119.1 | 0.1864 | 1.7183 | 1.9047 | 100.78 | 953.3 | 1054.1 | **2.4** |
| **2.6** | 135.94 | 0.01627 | 135.78 | 103.83 | 1016.5 | 1120.3 | 0.1916 | 1.7065 | 1.8981 | 103.82 | 951.2 | 1055.0 | **2.6** |
| **2.8** | 138.79 | 0.01629 | 126.65 | 106.68 | 1014.8 | 1121.5 | 0.1963 | 1.6957 | 1.8920 | 106.67 | 949.2 | 1055.9 | **2.8** |

# Table 2. Saturation: Pressures

| Abs. Press. Lb./Sq. In. | Temp. Fahr. | Specific Volume | | Enthalpy | | | Entropy | | | Internal Energy | | | Abs. Press. Lb./Sq. In. |
|---|---|---|---|---|---|---|---|---|---|---|---|---|---|
| | | Sat. Liquid | Sat. Vapor | Sat. Liquid | Evap. | Sat. Vapor | Sat. Liquid | Evap. | Sat. Vapor | Sat. Liquid | Evap. | Sat. Vapor | |
| $p$ | $t$ | $v_f$ | $v_g$ | $h_f$ | $h_{fg}$ | $h_g$ | $s_f$ | $s_{fg}$ | $s_g$ | $u_f$ | $u_{fg}$ | $u_g$ | $p$ |
| **3.0** | 141.48 | 0.01630 | 118.71 | 109.37 | 1013.2 | 1122.6 | 0.2008 | 1.6855 | 1.8863 | 109.36 | 947.3 | 1056.7 | **3.0** |
| **3.5** | 147.57 | 0.01633 | 102.72 | 115.46 | 1009.6 | 1125.1 | 0.2109 | 1.6626 | 1.8735 | 115.45 | 943.1 | 1058.6 | **3.5** |
| **4.0** | 152.97 | 0.01636 | 90.63 | 120.86 | 1006.4 | 1127.3 | 0.2198 | 1.6427 | 1.8625 | 120.85 | 939.3 | 1060.2 | **4.0** |
| **4.5** | 157.83 | 0.01638 | 81.16 | 125.71 | 1003.6 | 1129.3 | 0.2276 | 1.6252 | 1.8528 | 125.70 | 936.0 | 1061.7 | **4.5** |
| **5.0** | 162.24 | 0.01640 | 73.52 | 130.13 | 1001.0 | 1131.1 | 0.2347 | 1.6094 | 1.8441 | 130.12 | 933.0 | 1063.1 | **5.0** |
| **5.5** | 166.30 | 0.01643 | 67.24 | 134.19 | 998.5 | 1132.7 | 0.2411 | 1.5951 | 1.8363 | 134.17 | 930.1 | 1064.3 | **5.5** |
| **6.0** | 170.06 | 0.01645 | 61.98 | 137.96 | 996.2 | 1134.2 | 0.2472 | 1.5820 | 1.8292 | 137.94 | 927.5 | 1065.4 | **6.0** |
| **6.5** | 173.56 | 0.01647 | 57.50 | 141.47 | 994.1 | 1135.6 | 0.2528 | 1.5699 | 1.8227 | 141.45 | 925.0 | 1066.4 | **6.5** |
| **7.0** | 176.85 | 0.01649 | 53.64 | 144.76 | 992.1 | 1136.9 | 0.2581 | 1.5586 | 1.8167 | 144.74 | 922.7 | 1067.4 | **7.0** |
| **7.5** | 179.94 | 0.01651 | 50.29 | 147.86 | 990.2 | 1138.1 | 0.2629 | 1.5481 | 1.8110 | 147.84 | 920.5 | 1068.3 | **7.5** |
| **8.0** | 182.86 | 0.01653 | 47.34 | 150.79 | 988.5 | 1139.3 | 0.2674 | 1.5383 | 1.8057 | 150.77 | 918.4 | 1069.2 | **8.0** |
| **8.5** | 185.64 | 0.01654 | 44.73 | 153.57 | 986.8 | 1140.4 | 0.2718 | 1.5290 | 1.8008 | 153.54 | 916.5 | 1070.0 | **8.5** |
| **9.0** | 188.28 | 0.01656 | 42.40 | 156.22 | 985.2 | 1141.4 | 0.2759 | 1.5203 | 1.7962 | 156.19 | 914.6 | 1070.8 | **9.0** |
| **9.5** | 190.80 | 0.01658 | 40.31 | 158.75 | 983.6 | 1142.3 | 0.2798 | 1.5120 | 1.7918 | 158.72 | 912.8 | 1071.5 | **9.5** |
| **10** | 193.21 | 0.01659 | 38.42 | 161.17 | 982.1 | 1143.3 | 0.2835 | 1.5041 | 1.7876 | 161.14 | 911.1 | 1072.2 | **10** |
| **11** | 197.75 | 0.01662 | 35.14 | 165.73 | 979.3 | 1145.0 | 0.2903 | 1.4897 | 1.7800 | 165.70 | 907.8 | 1073.5 | **11** |
| **12** | 201.96 | 0.01665 | 32.40 | 169.96 | 976.6 | 1146.6 | 0.2967 | 1.4763 | 1.7730 | 169.92 | 904.8 | 1074.7 | **12** |
| **13** | 205.88 | 0.01667 | 30.06 | 173.91 | 974.2 | 1148.1 | 0.3027 | 1.4638 | 1.7665 | 173.87 | 901.9 | 1075.8 | **13** |
| **14** | 209.56 | 0.01670 | 28.04 | 177.61 | 971.9 | 1149.5 | 0.3083 | 1.4522 | 1.7605 | 177.57 | 899.3 | 1076.9 | **14** |
| **14.696** | 212.00 | 0.01672 | 26.80 | 180.07 | 970.3 | 1150.4 | 0.3120 | 1.4446 | 1.7566 | 180.02 | 897.5 | 1077.5 | **14.696** |
| **15** | 213.03 | 0.01672 | 26.29 | 181.11 | 969.7 | 1150.8 | 0.3135 | 1.4415 | 1.7549 | 181.06 | 896.7 | 1077.8 | **15** |
| **16** | 216.32 | 0.01674 | 24.75 | 184.42 | 967.6 | 1152.0 | 0.3184 | 1.4313 | 1.7497 | 184.37 | 894.3 | 1078.7 | **16** |
| **17** | 219.44 | 0.01677 | 23.39 | 187.56 | 965.5 | 1153.1 | 0.3231 | 1.4218 | 1.7449 | 187.51 | 892.0 | 1079.5 | **17** |
| **18** | 222.41 | 0.01679 | 22.17 | 190.56 | 963.6 | 1154.2 | 0.3275 | 1.4128 | 1.7403 | 190.50 | 889.9 | 1080.4 | **18** |
| **19** | 225.24 | 0.01681 | 21.08 | 193.42 | 961.9 | 1155.3 | 0.3317 | 1.4043 | 1.7360 | 193.36 | 887.8 | 1081.2 | **19** |
| **20** | 227.96 | 0.01683 | 20.089 | 196.16 | 960.1 | 1156.3 | 0.3356 | 1.3962 | 1.7319 | 196.10 | 885.8 | 1081.9 | **20** |
| **21** | 230.57 | 0.01685 | 19.192 | 198.79 | 958.4 | 1157.2 | 0.3395 | 1.3885 | 1.7280 | 198.73 | 883.9 | 1082.6 | **21** |
| **22** | 233.07 | 0.01687 | 18.375 | 201.33 | 956.8 | 1158.1 | 0.3431 | 1.3811 | 1.7242 | 201.26 | 882.0 | 1083.3 | **22** |
| **23** | 235.49 | 0.01689 | 17.627 | 203.78 | 955.2 | 1159.0 | 0.3466 | 1.3740 | 1.7206 | 203.71 | 880.2 | 1083.9 | **23** |
| **24** | 237.82 | 0.01691 | 16.938 | 206.14 | 953.7 | 1159.8 | 0.3500 | 1.3672 | 1.7172 | 206.07 | 878.5 | 1084.6 | **24** |
| **25** | 240.07 | 0.01692 | 16.303 | 208.42 | 952.1 | 1160.6 | 0.3533 | 1.3606 | 1.7139 | 208.34 | 876.8 | 1085.1 | **25** |
| **26** | 242.25 | 0.01694 | 15.715 | 210.62 | 950.7 | 1161.3 | 0.3564 | 1.3544 | 1.7108 | 210.54 | 875.2 | 1085.7 | **26** |
| **27** | 244.36 | 0.01696 | 15.170 | 212.75 | 949.3 | 1162.0 | 0.3594 | 1.3484 | 1.7078 | 212.67 | 873.6 | 1086.3 | **27** |
| **28** | 246.41 | 0.01698 | 14.663 | 214.83 | 947.9 | 1162.7 | 0.3623 | 1.3425 | 1.7048 | 214.74 | 872.1 | 1086.8 | **28** |
| **29** | 248.40 | 0.01699 | 14.189 | 216.86 | 946.5 | 1163.4 | 0.3652 | 1.3368 | 1.7020 | 216.77 | 870.5 | 1087.3 | **29** |
| **30** | 250.33 | 0.01701 | 13.746 | 218.82 | 945.3 | 1164.1 | 0.3680 | 1.3313 | 1.6993 | 218.73 | 869.1 | 1087.8 | **30** |
| **31** | 252.22 | 0.01702 | 13.330 | 220.73 | 944.0 | 1164.7 | 0.3707 | 1.3260 | 1.6967 | 220.63 | 867.7 | 1088.3 | **31** |
| **32** | 254.05 | 0.01704 | 12.940 | 222.59 | 942.8 | 1165.4 | 0.3733 | 1.3209 | 1.6941 | 222.49 | 866.3 | 1088.7 | **32** |
| **33** | 255.84 | 0.01705 | 12.572 | 224.41 | 941.6 | 1166.0 | 0.3758 | 1.3159 | 1.6917 | 224.31 | 864.9 | 1089.2 | **33** |
| **34** | 257.58 | 0.01707 | 12.226 | 226.18 | 940.3 | 1166.5 | 0.3783 | 1.3110 | 1.6893 | 226.07 | 863.5 | 1089.6 | **34** |
| **35** | 259.28 | 0.01708 | 11.898 | 227.91 | 939.2 | 1167.1 | 0.3807 | 1.3063 | 1.6870 | 227.80 | 862.3 | 1090.1 | **35** |
| **36** | 260.95 | 0.01709 | 11.588 | 229.60 | 938.0 | 1167.6 | 0.3831 | 1.3017 | 1.6848 | 229.49 | 861.0 | 1090.5 | **36** |
| **37** | 262.57 | 0.01711 | 11.294 | 231.26 | 936.9 | 1168.2 | 0.3854 | 1.2972 | 1.6826 | 231.14 | 859.8 | 1090.9 | **37** |
| **38** | 264.16 | 0.01712 | 11.015 | 232.89 | 935.8 | 1168.7 | 0.3876 | 1.2929 | 1.6805 | 232.77 | 858.5 | 1091.3 | **38** |
| **39** | 265.72 | 0.01714 | 10.750 | 234.48 | 934.7 | 1169.2 | 0.3898 | 1.2886 | 1.6784 | 234.36 | 857.2 | 1091.6 | **39** |
| **40** | 267.25 | 0.01715 | 10.498 | 236.03 | 933.7 | 1169.7 | 0.3919 | 1.2844 | 1.6763 | 235.90 | 856.1 | 1092.0 | **40** |
| **41** | 268.74 | 0.01716 | 10.258 | 237.55 | 932.6 | 1170.2 | 0.3940 | 1.2803 | 1.6743 | 237.42 | 855.0 | 1092.4 | **41** |
| **42** | 270.21 | 0.01717 | 10.029 | 239.04 | 931.6 | 1170.7 | 0.3960 | 1.2764 | 1.6724 | 238.91 | 853.8 | 1092.7 | **42** |
| **43** | 271.64 | 0.01719 | 9.810 | 240.51 | 930.6 | 1171.1 | 0.3980 | 1.2726 | 1.6706 | 240.37 | 852.7 | 1093.1 | **43** |
| **44** | 273.05 | 0.01720 | 9.601 | 241.95 | 929.6 | 1171.6 | 0.4000 | 1.2687 | 1.6687 | 241.81 | 851.6 | 1093.4 | **44** |
| **45** | 274.44 | 0.01721 | 9.401 | 243.36 | 928.6 | 1172.0 | 0.4019 | 1.2650 | 1.6669 | 243.22 | 850.5 | 1093.7 | **45** |
| **46** | 275.80 | 0.01722 | 9.209 | 244.75 | 927.7 | 1172.4 | 0.4038 | 1.2613 | 1.6652 | 244.60 | 849.5 | 1094.1 | **46** |
| **47** | 277.13 | 0.01723 | 9.025 | 246.12 | 926.7 | 1172.9 | 0.4057 | 1.2577 | 1.6634 | 245.97 | 848.4 | 1094.4 | **47** |
| **48** | 278.45 | 0.01725 | 8.848 | 247.47 | 925.8 | 1173.3 | 0.4075 | 1.2542 | 1.6617 | 247.32 | 847.4 | 1094.7 | **48** |
| **49** | 279.74 | 0.01726 | 8.678 | 248.79 | 924.9 | 1173.7 | 0.4093 | 1.2508 | 1.6601 | 248.63 | 846.4 | 1095.0 | **49** |

## Table 2. Saturation: Pressures

| Abs. Press. Lb./Sq. In. | Temp. Fahr. | Specific Volume Sat. Liquid | Specific Volume Sat. Vapor | Enthalpy Sat. Liquid | Enthalpy Evap. | Enthalpy Sat. Vapor | Entropy Sat. Liquid | Entropy Evap. | Entropy Sat. Vapor | Internal Energy Sat. Liquid | Internal Energy Evap. | Internal Energy Sat. Vapor | Abs. Press. Lb./Sq. In. |
|---|---|---|---|---|---|---|---|---|---|---|---|---|---|
| p | t | $v_f$ | $v_g$ | $h_f$ | $h_{fg}$ | $h_g$ | $s_f$ | $s_{fg}$ | $s_g$ | $u_f$ | $u_{fg}$ | $u_g$ | p |
| 50 | 281.01 | 0.01727 | 8.515 | 250.09 | 924.0 | 1174.1 | 0.4110 | 1.2474 | 1.6585 | 249.93 | 845.4 | 1095.3 | 50 |
| 51 | 282.26 | 0.01728 | 8.359 | 251.37 | 923.0 | 1174.4 | 0.4127 | 1.2442 | 1.6569 | 251.21 | 844.3 | 1095.5 | 51 |
| 52 | 283.49 | 0.01729 | 8.208 | 252.63 | 922.2 | 1174.8 | 0.4144 | 1.2409 | 1.6553 | 252.46 | 843.3 | 1095.8 | 52 |
| 53 | 284.70 | 0.01730 | 8.062 | 253.87 | 921.3 | 1175.2 | 0.4161 | 1.2377 | 1.6538 | 253.70 | 842.4 | 1096.1 | 53 |
| 54 | 285.90 | 0.01731 | 7.922 | 255.09 | 920.5 | 1175.6 | 0.4177 | 1.2346 | 1.6523 | 254.92 | 841.5 | 1096.4 | 54 |
| 55 | 287.07 | 0.01732 | 7.787 | 256.30 | 919.6 | 1175.9 | 0.4193 | 1.2316 | 1.6509 | 256.12 | 840.6 | 1096.7 | 55 |
| 56 | 288.23 | 0.01733 | 7.656 | 257.50 | 918.8 | 1176.3 | 0.4209 | 1.2285 | 1.6494 | 257.32 | 839.7 | 1097.0 | 56 |
| 57 | 289.37 | 0.01734 | 7.529 | 258.67 | 917.9 | 1176.6 | 0.4225 | 1.2255 | 1.6480 | 258.49 | 838.7 | 1097.2 | 57 |
| 58 | 290.50 | 0.01736 | 7.407 | 259.82 | 917.1 | 1176.9 | 0.4240 | 1.2226 | 1.6466 | 259.63 | 837.8 | 1097.4 | 58 |
| 59 | 291.61 | 0.01737 | 7.289 | 260.96 | 916.3 | 1177.3 | 0.4255 | 1.2197 | 1.6452 | 260.77 | 836.9 | 1097.7 | 59 |
| 60 | 292.71 | 0.01738 | 7.175 | 262.09 | 915.5 | 1177.6 | 0.4270 | 1.2168 | 1.6438 | 261.90 | 836.0 | 1097.9 | 60 |
| 61 | 293.79 | 0.01739 | 7.064 | 263.20 | 914.7 | 1177.9 | 0.4285 | 1.2140 | 1.6425 | 263.00 | 835.2 | 1098.2 | 61 |
| 62 | 294.85 | 0.01740 | 6.957 | 264.30 | 913.9 | 1178.2 | 0.4300 | 1.2112 | 1.6412 | 264.10 | 834.3 | 1098.4 | 62 |
| 63 | 295.90 | 0.01741 | 6.853 | 265.38 | 913.1 | 1178.5 | 0.4314 | 1.2085 | 1.6399 | 265.18 | 833.4 | 1098.6 | 63 |
| 64 | 296.94 | 0.01742 | 6.752 | 266.45 | 912.3 | 1178.8 | 0.4328 | 1.2059 | 1.6387 | 266.24 | 832.6 | 1098.8 | 64 |
| 65 | 297.97 | 0.01743 | 6.655 | 267.50 | 911.6 | 1179.1 | 0.4342 | 1.2032 | 1.6374 | 267.29 | 831.8 | 1099.1 | 65 |
| 66 | 298.99 | 0.01744 | 6.560 | 268.55 | 910.8 | 1179.4 | 0.4356 | 1.2006 | 1.6362 | 268.34 | 831.0 | 1099.3 | 66 |
| 67 | 299.99 | 0.01745 | 6.468 | 269.58 | 910.1 | 1179.7 | 0.4369 | 1.1981 | 1.6350 | 269.36 | 830.2 | 1099.5 | 67 |
| 68 | 300.98 | 0.01746 | 6.378 | 270.60 | 909.4 | 1180.0 | 0.4383 | 1.1955 | 1.6338 | 270.38 | 829.4 | 1099.8 | 68 |
| 69 | 301.96 | 0.01747 | 6.291 | 271.61 | 908.7 | 1180.3 | 0.4396 | 1.1930 | 1.6326 | 271.39 | 828.6 | 1100.0 | 69 |
| 70 | 302.92 | 0.01748 | 6.206 | 272.61 | 907.9 | 1180.6 | 0.4409 | 1.1906 | 1.6315 | 272.38 | 827.8 | 1100.2 | 70 |
| 71 | 303.88 | 0.01749 | 6.124 | 273.60 | 907.2 | 1180.8 | 0.4422 | 1.1881 | 1.6303 | 273.37 | 827.0 | 1100.4 | 71 |
| 72 | 304.83 | 0.01750 | 6.044 | 274.57 | 906.5 | 1181.1 | 0.4435 | 1.1857 | 1.6292 | 274.34 | 826.3 | 1100.6 | 72 |
| 73 | 305.76 | 0.01751 | 5.966 | 275.54 | 905.8 | 1181.3 | 0.4447 | 1.1834 | 1.6281 | 275.30 | 825.5 | 1100.8 | 73 |
| 74 | 306.68 | 0.01752 | 5.890 | 276.49 | 905.1 | 1181.6 | 0.4460 | 1.1810 | 1.6270 | 276.25 | 824.7 | 1101.0 | 74 |
| 75 | 307.60 | 0.01753 | 5.816 | 277.43 | 904.5 | 1181.9 | 0.4472 | 1.1787 | 1.6259 | 277.19 | 824.0 | 1101.2 | 75 |
| 76 | 308.50 | 0.01754 | 5.743 | 278.37 | 903.7 | 1182.1 | 0.4484 | 1.1764 | 1.6248 | 278.12 | 823.3 | 1101.4 | 76 |
| 77 | 309.40 | 0.01754 | 5.673 | 279.30 | 903.1 | 1182.4 | 0.4496 | 1.1742 | 1.6238 | 279.05 | 822.5 | 1101.6 | 77 |
| 78 | 310.29 | 0.01755 | 5.604 | 280.21 | 902.4 | 1182.6 | 0.4508 | 1.1720 | 1.6228 | 279.96 | 821.7 | 1101.7 | 78 |
| 79 | 311.16 | 0.01756 | 5.537 | 281.12 | 901.7 | 1182.8 | 0.4520 | 1.1698 | 1.6217 | 280.86 | 821.0 | 1101.9 | 79 |
| 80 | 312.03 | 0.01757 | 5.472 | 282.02 | 901.1 | 1183.1 | 0.4531 | 1.1676 | 1.6207 | 281.76 | 820.3 | 1102.1 | 80 |
| 81 | 312.89 | 0.01758 | 5.408 | 282.91 | 900.4 | 1183.3 | 0.4543 | 1.1654 | 1.6197 | 282.65 | 819.6 | 1102.2 | 81 |
| 82 | 313.74 | 0.01759 | 5.346 | 283.79 | 899.7 | 1183.5 | 0.4554 | 1.1633 | 1.6187 | 283.52 | 818.9 | 1102.4 | 82 |
| 83 | 314.59 | 0.01760 | 5.285 | 284.66 | 899.1 | 1183.8 | 0.4565 | 1.1612 | 1.6177 | 284.39 | 818.2 | 1102.6 | 83 |
| 84 | 315.42 | 0.01761 | 5.226 | 285.53 | 898.5 | 1184.0 | 0.4576 | 1.1592 | 1.6168 | 285.26 | 817.5 | 1102.8 | 84 |
| 85 | 316.25 | 0.01761 | 5.168 | 286.39 | 897.8 | 1184.2 | 0.4587 | 1.1571 | 1.6158 | 286.11 | 816.8 | 1102.9 | 85 |
| 86 | 317.07 | 0.01762 | 5.111 | 287.24 | 897.2 | 1184.4 | 0.4598 | 1.1551 | 1.6149 | 286.96 | 816.1 | 1103.1 | 86 |
| 87 | 317.88 | 0.01763 | 5.055 | 288.08 | 896.5 | 1184.6 | 0.4609 | 1.1530 | 1.6139 | 287.80 | 815.4 | 1103.2 | 87 |
| 88 | 318.68 | 0.01764 | 5.001 | 288.91 | 895.9 | 1184.8 | 0.4620 | 1.1510 | 1.6130 | 288.63 | 814.8 | 1103.4 | 88 |
| 89 | 319.48 | 0.01765 | 4.948 | 289.74 | 895.3 | 1185.1 | 0.4630 | 1.1491 | 1.6121 | 289.45 | 814.1 | 1103.6 | 89 |
| 90 | 320.27 | 0.01766 | 4.896 | 290.56 | 894.7 | 1185.3 | 0.4641 | 1.1471 | 1.6112 | 290.27 | 813.4 | 1103.7 | 90 |
| 91 | 321.06 | 0.01767 | 4.845 | 291.38 | 894.1 | 1185.5 | 0.4651 | 1.1452 | 1.6103 | 291.08 | 812.8 | 1103.9 | 91 |
| 92 | 321.83 | 0.01768 | 4.796 | 292.18 | 893.5 | 1185.7 | 0.4661 | 1.1433 | 1.6094 | 291.88 | 812.2 | 1104.1 | 92 |
| 93 | 322.60 | 0.01768 | 4.747 | 292.98 | 892.9 | 1185.9 | 0.4672 | 1.1413 | 1.6085 | 292.68 | 811.5 | 1104.2 | 93 |
| 94 | 323.36 | 0.01769 | 4.699 | 293.78 | 892.3 | 1186.1 | 0.4682 | 1.1394 | 1.6076 | 293.47 | 810.9 | 1104.4 | 94 |
| 95 | 324.12 | 0.01770 | 4.652 | 294.56 | 891.7 | 1186.2 | 0.4692 | 1.1376 | 1.6068 | 294.25 | 810.2 | 1104.5 | 95 |
| 96 | 324.87 | 0.01771 | 4.606 | 295.34 | 891.1 | 1186.4 | 0.4702 | 1.1358 | 1.6060 | 295.03 | 809.6 | 1104.6 | 96 |
| 97 | 325.61 | 0.01772 | 4.561 | 296.12 | 890.5 | 1186.6 | 0.4711 | 1.1340 | 1.6051 | 295.80 | 808.9 | 1104.7 | 97 |
| 98 | 326.35 | 0.01772 | 4.517 | 296.89 | 889.9 | 1186.8 | 0.4721 | 1.1322 | 1.6043 | 296.57 | 808.3 | 1104.9 | 98 |
| 99 | 327.08 | 0.01773 | 4.474 | 297.65 | 889.4 | 1187.0 | 0.4731 | 1.1304 | 1.6035 | 297.33 | 807.7 | 1105.0 | 99 |
| 100 | 327.81 | 0.01774 | 4.432 | 298.40 | 888.8 | 1187.2 | 0.4740 | 1.1286 | 1.6026 | 298.08 | 807.1 | 1105.2 | 100 |
| 101 | 328.53 | 0.01775 | 4.391 | 299.15 | 888.2 | 1187.4 | 0.4750 | 1.1268 | 1.6018 | 298.82 | 806.5 | 1105.3 | 101 |
| 102 | 329.25 | 0.01775 | 4.350 | 299.90 | 887.6 | 1187.5 | 0.4759 | 1.1251 | 1.6010 | 299.57 | 805.9 | 1105.4 | 102 |
| 103 | 329.96 | 0.01776 | 4.310 | 300.64 | 887.1 | 1187.7 | 0.4768 | 1.1234 | 1.6002 | 300.30 | 805.3 | 1105.6 | 103 |
| 104 | 330.66 | 0.01777 | 4.271 | 301.37 | 886.5 | 1187.9 | 0.4778 | 1.1216 | 1.5994 | 301.03 | 804.7 | 1105.7 | 104 |

## Table 2. Saturation: Pressures

| Abs. Press. Lb./Sq. In. | Temp. Fahr. | Specific Volume | | Enthalpy | | | Entropy | | | Internal Energy | | | Abs. Press. Lb./Sq. In. |
|---|---|---|---|---|---|---|---|---|---|---|---|---|---|
| | | Sat. Liquid | Sat. Vapor | Sat. Liquid | Evap. | Sat. Vapor | Sat. Liquid | Evap. | Sat. Vapor | Sat. Liquid | Evap. | Sat. Vapor | |
| p | t | $v_f$ | $v_g$ | $h_f$ | $h_{fg}$ | $h_g$ | $s_f$ | $s_{fg}$ | $s_g$ | $u_f$ | $u_{fg}$ | $u_g$ | p |
| **105** | 331.36 | 0.01778 | 4.232 | 302.10 | 886.0 | 1188.1 | 0.4787 | 1.1199 | 1.5986 | 301.75 | 804.1 | 1105.9 | **105** |
| **106** | 332.05 | 0.01778 | 4.194 | 302.82 | 885.4 | 1188.2 | 0.4796 | 1.1182 | 1.5978 | 302.47 | 803.5 | 1106.0 | **106** |
| **107** | 332.74 | 0.01779 | 4.157 | 303.54 | 884.9 | 1188.4 | 0.4805 | 1.1166 | 1.5971 | 303.19 | 802.9 | 1106.1 | **107** |
| **108** | 333.42 | 0.01780 | 4.120 | 304.26 | 884.3 | 1188.6 | 0.4814 | 1.1149 | 1.5963 | 303.90 | 802.4 | 1106.3 | **108** |
| **109** | 334.10 | 0.01781 | 4.084 | 304.97 | 883.7 | 1188.7 | 0.4823 | 1.1133 | 1.5956 | 304.61 | 801.8 | 1106.4 | **109** |
| **110** | 334.77 | 0.01782 | 4.049 | 305.66 | 883.2 | 1188.9 | 0.4832 | 1.1117 | 1.5948 | 305.30 | 801.2 | 1106.5 | **110** |
| **111** | 335.44 | 0.01782 | 4.015 | 306.37 | 882.6 | 1189.0 | 0.4840 | 1.1101 | 1.5941 | 306.00 | 800.6 | 1106.6 | **111** |
| **112** | 336.11 | 0.01783 | 3.981 | 307.06 | 882.1 | 1189.2 | 0.4849 | 1.1085 | 1.5934 | 306.69 | 800.0 | 1106.7 | **112** |
| **113** | 336.77 | 0.01784 | 3.947 | 307.75 | 881.6 | 1189.4 | 0.4858 | 1.1069 | 1.5927 | 307.38 | 799.4 | 1106.8 | **113** |
| **114** | 337.42 | 0.01784 | 3.914 | 308.43 | 881.1 | 1189.5 | 0.4866 | 1.1053 | 1.5919 | 308.05 | 798.9 | 1106.9 | **114** |
| **115** | 338.07 | 0.01785 | 3.882 | 309.11 | 880.6 | 1189.7 | 0.4875 | 1.1037 | 1.5912 | 308.73 | 798.4 | 1107.1 | **115** |
| **116** | 338.72 | 0.01786 | 3.850 | 309.79 | 880.0 | 1189.8 | 0.4883 | 1.1022 | 1.5905 | 309.41 | 797.8 | 1107.2 | **116** |
| **117** | 339.36 | 0.01787 | 3.819 | 310.46 | 879.5 | 1190.0 | 0.4891 | 1.1007 | 1.5898 | 310.07 | 797.2 | 1107.3 | **117** |
| **118** | 339.99 | 0.01787 | 3.788 | 311.12 | 879.0 | 1190.1 | 0.4900 | 1.0992 | 1.5891 | 310.73 | 796.7 | 1107.4 | **118** |
| **119** | 340.62 | 0.01788 | 3.758 | 311.78 | 878.4 | 1190.2 | 0.4908 | 1.0977 | 1.5885 | 311.39 | 796.1 | 1107.5 | **119** |
| **120** | 341.25 | 0.01789 | 3.728 | 312.44 | 877.9 | 1190.4 | 0.4916 | 1.0962 | 1.5878 | 312.05 | 795.6 | 1107.6 | **120** |
| **121** | 341.88 | 0.01790 | 3.699 | 313.10 | 877.4 | 1190.5 | 0.4924 | 1.0947 | 1.5871 | 312.70 | 795.0 | 1107.7 | **121** |
| **122** | 342.50 | 0.01791 | 3.670 | 313.75 | 876.9 | 1190.7 | 0.4932 | 1.0933 | 1.5865 | 313.35 | 794.5 | 1107.8 | **122** |
| **123** | 343.11 | 0.01791 | 3.642 | 314.40 | 876.4 | 1190.8 | 0.4940 | 1.0918 | 1.5858 | 313.99 | 793.9 | 1107.9 | **123** |
| **124** | 343.72 | 0.01792 | 3.614 | 315.04 | 875.9 | 1190.9 | 0.4948 | 1.0903 | 1.5851 | 314.63 | 793.4 | 1108.0 | **124** |
| **125** | 344.33 | 0.01792 | 3.587 | 315.68 | 875.4 | 1191.1 | 0.4956 | 1.0888 | 1.5844 | 315.26 | 792.8 | 1108.1 | **125** |
| **126** | 344.94 | 0.01793 | 3.560 | 316.31 | 874.9 | 1191.2 | 0.4964 | 1.0874 | 1.5838 | 315.89 | 792.3 | 1108.2 | **126** |
| **127** | 345.54 | 0.01794 | 3.533 | 316.94 | 874.4 | 1191.3 | 0.4972 | 1.0859 | 1.5831 | 316.52 | 791.8 | 1108.3 | **127** |
| **128** | 346.13 | 0.01794 | 3.507 | 317.57 | 873.9 | 1191.5 | 0.4980 | 1.0845 | 1.5825 | 317.15 | 791.3 | 1108.4 | **128** |
| **129** | 346.73 | 0.01795 | 3.481 | 318.19 | 873.4 | 1191.6 | 0.4987 | 1.0832 | 1.5819 | 317.77 | 790.7 | 1108.5 | **129** |
| **130** | 347.32 | 0.01796 | 3.455 | 318.81 | 872.9 | 1191.7 | 0.4995 | 1.0817 | 1.5812 | 318.38 | 790.2 | 1108.6 | **130** |
| **131** | 347.90 | 0.01797 | 3.430 | 319.43 | 872.5 | 1191.9 | 0.5002 | 1.0804 | 1.5806 | 318.99 | 789.7 | 1108.7 | **131** |
| **132** | 348.48 | 0.01797 | 3.405 | 320.04 | 872.0 | 1192.0 | 0.5010 | 1.0790 | 1.5800 | 319.60 | 789.2 | 1108.8 | **132** |
| **133** | 349.06 | 0.01798 | 3.381 | 320.65 | 871.5 | 1192.1 | 0.5018 | 1.0776 | 1.5793 | 320.21 | 788.7 | 1108.9 | **133** |
| **134** | 349.64 | 0.01799 | 3.357 | 321.25 | 871.0 | 1192.2 | 0.5025 | 1.0762 | 1.5787 | 320.80 | 788.2 | 1109.0 | **134** |
| **135** | 350.21 | 0.01800 | 3.333 | 321.85 | 870.6 | 1192.4 | 0.5032 | 1.0749 | 1.5781 | 321.40 | 787.7 | 1109.1 | **135** |
| **136** | 350.78 | 0.01800 | 3.310 | 322.45 | 870.1 | 1192.5 | 0.5040 | 1.0735 | 1.5775 | 322.00 | 787.2 | 1109.2 | **136** |
| **137** | 351.35 | 0.01801 | 3.287 | 323.05 | 869.6 | 1192.6 | 0.5047 | 1.0722 | 1.5769 | 322.59 | 786.7 | 1109.3 | **137** |
| **138** | 351.91 | 0.01801 | 3.264 | 323.64 | 869.1 | 1192.7 | 0.5054 | 1.0709 | 1.5763 | 323.18 | 786.2 | 1109.4 | **138** |
| **139** | 352.47 | 0.01802 | 3.242 | 324.23 | 868.7 | 1192.9 | 0.5061 | 1.0696 | 1.5757 | 323.77 | 785.7 | 1109.5 | **139** |
| **140** | 353.02 | 0.01802 | 3.220 | 324.82 | 868.2 | 1193.0 | 0.5069 | 1.0682 | 1.5751 | 324.35 | 785.2 | 1109.6 | **140** |
| **141** | 353.57 | 0.01803 | 3.198 | 325.40 | 867.7 | 1193.1 | 0.5076 | 1.0669 | 1.5745 | 324.93 | 784.8 | 1109.7 | **141** |
| **142** | 354.12 | 0.01804 | 3.177 | 325.98 | 867.2 | 1193.2 | 0.5083 | 1.0657 | 1.5740 | 325.51 | 784.3 | 1109.8 | **142** |
| **143** | 354.67 | 0.01804 | 3.155 | 326.56 | 866.7 | 1193.3 | 0.5090 | 1.0644 | 1.5734 | 326.08 | 783.8 | 1109.8 | **143** |
| **144** | 355.21 | 0.01805 | 3.134 | 327.13 | 866.3 | 1193.4 | 0.5097 | 1.0631 | 1.5728 | 326.65 | 783.3 | 1109.9 | **144** |
| **145** | 355.76 | 0.01806 | 3.114 | 327.70 | 865.8 | 1193.5 | 0.5104 | 1.0618 | 1.5722 | 327.22 | 782.8 | 1110.0 | **145** |
| **146** | 356.29 | 0.01806 | 3.094 | 328.27 | 865.3 | 1193.6 | 0.5111 | 1.0605 | 1.5716 | 327.78 | 782.3 | 1110.1 | **146** |
| **147** | 356.83 | 0.01807 | 3.074 | 328.83 | 864.9 | 1193.8 | 0.5118 | 1.0592 | 1.5710 | 328.34 | 781.9 | 1110.2 | **147** |
| **148** | 357.36 | 0.01808 | 3.054 | 329.39 | 864.5 | 1193.9 | 0.5124 | 1.0580 | 1.5705 | 328.90 | 781.4 | 1110.3 | **148** |
| **149** | 357.89 | 0.01808 | 3.034 | 329.95 | 864.0 | 1194.0 | 0.5131 | 1.0568 | 1.5699 | 329.45 | 780.9 | 1110.4 | **149** |
| **150** | 358.42 | 0.01809 | 3.015 | 330.51 | 863.6 | 1194.1 | 0.5138 | 1.0556 | 1.5694 | 330.01 | 780.5 | 1110.5 | **150** |
| **152** | 359.46 | 0.01810 | 2.977 | 331.61 | 862.7 | 1194.3 | 0.5151 | 1.0532 | 1.5683 | 331.10 | 779.5 | 1110.6 | **152** |
| **154** | 360.49 | 0.01812 | 2.940 | 332.70 | 861.8 | 1194.5 | 0.5165 | 1.0507 | 1.5672 | 332.18 | 778.5 | 1110.7 | **154** |
| **156** | 361.52 | 0.01813 | 2.904 | 333.79 | 860.9 | 1194.7 | 0.5178 | 1.0483 | 1.5661 | 333.26 | 777.6 | 1110.9 | **156** |
| **158** | 362.53 | 0.01814 | 2.869 | 334.86 | 860.0 | 1194.9 | 0.5191 | 1.0459 | 1.5650 | 334.23 | 776.8 | 1111.0 | **158** |
| **160** | 363.53 | 0.01815 | 2.834 | 335.93 | 859.2 | 1195.1 | 0.5204 | 1.0436 | 1.5640 | 335.39 | 775.8 | 1111.2 | **160** |
| **162** | 364.53 | 0.01817 | 2.801 | 336.98 | 858.3 | 1195.3 | 0.5216 | 1.0414 | 1.5630 | 336.44 | 775.0 | 1111.4 | **162** |
| **164** | 365.51 | 0.01818 | 2.768 | 338.02 | 857.5 | 1195.5 | 0.5229 | 1.0391 | 1.5620 | 337.47 | 774.1 | 1111.5 | **164** |
| **166** | 366.48 | 0.01819 | 2.736 | 339.05 | 856.6 | 1195.7 | 0.5241 | 1.0369 | 1.5610 | 338.49 | 773.2 | 1111.7 | **166** |
| **168** | 367.45 | 0.01820 | 2.705 | 340.07 | 855.7 | 1195.8 | 0.5254 | 1.0346 | 1.5600 | 339.51 | 772.3 | 1111.8 | **168** |

## Table 2. Saturation: Pressures

| Abs. Press. Lb./Sq. In. | Temp. Fahr. | Specific Volume | | Enthalpy | | | Entropy | | | Internal Energy | | | Abs. Press. Lb./Sq. In. |
|---|---|---|---|---|---|---|---|---|---|---|---|---|---|
| | | Sat. Liquid | Sat. Vapor | Sat. Liquid | Evap. | Sat. Vapor | Sat. Liquid | Evap. | Sat. Vapor | Sat. Liquid | Evap. | Sat. Vapor | |
| $p$ | $t$ | $v_f$ | $v_g$ | $h_f$ | $h_{fg}$ | $h_g$ | $s_f$ | $s_{fg}$ | $s_g$ | $u_f$ | $u_{fg}$ | $u_g$ | $p$ |
| **170** | 368.41 | 0.01822 | 2.675 | 341.09 | 854.9 | 1196.0 | 0.5266 | 1.0324 | 1.5590 | 340.52 | 771.4 | 1111.9 | **170** |
| **172** | 369.35 | 0.01823 | 2.645 | 342.10 | 854.1 | 1196.2 | 0.5278 | 1.0302 | 1.5580 | 341.52 | 770.5 | 1112.0 | **172** |
| **174** | 370.29 | 0.01824 | 2.616 | 343.10 | 853.3 | 1196.4 | 0.5290 | 1.0280 | 1.5570 | 342.51 | 769.7 | 1112.2 | **174** |
| **176** | 371.22 | 0.01825 | 2.587 | 344.09 | 852.4 | 1196.5 | 0.5302 | 1.0259 | 1.5561 | 343.50 | 768.8 | 1112.3 | **176** |
| **178** | 372.14 | 0.01826 | 2.559 | 345.06 | 851.6 | 1196.7 | 0.5313 | 1.0238 | 1.5551 | 344.46 | 767.9 | 1112.4 | **178** |
| **180** | 373.06 | 0.01827 | 2.532 | 346.03 | 850.8 | 1196.9 | 0.5325 | 1.0217 | 1.5542 | 345.42 | 767.1 | 1112.5 | **180** |
| **182** | 373.96 | 0.01829 | 2.505 | 347.00 | 850.0 | 1197.0 | 0.5336 | 1.0196 | 1.5532 | 346.38 | 766.2 | 1112.6 | **182** |
| **184** | 374.86 | 0.01830 | 2.479 | 347.96 | 849.2 | 1197.2 | 0.5348 | 1.0175 | 1.5523 | 347.34 | 765.4 | 1112.8 | **184** |
| **186** | 375.75 | 0.01831 | 2.454 | 348.92 | 848.4 | 1197.3 | 0.5359 | 1.0155 | 1.5514 | 348.29 | 764.6 | 1112.9 | **186** |
| **188** | 376.64 | 0.01832 | 2.429 | 349.86 | 847.6 | 1197.5 | 0.5370 | 1.0136 | 1.5506 | 349.22 | 763.8 | 1113.0 | **188** |
| **190** | 377.51 | 0.01833 | 2.404 | 350.79 | 846.8 | 1197.6 | 0.5381 | 1.0116 | 1.5497 | 350.15 | 763.0 | 1113.1 | **190** |
| **192** | 378.38 | 0.01834 | 2.380 | 351.72 | 846.1 | 1197.8 | 0.5392 | 1.0096 | 1.5488 | 351.07 | 762.1 | 1113.2 | **192** |
| **194** | 379.24 | 0.01835 | 2.356 | 352.64 | 845.3 | 1197.9 | 0.5403 | 1.0076 | 1.5479 | 351.98 | 761.3 | 1113.3 | **194** |
| **196** | 380.10 | 0.01836 | 2.333 | 353.55 | 844.5 | 1198.1 | 0.5414 | 1.0056 | 1.5470 | 352.89 | 760.6 | 1113.5 | **196** |
| **198** | 380.95 | 0.01838 | 2.310 | 354.46 | 843.7 | 1198.2 | 0.5425 | 1.0037 | 1.5462 | 353.79 | 759.8 | 1113.6 | **198** |
| **200** | 381.79 | 0.01839 | 2.288 | 355.36 | 843.0 | 1198.4 | 0.5435 | 1.0018 | 1.5453 | 354.68 | 759.0 | 1113.7 | **200** |
| **205** | 383.86 | 0.01842 | 2.234 | 357.58 | 841.1 | 1198.7 | 0.5461 | 0.9971 | 1.5432 | 356.88 | 757.1 | 1114.0 | **205** |
| **210** | 385.90 | 0.01844 | 2.183 | 359.77 | 839.2 | 1199.0 | 0.5487 | 0.9925 | 1.5412 | 359.05 | 755.2 | 1114.2 | **210** |
| **215** | 387.89 | 0.01847 | 2.134 | 361.91 | 837.4 | 1199.3 | 0.5512 | 0.9880 | 1.5392 | 361.18 | 753.2 | 1114.4 | **215** |
| **220** | 389.86 | 0.01850 | 2.087 | 364.02 | 835.6 | 1199.6 | 0.5537 | 0.9835 | 1.5372 | 363.27 | 751.3 | 1114.6 | **220** |
| **225** | 391.79 | 0.01852 | 2.0422 | 366.09 | 833.8 | 1199.9 | 0.5561 | 0.9792 | 1.5353 | 365.32 | 749.5 | 1114.8 | **225** |
| **230** | 393.68 | 0.01854 | 1.9992 | 368.13 | 832.0 | 1200.1 | 0.5585 | 0.9750 | 1.5334 | 367.34 | 747.7 | 1115.0 | **230** |
| **235** | 395.54 | 0.01857 | 1.9579 | 370.14 | 830.3 | 1200.4 | 0.5608 | 0.9708 | 1.5316 | 369.33 | 745.9 | 1115.3 | **235** |
| **240** | 397.37 | 0.01860 | 1.9183 | 372.12 | 828.5 | 1200.6 | 0.5631 | 0.9667 | 1.5298 | 371.29 | 744.1 | 1115.4 | **240** |
| **245** | 399.18 | 0.01863 | 1.8803 | 374.08 | 826.8 | 1200.9 | 0.5653 | 0.9627 | 1.5280 | 373.23 | 742.4 | 1115.6 | **245** |
| **250** | 400.95 | 0.01865 | 1.8438 | 376.00 | 825.1 | 1201.1 | 0.5675 | 0.9588 | 1.5263 | 375.14 | 740.7 | 1115.8 | **250** |
| **255** | 402.70 | 0.01868 | 1.8086 | 377.89 | 823.4 | 1201.3 | 0.5697 | 0.9549 | 1.5246 | 377.01 | 739.0 | 1116.0 | **255** |
| **260** | 404.42 | 0.01870 | 1.7748 | 379.76 | 821.8 | 1201.5 | 0.5719 | 0.9510 | 1.5229 | 378.86 | 737.3 | 1116.1 | **260** |
| **265** | 406.11 | 0.01873 | 1.7422 | 381.60 | 820.1 | 1201.7 | 0.5740 | 0.9472 | 1.5212 | 380.68 | 735.6 | 1116.3 | **265** |
| **270** | 407.78 | 0.01875 | 1.7107 | 383.42 | 818.5 | 1201.9 | 0.5760 | 0.9436 | 1.5196 | 382.48 | 733.9 | 1116.4 | **270** |
| **275** | 409.43 | 0.01878 | 1.6804 | 385.21 | 816.9 | 1202.1 | 0.5781 | 0.9399 | 1.5180 | 384.26 | 732.3 | 1116.6 | **275** |
| **280** | 411.05 | 0.01880 | 1.6511 | 386.98 | 815.3 | 1202.3 | 0.5801 | 0.9363 | 1.5164 | 386.01 | 730.7 | 1116.7 | **280** |
| **285** | 412.65 | 0.01883 | 1.6228 | 388.73 | 813.7 | 1202.4 | 0.5821 | 0.9327 | 1.5149 | 387.74 | 729.1 | 1116.8 | **285** |
| **290** | 414.23 | 0.01885 | 1.5954 | 390.46 | 812.1 | 1202.6 | 0.5841 | 0.9292 | 1.5133 | 389.45 | 727.5 | 1116.9 | **290** |
| **295** | 415.79 | 0.01887 | 1.5689 | 392.16 | 810.5 | 1202.7 | 0.5860 | 0.9258 | 1.5118 | 391.13 | 725.9 | 1117.0 | **295** |
| **300** | 417.33 | 0.01890 | 1.5433 | 393.84 | 809.0 | 1202.8 | 0.5879 | 0.9225 | 1.5104 | 392.79 | 724.3 | 1117.1 | **300** |
| **310** | 420.35 | 0.01894 | 1.4944 | 397.15 | 806.0 | 1203.1 | 0.5916 | 0.9159 | 1.5075 | 396.06 | 721.3 | 1117.4 | **310** |
| **320** | 423.29 | 0.01899 | 1.4485 | 400.39 | 803.0 | 1203.4 | 0.5952 | 0.9094 | 1.5046 | 399.26 | 718.3 | 1117.6 | **320** |
| **330** | 426.16 | 0.01904 | 1.4053 | 403.56 | 800.0 | 1203.6 | 0.5988 | 0.9031 | 1.5019 | 402.40 | 715.4 | 1117.8 | **330** |
| **340** | 428.97 | 0.01908 | 1.3645 | 406.66 | 797.1 | 1203.7 | 0.6022 | 0.8970 | 1.4992 | 405.46 | 712.4 | 1117.9 | **340** |
| **350** | 431.72 | 0.01913 | 1.3260 | 409.69 | 794.2 | 1203.9 | 0.6056 | 0.8910 | 1.4966 | 408.45 | 709.6 | 1118.0 | **350** |
| **360** | 434.40 | 0.01917 | 1.2895 | 412.67 | 791.4 | 1204.1 | 0.6090 | 0.8851 | 1.4941 | 411.39 | 706.8 | 1118.2 | **360** |
| **370** | 437.03 | 0.01921 | 1.2550 | 415.59 | 788.6 | 1204.2 | 0.6122 | 0.8794 | 1.4916 | 414.27 | 704.0 | 1118.3 | **370** |
| **380** | 439.60 | 0.01925 | 1.2222 | 418.45 | 785.8 | 1204.3 | 0.6153 | 0.8738 | 1.4891 | 417.10 | 701.3 | 1118.4 | **380** |
| **390** | 442.12 | 0.01930 | 1.1910 | 421.27 | 783.1 | 1204.4 | 0.6184 | 0.8683 | 1.4867 | 419.88 | 698.6 | 1118.5 | **390** |
| **400** | 444.59 | 0.0193 | 1.1613 | 424.0 | 780.5 | 1204.5 | 0.6214 | 0.8630 | 1.4844 | 422.6 | 695.9 | 1118.5 | **400** |
| **410** | 447.01 | 0.0194 | 1.1330 | 426.8 | 777.7 | 1204.5 | 0.6243 | 0.8578 | 1.4821 | 425.3 | 693.3 | 1118.6 | **410** |
| **420** | 449.39 | 0.0194 | 1.1061 | 429.4 | 775.2 | 1204.6 | 0.6272 | 0.8527 | 1.4799 | 427.9 | 690.8 | 1118.7 | **420** |
| **430** | 451.73 | 0.0194 | 1.0803 | 432.1 | 772.5 | 1204.6 | 0.6301 | 0.8476 | 1.4777 | 430.5 | 688.2 | 1118.7 | **430** |
| **440** | 454.02 | 0.0195 | 1.0556 | 434.6 | 770.0 | 1204.6 | 0.6329 | 0.8426 | 1.4755 | 433.0 | 685.7 | 1118.7 | **440** |
| **450** | 456.28 | 0.0195 | 1.0320 | 437.2 | 767.4 | 1204.6 | 0.6356 | 0.8378 | 1.4734 | 435.5 | 683.2 | 1118.7 | **450** |
| **460** | 458.50 | 0.0196 | 1.0094 | 439.7 | 764.9 | 1204.6 | 0.6383 | 0.8330 | 1.4713 | 438.0 | 680.7 | 1118.7 | **460** |
| **470** | 460.68 | 0.0196 | 0.9878 | 442.2 | 762.4 | 1204.6 | 0.6410 | 0.8283 | 1.4693 | 440.5 | 678.2 | 1118.7 | **470** |
| **480** | 462.82 | 0.0197 | 0.9670 | 444.6 | 759.9 | 1204.5 | 0.6436 | 0.8237 | 1.4673 | 442.9 | 675.7 | 1118.6 | **480** |
| **490** | 464.93 | 0.0197 | 0.9470 | 447.0 | 757.5 | 1204.5 | 0.6462 | 0.8191 | 1.4653 | 445.2 | 673.4 | 1118.6 | **490** |

## Table 2. Saturation: Pressures

| Abs. Press. Lb./Sq. In. p | Temp. Fahr. t | Specific Volume Sat. Liquid $v_f$ | Specific Volume Evap. $v_{fg}$ | Specific Volume Sat. Vapor $v_g$ | Enthalpy Sat. Liquid $h_f$ | Enthalpy Evap. $h_{fg}$ | Enthalpy Sat. Vapor $h_g$ | Entropy Sat. Liquid $s_f$ | Entropy Evap. $s_{fg}$ | Entropy Sat. Vapor $s_g$ | Internal Energy Sat. Liquid $u_f$ | Internal Energy Sat. Vapor $u_g$ | Abs. Press. Lb./Sq. In. p |
|---|---|---|---|---|---|---|---|---|---|---|---|---|---|
| **500** | 467.01 | 0.0197 | 0.9081 | 0.9278 | 449.4 | 755.0 | 1204.4 | 0.6487 | 0.8147 | 1.4634 | 447.6 | 1118.6 | **500** |
| **520** | 471.07 | 0.0198 | 0.8717 | 0.8915 | 454.1 | 750.1 | 1204.2 | 0.6536 | 0.8060 | 1.4596 | 452.2 | 1118.4 | **520** |
| **540** | 475.01 | 0.0199 | 0.8379 | 0.8578 | 458.6 | 745.4 | 1204.0 | 0.6584 | 0.7976 | 1.4560 | 456.6 | 1118.3 | **540** |
| **560** | 478.85 | 0.0200 | 0.8065 | 0.8265 | 463.0 | 740.8 | 1203.8 | 0.6631 | 0.7893 | 1.4524 | 460.9 | 1118.2 | **560** |
| **580** | 482.58 | 0.0201 | 0.7772 | 0.7973 | 467.4 | 736.1 | 1203.5 | 0.6676 | 0.7813 | 1.4489 | 465.2 | 1118.0 | **580** |
| **600** | 486.21 | 0.0201 | 0.7497 | 0.7698 | 471.6 | 731.6 | 1203.2 | 0.6720 | 0.7734 | 1.4454 | 469.4 | 1117.7 | **600** |
| **620** | 489.75 | 0.0202 | 0.7238 | 0.7440 | 475.7 | 727.2 | 1202.9 | 0.6763 | 0.7658 | 1.4421 | 473.4 | 1117.5 | **620** |
| **640** | 493.21 | 0.0203 | 0.6995 | 0.7198 | 479.8 | 722.7 | 1202.5 | 0.6805 | 0.7584 | 1.4389 | 477.4 | 1117.3 | **640** |
| **660** | 496.58 | 0.0204 | 0.6767 | 0.6971 | 483.8 | 718.3 | 1202.1 | 0.6846 | 0.7512 | 1.4358 | 481.3 | 1117.0 | **660** |
| **680** | 499.88 | 0.0204 | 0.6553 | 0.6757 | 487.7 | 714.0 | 1201.7 | 0.6886 | 0.7441 | 1.4327 | 485.1 | 1116.7 | **680** |
| **700** | 503.10 | 0.0205 | 0.6349 | 0.6554 | 491.5 | 709.7 | 1201.2 | 0.6925 | 0.7371 | 1.4296 | 488.8 | 1116.3 | **700** |
| **720** | 506.25 | 0.0206 | 0.6156 | 0.6362 | 495.3 | 705.4 | 1200.7 | 0.6963 | 0.7303 | 1.4266 | 492.5 | 1116.0 | **720** |
| **740** | 509.34 | 0.0207 | 0.5973 | 0.6180 | 499.0 | 701.2 | 1200.2 | 0.7001 | 0.7237 | 1.4237 | 496.2 | 1115.6 | **740** |
| **760** | 512.36 | 0.0207 | 0.5800 | 0.6007 | 502.6 | 697.1 | 1199.7 | 0.7037 | 0.7172 | 1.4209 | 499.7 | 1115.2 | **760** |
| **780** | 515.33 | 0.0208 | 0.5635 | 0.5843 | 506.2 | 692.9 | 1199.1 | 0.7073 | 0.7108 | 1.4181 | 503.2 | 1114.8 | **780** |
| **800** | 518.23 | 0.0209 | 0.5478 | 0.5687 | 509.7 | 688.9 | 1198.6 | 0.7108 | 0.7045 | 1.4153 | 506.6 | 1114.4 | **800** |
| **820** | 521.08 | 0.0209 | 0.5329 | 0.5538 | 513.2 | 684.8 | 1198.0 | 0.7143 | 0.6983 | 1.4126 | 510.0 | 1114.0 | **820** |
| **840** | 523.88 | 0.0210 | 0.5186 | 0.5396 | 516.6 | 680.8 | 1197.4 | 0.7177 | 0.6922 | 1.4099 | 513.3 | 1113.6 | **840** |
| **860** | 526.63 | 0.0211 | 0.5049 | 0.5260 | 520.0 | 676.8 | 1196.8 | 0.7210 | 0.6862 | 1.4072 | 516.6 | 1113.1 | **860** |
| **880** | 529.33 | 0.0212 | 0.4918 | 0.5130 | 523.3 | 672.8 | 1196.1 | 0.7243 | 0.6803 | 1.4046 | 519.9 | 1112.6 | **880** |
| **900** | 531.98 | 0.0212 | 0.4794 | 0.5006 | 526.6 | 668.8 | 1195.4 | 0.7275 | 0.6744 | 1.4020 | 523.1 | 1112.1 | **900** |
| **920** | 534.59 | 0.0213 | 0.4673 | 0.4886 | 529.8 | 664.9 | 1194.7 | 0.7307 | 0.6687 | 1.3995 | 526.2 | 1111.5 | **920** |
| **940** | 537.16 | 0.0214 | 0.4558 | 0.4772 | 533.0 | 661.0 | 1194.0 | 0.7339 | 0.6631 | 1.3970 | 529.3 | 1111.0 | **940** |
| **960** | 539.68 | 0.0214 | 0.4449 | 0.4663 | 536.2 | 657.1 | 1193.3 | 0.7370 | 0.6576 | 1.3945 | 532.4 | 1110.5 | **960** |
| **980** | 542.17 | 0.0215 | 0.4342 | 0.4557 | 539.3 | 653.3 | 1192.6 | 0.7400 | 0.6521 | 1.3921 | 535.4 | 1110.0 | **980** |
| **1000** | 544.61 | 0.0216 | 0.4240 | 0.4456 | 542.4 | 649.4 | 1191.8 | 0.7430 | 0.6467 | 1.3897 | 538.4 | 1109.4 | **1000** |
| **1050** | 550.57 | 0.0218 | 0.4000 | 0.4218 | 550.0 | 639.9 | 1189.9 | 0.7504 | 0.6334 | 1.3838 | 545.8 | 1108.0 | **1050** |
| **1100** | 556.31 | 0.0220 | 0.3781 | 0.4001 | 557.4 | 630.4 | 1187.8 | 0.7575 | 0.6205 | 1.3780 | 552.9 | 1106.4 | **1100** |
| **1150** | 561.86 | 0.0221 | 0.3581 | 0.3802 | 564.6 | 621.0 | 1185.6 | 0.7644 | 0.6079 | 1.3723 | 559.9 | 1104.7 | **1150** |
| **1200** | 567.22 | 0.0223 | 0.3396 | 0.3619 | 571.7 | 611.7 | 1183.4 | 0.7711 | 0.5956 | 1.3667 | 566.7 | 1103.0 | **1200** |
| **1250** | 572.42 | 0.0225 | 0.3225 | 0.3450 | 578.6 | 602.4 | 1181.0 | 0.7776 | 0.5836 | 1.3612 | 573.4 | 1101.2 | **1250** |
| **1300** | 577.46 | 0.0227 | 0.3066 | 0.3293 | 585.4 | 593.2 | 1178.6 | 0.7840 | 0.5719 | 1.3559 | 580.0 | 1099.4 | **1300** |
| **1350** | 582.35 | 0.0229 | 0.2919 | 0.3148 | 592.1 | 584.0 | 1176.1 | 0.7902 | 0.5604 | 1.3506 | 586.4 | 1097.5 | **1350** |
| **1400** | 587.10 | 0.0231 | 0.2781 | 0.3012 | 598.7 | 574.7 | 1173.4 | 0.7963 | 0.5491 | 1.3454 | 592.7 | 1095.4 | **1400** |
| **1450** | 591.73 | 0.0233 | 0.2651 | 0.2884 | 605.2 | 565.5 | 1170.7 | 0.8023 | 0.5379 | 1.3402 | 599.0 | 1093.3 | **1450** |
| **1500** | 596.23 | 0.0235 | 0.2530 | 0.2765 | 611.6 | 556.3 | 1167.9 | 0.8082 | 0.5269 | 1.3351 | 605.1 | 1091.2 | **1500** |
| **1600** | 604.90 | 0.0239 | 0.2309 | 0.2548 | 624.1 | 538.0 | 1162.1 | 0.8196 | 0.5053 | 1.3249 | 617.0 | 1086.7 | **1600** |
| **1700** | 613.15 | 0.0243 | 0.2111 | 0.2354 | 636.3 | 519.6 | 1155.9 | 0.8306 | 0.4843 | 1.3149 | 628.7 | 1081.8 | **1700** |
| **1800** | 621.03 | 0.0247 | 0.1932 | 0.2179 | 648.3 | 501.1 | 1149.4 | 0.8412 | 0.4637 | 1.3049 | 640.1 | 1076.8 | **1800** |
| **1900** | 628.58 | 0.0252 | 0.1769 | 0.2021 | 660.1 | 482.4 | 1142.4 | 0.8516 | 0.4433 | 1.2949 | 651.2 | 1071.4 | **1900** |
| **2000** | 635.82 | 0.0257 | 0.1621 | 0.1878 | 671.7 | 463.4 | 1135.1 | 0.8619 | 0.4230 | 1.2849 | 662.2 | 1065.6 | **2000** |
| **2100** | 642.77 | 0.0262 | 0.1484 | 0.1746 | 683.3 | 444.1 | 1127.4 | 0.8721 | 0.4027 | 1.2748 | 673.1 | 1059.6 | **2100** |
| **2200** | 649.46 | 0.0268 | 0.1358 | 0.1625 | 694.8 | 424.4 | 1119.2 | 0.8820 | 0.3826 | 1.2646 | 683.9 | 1053.1 | **2200** |
| **2300** | 655.91 | 0.0274 | 0.1239 | 0.1513 | 706.5 | 403.9 | 1110.4 | 0.8921 | 0.3621 | 1.2541 | 694.8 | 1046.0 | **2300** |
| **2400** | 662.12 | 0.0280 | 0.1128 | 0.1407 | 718.4 | 382.7 | 1101.1 | 0.9023 | 0.3411 | 1.2434 | 706.0 | 1038.6 | **2400** |
| **2500** | 668.13 | 0.0287 | 0.1021 | 0.1307 | 730.6 | 360.5 | 1091.1 | 0.9126 | 0.3197 | 1.2322 | 717.3 | 1030.6 | **2500** |
| **2600** | 673.94 | 0.0295 | 0.0918 | 0.1213 | 743.0 | 337.2 | 1080.2 | 0.9232 | 0.2973 | 1.2205 | 728.8 | 1021.9 | **2600** |
| **2700** | 679.55 | 0.0305 | 0.0818 | 0.1123 | 756.2 | 312.1 | 1068.3 | 0.9342 | 0.2740 | 1.2082 | 741.0 | 1012.3 | **2700** |
| **2800** | 684.99 | 0.0315 | 0.0719 | 0.1035 | 770.1 | 284.7 | 1054.8 | 0.9459 | 0.2487 | 1.1946 | 753.8 | 1001.2 | **2800** |
| **2900** | 690.26 | 0.0329 | 0.0618 | 0.0947 | 785.4 | 253.6 | 1039.0 | 0.9587 | 0.2205 | 1.1792 | 767.7 | 988.2 | **2900** |
| **3000** | 695.36 | 0.0346 | 0.0512 | 0.0858 | 802.5 | 217.8 | 1020.3 | 0.9731 | 0.1885 | 1.1615 | 783.4 | 972.7 | **3000** |
| **3100** | 700.31 | 0.0371 | 0.0382 | 0.0753 | 825.0 | 168.1 | 993.1 | 0.9919 | 0.1449 | 1.1368 | 803.7 | 949.9 | **3100** |
| **3200** | 705.11 | 0.0444 | 0.0136 | 0.0580 | 872.4 | 62.0 | 934.4 | 1.0320 | 0.0532 | 1.0852 | 846.0 | 898.4 | **3200** |
| **3206.2** | 705.40 | 0.0503 | 0 | 0.0503 | 902.7 | 0 | 902.7 | 1.0580 | 0 | 1.0580 | 872.9 | 872.9 | **3206.2** |

# Table 3. Superheated Vapor

| Abs. Press. Lb./Sq. In. (Sat. Temp.) | | Sat. Liquid | Sat. Vapor | 120° | 140° | 160° | 180° | 200° | 220° | 240° | 260° | 280° | 300° | 320° | 340° | 360° | 380° |
|---|---|---|---|---|---|---|---|---|---|---|---|---|---|---|---|---|---|
| | v | 0.02 | 333.6 | 344.6 | 356.6 | 368.6 | 380.6 | 392.6 | 404.5 | 416.5 | 428.4 | 440.4 | 452.3 | 464.2 | 476.2 | 488.1 | 500.0 |
| 1 | h | 69.7 | 1106.0 | 1114.3 | 1123.3 | 1132.4 | 1141.4 | 1150.4 | 1159.5 | 1168.5 | 1177.6 | 1186.7 | 1195.8 | 1204.9 | 1214.1 | 1223.3 | 1232.5 |
| (101.74) | s | 0.1326 | 1.9782 | 1.9928 | 2.0081 | 2.0230 | 2.0373 | 2.0512 | 2.0647 | 2.0779 | 2.0907 | 2.1031 | 2.1153 | 2.1271 | 2.1387 | 2.1501 | 2.1612 |
| | v | 0.02 | 173.7 | | 177.96 | 184.01 | 190.04 | 196.06 | 202.1 | 208.1 | 214.1 | 220.0 | 226.0 | 232.0 | 238.0 | 244.0 | 249.9 |
| 2 | h | 94.0 | 1116.2 | | 1122.6 | 1131.8 | 1140.9 | 1150.0 | 1159.1 | 1168.2 | 1177.3 | 1186.5 | 1195.6 | 1204.8 | 1213.9 | 1223.1 | 1232.3 |
| (126.08) | s | 0.1749 | 1.9200 | | 1.9308 | 1.9458 | 1.9603 | 1.9743 | 1.9879 | 2.0011 | 2.0140 | 2.0265 | 2.0387 | 2.0506 | 2.0622 | 2.0735 | 2.0846 |
| | v | 0.02 | 118.7 | | | 122.48 | 126.53 | 130.56 | 134.58 | 138.59 | 142.60 | 146.60 | 150.60 | 154.59 | 158.58 | 162.57 | 166.56 |
| 3 | h | 109.4 | 1122.6 | | | 1131.2 | 1140.4 | 1149.6 | 1158.8 | 1167.9 | 1177.1 | 1186.2 | 1195.4 | 1204.6 | 1213.8 | 1223.0 | 1232.2 |
| (141.48) | s | 0.2008 | 1.8863 | | | 1.9003 | 1.9150 | 1.9291 | 1.9428 | 1.9561 | 1.9690 | 1.9815 | 1.9937 | 2.0057 | 2.0173 | 2.0287 | 2.0398 |
| | v | 0.02 | 90.63 | | | 91.71 | 94.77 | 97.81 | 100.84 | 103.86 | 106.87 | 109.88 | 112.88 | 115.88 | 118.88 | 121.88 | 124.87 |
| 4 | h | 120.9 | 1127.3 | | | 1130.6 | 1139.9 | 1149.2 | 1158.5 | 1167.6 | 1176.8 | 1186.0 | 1195.2 | 1204.4 | 1213.6 | 1222.8 | 1232.1 |
| (152.97) | s | 0.2198 | 1.8625 | | | 1.8678 | 1.8827 | 1.8969 | 1.9107 | 1.9240 | 1.9370 | 1.9496 | 1.9618 | 1.9738 | 1.9854 | 1.9968 | 2.0080 |
| | v | 0.02 | 73.52 | | | | 75.71 | 78.16 | 80.59 | 83.01 | 85.43 | 87.85 | 90.25 | 92.66 | 95.06 | 97.46 | 99.86 |
| 5 | h | 130.1 | 1131.1 | | | | 1139.4 | 1148.8 | 1158.1 | 1167.3 | 1176.5 | 1185.7 | 1195.0 | 1204.2 | 1213.4 | 1222.7 | 1231.9 |
| (162.24) | s | 0.2347 | 1.8441 | | | | 1.8574 | 1.8718 | 1.8857 | 1.8991 | 1.9121 | 1.9247 | 1.9370 | 1.9490 | 1.9607 | 1.9721 | 1.9833 |
| | v | 0.02 | 61.98 | | | | 63.00 | 65.05 | 67.09 | 69.12 | 71.14 | 73.16 | 75.17 | 77.18 | 79.18 | 81.18 | 83.18 |
| 6 | h | 138.0 | 1134.2 | | | | 1138.9 | 1148.3 | 1157.7 | 1167.0 | 1176.3 | 1185.5 | 1194.7 | 1204.0 | 1213.2 | 1222.5 | 1231.8 |
| (170.06) | s | 0.2472 | 1.8292 | | | | 1.8367 | 1.8512 | 1.8651 | 1.8786 | 1.8917 | 1.9044 | 1.9167 | 1.9287 | 1.9404 | 1.9518 | 1.9630 |
| | v | 0.02 | 53.64 | | | | 53.93 | 55.69 | 57.45 | 59.19 | 60.93 | 62.67 | 64.39 | 66.12 | 67.84 | 69.56 | 71.27 |
| 7 | h | 144.8 | 1136.9 | | | | 1138.4 | 1147.9 | 1157.4 | 1166.7 | 1176.0 | 1185.3 | 1194.5 | 1203.8 | 1213.1 | 1222.3 | 1231.6 |
| (176.85) | s | 0.2581 | 1.8167 | | | | 1.8190 | 1.8336 | 1.8477 | 1.8613 | 1.8744 | 1.8871 | 1.8995 | 1.9115 | 1.9232 | 1.9347 | 1.9459 |
| | v | 0.02 | 47.34 | | | | | 48.67 | 50.22 | 51.75 | 53.28 | 54.80 | 56.31 | 57.82 | 59.33 | 60.84 | 62.34 |
| 8 | h | 150.8 | 1139.3 | | | | | 1147.5 | 1157.0 | 1166.3 | 1175.7 | 1185.0 | 1194.3 | 1203.6 | 1212.9 | 1222.2 | 1231.5 |
| (182.86) | s | 0.2674 | 1.8057 | | | | | 1.8184 | 1.8325 | 1.8462 | 1.8594 | 1.8721 | 1.8845 | 1.8966 | 1.9084 | 1.9199 | 1.9311 |
| | v | 0.02 | 42.40 | | | | | 43.21 | 44.59 | 45.96 | 47.32 | 48.67 | 50.03 | 51.37 | 52.71 | 54.06 | 55.39 |
| 9 | h | 156.2 | 1141.4 | | | | | 1147.0 | 1156.6 | 1166.0 | 1175.4 | 1184.8 | 1194.1 | 1203.4 | 1212.7 | 1222.0 | 1231.4 |
| (188.28) | s | 0.2759 | 1.7962 | | | | | 1.8049 | 1.8191 | 1.8328 | 1.8461 | 1.8589 | 1.8713 | 1.8834 | 1.8952 | 1.9067 | 1.9179 |
| | v | 0.02 | 38.42 | | | | | 38.85 | 40.09 | 41.33 | 42.56 | 43.78 | 45.00 | 46.21 | 47.42 | 48.63 | 49.84 |
| 10 | h | 161.2 | 1143.3 | | | | | 1146.6 | 1156.2 | 1165.7 | 1175.1 | 1184.5 | 1193.9 | 1203.2 | 1212.5 | 1221.9 | 1231.2 |
| (193.21) | s | 0.2835 | 1.7876 | | | | | 1.7927 | 1.8071 | 1.8208 | 1.8341 | 1.8470 | 1.8595 | 1.8716 | 1.8834 | 1.8950 | 1.9062 |
| | v | 0.02 | 35.14 | | | | | 35.27 | 36.41 | 37.54 | 38.66 | 39.77 | 40.88 | 41.99 | 43.09 | 44.19 | 45.29 |
| 11 | h | 165.7 | 1145.0 | | | | | 1146.1 | 1155.8 | 1165.4 | 1174.9 | 1184.3 | 1193.6 | 1203.0 | 1212.4 | 1221.7 | 1231.1 |
| (197.75) | s | 0.2903 | 1.7800 | | | | | 1.7816 | 1.7961 | 1.8100 | 1.8233 | 1.8362 | 1.8487 | 1.8609 | 1.8728 | 1.8843 | 1.8956 |
| | v | 0.02 | 32.40 | | | | | | 33.34 | 34.38 | 35.41 | 36.43 | 37.45 | 38.47 | 39.48 | 40.49 | 41.50 |
| 12 | h | 170.0 | 1146.6 | | | | | | 1155.4 | 1165.0 | 1174.6 | 1184.0 | 1193.4 | 1202.8 | 1212.2 | 1221.6 | 1230.9 |
| (201.96) | s | 0.2967 | 1.7730 | | | | | | 1.7860 | 1.8000 | 1.8134 | 1.8264 | 1.8389 | 1.8511 | 1.8630 | 1.8745 | 1.8858 |
| | v | 0.02 | 30.06 | | | | | | 30.74 | 31.71 | 32.66 | 33.61 | 34.55 | 35.49 | 36.43 | 37.36 | 38.29 |
| 13 | h | 173.9 | 1148.1 | | | | | | 1155.1 | 1164.7 | 1174.3 | 1183.8 | 1193.2 | 1202.6 | 1212.0 | 1221.4 | 1230.8 |
| (205.88) | s | 0.3027 | 1.7665 | | | | | | 1.7768 | 1.7908 | 1.8043 | 1.8173 | 1.8299 | 1.8421 | 1.8540 | 1.8656 | 1.8769 |
| | v | 0.02 | 28.04 | | | | | | 28.52 | 29.41 | 30.30 | 31.19 | 32.06 | 32.94 | 33.81 | 34.68 | 35.54 |
| 14 | h | 177.6 | 1149.5 | | | | | | 1154.6 | 1164.4 | 1174.0 | 1183.5 | 1193.0 | 1202.4 | 1211.8 | 1221.2 | 1230.6 |
| (209.56) | s | 0.3083 | 1.7605 | | | | | | 1.7681 | 1.7822 | 1.7958 | 1.8088 | 1.8215 | 1.8337 | 1.8457 | 1.8573 | 1.8686 |
| | v | 0.02 | 26.80 | | | | | | 27.15 | 28.00 | 28.85 | 29.70 | 30.53 | 31.37 | 32.20 | 33.03 | 33.85 |
| 14.696 | h | 180.1 | 1150.4 | | | | | | 1154.4 | 1164.2 | 1173.8 | 1183.3 | 1192.8 | 1202.3 | 1211.7 | 1221.1 | 1230.5 |
| (212.00) | s | 0.3120 | 1.7566 | | | | | | 1.7624 | 1.7766 | 1.7902 | 1.8033 | 1.8160 | 1.8283 | 1.8402 | 1.8518 | 1.8631 |
| | v | 0.02 | 26.29 | | | | | | 26.59 | 27.43 | 28.26 | 29.09 | 29.91 | 30.73 | 31.54 | 32.35 | 33.16 |
| 15 | h | 181.1 | 1150.8 | | | | | | 1154.3 | 1164.1 | 1173.7 | 1183.2 | 1192.8 | 1202.2 | 1211.7 | 1221.1 | 1230.5 |
| (213.03) | s | 0.3135 | 1.7549 | | | | | | 1.7601 | 1.7742 | 1.7879 | 1.8010 | 1.8136 | 1.8259 | 1.8379 | 1.8495 | 1.8609 |
| | v | 0.02 | 24.75 | | | | | | 24.90 | 25.69 | 26.47 | 27.25 | 28.02 | 28.79 | 29.56 | 30.32 | 31.08 |
| 16 | h | 184.4 | 1152.0 | | | | | | 1153.8 | 1163.7 | 1173.4 | 1183.0 | 1192.5 | 1202.0 | 1211.5 | 1220.9 | 1230.3 |
| (216.32) | s | 0.3184 | 1.7497 | | | | | | 1.7524 | 1.7667 | 1.7804 | 1.7936 | 1.8063 | 1.8186 | 1.8306 | 1.8422 | 1.8536 |

Temperature—Degrees Fahrenheit

Temperature—Degrees Fahrenheit

| 400° | 420° | 440° | 460° | 480° | 500° | 600° | 700° | 800° | 900° | 1000° | 1100° | 1200° | 1300° | 1400° | 1600° | | Abs. Press. Lb./Sq. In. (Sat. Temp.) |
|---|---|---|---|---|---|---|---|---|---|---|---|---|---|---|---|---|---|
| 512.0 | 523.9 | 535.8 | 547.7 | 559.7 | 571.6 | 631.2 | 690.8 | 750.4 | 809.9 | 869.5 | 929.1 | 988.7 | 1048.3 | 1107.8 | 1227.0 | v | |
| 1241.7 | 1251.0 | 1260.3 | 1269.6 | 1278.9 | 1288.3 | 1335.7 | 1383.8 | 1432.8 | 1482.7 | 1533.5 | 1585.2 | 1637.7 | 1691.2 | 1745.7 | 1857.5 | h | **1** |
| 2.1720 | 2.1827 | 2.1931 | 2.2034 | 2.2134 | 2.2233 | 2.2702 | 2.3137 | 2.3542 | 2.3923 | 2.4283 | 2.4625 | 2.4952 | 2.5265 | 2.5566 | 2.6137 | s | (101.74) |
| 255.9 | 261.9 | 267.8 | 273.8 | 279.8 | 285.7 | 315.5 | 345.4 | 375.1 | 404.9 | 434.7 | 464.5 | 494.3 | 524.1 | 553.9 | 613.5 | v | |
| 1241.6 | 1250.9 | 1260.2 | 1269.5 | 1278.8 | 1288.2 | 1335.6 | 1383.8 | 1432.8 | 1482.7 | 1533.5 | 1585.1 | 1637.7 | 1691.2 | 1745.7 | 1857.4 | h | **2** |
| 2.0955 | 2.1062 | 2.1166 | 2.1269 | 2.1369 | 2.1468 | 2.1938 | 2.2372 | 2.2778 | 2.3159 | 2.3519 | 2.3861 | 2.4188 | 2.4501 | 2.4802 | 2.5373 | s | (126.08) |
| 170.54 | 174.52 | 178.51 | 182.48 | 186.47 | 190.44 | 210.3 | 230.2 | 250.1 | 269.9 | 289.8 | 309.7 | 329.5 | 349.4 | 369.3 | 409.0 | v | |
| 1241.5 | 1250.7 | 1260.0 | 1269.4 | 1278.7 | 1288.1 | 1335.5 | 1383.7 | 1432.8 | 1482.7 | 1533.4 | 1585.1 | 1637.7 | 1691.2 | 1745.7 | 1857.4 | h | **3** |
| 2.0507 | 2.0614 | 2.0719 | 2.0821 | 2.0921 | 2.1021 | 2.1490 | 2.1925 | 2.2330 | 2.2712 | 2.3072 | 2.3414 | 2.3741 | 2.4054 | 2.4355 | 2.4926 | s | (141.48) |
| 127.86 | 130.85 | 133.84 | 136.83 | 139.82 | 142.80 | 157.73 | 172.64 | 187.55 | 202.4 | 217.4 | 232.3 | 247.1 | 262.0 | 276.9 | 306.7 | v | |
| 1241.3 | 1250.6 | 1259.9 | 1269.3 | 1278.7 | 1288.1 | 1335.5 | 1383.7 | 1432.7 | 1482.6 | 1533.4 | 1585.1 | 1637.7 | 1691.2 | 1745.7 | 1857.4 | h | **4** |
| 2.0189 | 2.0296 | 2.0401 | 2.0503 | 2.0603 | 2.0703 | 2.1173 | 2.1608 | 2.2013 | 2.2394 | 2.2755 | 2.3097 | 2.3424 | 2.3737 | 2.4038 | 2.4609 | s | (152.97) |
| 102.26 | 104.65 | 107.04 | 109.44 | 111.83 | 114.22 | 126.16 | 138.10 | 150.03 | 161.95 | 173.87 | 185.79 | 197.71 | 209.6 | 221.6 | 245.4 | v | |
| 1241.2 | 1250.5 | 1259.8 | 1269.2 | 1278.6 | 1288.0 | 1335.4 | 1383.6 | 1432.7 | 1482.6 | 1533.4 | 1585.1 | 1637.7 | 1691.2 | 1745.7 | 1857.4 | h | **5** |
| 1.9942 | 2.0049 | 2.0154 | 2.0256 | 2.0356 | 2.0456 | 2.0927 | 2.1361 | 2.1767 | 2.2148 | 2.2509 | 2.2851 | 2.3178 | 2.3491 | 2.3792 | 2.4363 | s | (162.24) |
| 85.18 | 87.18 | 89.18 | 91.17 | 93.17 | 95.16 | 105.12 | 115.07 | 125.01 | 134.95 | 144.89 | 154.82 | 164.76 | 174.69 | 184.62 | 204.5 | v | |
| 1241.1 | 1250.4 | 1259.7 | 1269.1 | 1278.5 | 1287.9 | 1335.3 | 1383.6 | 1432.6 | 1482.6 | 1533.4 | 1585.0 | 1637.6 | 1691.1 | 1745.6 | 1857.4 | h | **6** |
| 1.9740 | 1.9847 | 1.9952 | 2.0055 | 2.0155 | 2.0255 | 2.0725 | 2.1160 | 2.1566 | 2.1947 | 2.2307 | 2.2650 | 2.2977 | 2.3290 | 2.3591 | 2.4162 | s | (170.06) |
| 72.99 | 74.70 | 76.42 | 78.13 | 79.84 | 81.55 | 90.09 | 98.62 | 107.14 | 115.67 | 124.18 | 132.70 | 141.22 | 149.73 | 158.25 | 175.27 | v | |
| 1240.9 | 1250.3 | 1259.6 | 1269.0 | 1278.4 | 1287.8 | 1335.3 | 1383.5 | 1432.6 | 1482.5 | 1533.3 | 1585.0 | 1637.6 | 1691.1 | 1745.6 | 1857.4 | h | **7** |
| 1.9569 | 1.9676 | 1.9781 | 1.9884 | 1.9984 | 2.0084 | 2.0555 | 2.0990 | 2.1396 | 2.1777 | 2.2137 | 2.2480 | 2.2807 | 2.3120 | 2.3421 | 2.3992 | s | (176.85) |
| 63.84 | 65.35 | 66.85 | 68.34 | 69.84 | 71.34 | 78.82 | 86.28 | 93.74 | 101.20 | 108.66 | 116.11 | 123.56 | 131.01 | 138.46 | 153.36 | v | |
| 1240.8 | 1250.1 | 1259.5 | 1268.9 | 1278.3 | 1287.7 | 1335.2 | 1383.5 | 1432.6 | 1482.5 | 1533.3 | 1585.0 | 1637.6 | 1691.1 | 1745.6 | 1857.4 | h | **8** |
| 1.9420 | 1.9528 | 1.9633 | 1.9736 | 1.9836 | 1.9936 | 2.0407 | 2.0842 | 2.1248 | 2.1630 | 2.1990 | 2.2333 | 2.2660 | 2.2973 | 2.3274 | 2.3844 | s | (182.86) |
| 56.73 | 58.07 | 59.40 | 60.74 | 62.07 | 63.40 | 70.05 | 76.69 | 83.32 | 89.95 | 96.58 | 103.20 | 109.83 | 116.45 | 123.08 | 136.32 | v | |
| 1240.7 | 1250.0 | 1259.4 | 1268.8 | 1278.2 | 1287.6 | 1335.2 | 1383.4 | 1432.5 | 1482.5 | 1533.3 | 1585.0 | 1637.6 | 1691.1 | 1745.6 | 1857.4 | h | **9** |
| 1.9289 | 1.9397 | 1.9502 | 1.9605 | 1.9706 | 1.9806 | 2.0277 | 2.0712 | 2.1118 | 2.1500 | 2.1860 | 2.2203 | 2.2530 | 2.2843 | 2.3144 | 2.3715 | s | (188.28) |
| 51.04 | 52.24 | 53.45 | 54.65 | 55.85 | 57.05 | 63.03 | 69.01 | 74.98 | 80.95 | 86.92 | 92.88 | 98.84 | 104.80 | 110.77 | 122.69 | v | |
| 1240.6 | 1249.9 | 1259.3 | 1268.7 | 1278.1 | 1287.5 | 1335.1 | 1383.4 | 1432.5 | 1482.4 | 1533.2 | 1585.0 | 1637.6 | 1691.1 | 1745.6 | 1857.3 | h | **10** |
| 1.9172 | 1.9280 | 1.9385 | 1.9488 | 1.9589 | 1.9689 | 2.0160 | 2.0596 | 2.1002 | 2.1383 | 2.1744 | 2.2086 | 2.2413 | 2.2727 | 2.3028 | 2.3598 | s | (193.21) |
| 46.39 | 47.48 | 48.58 | 49.67 | 50.76 | 51.85 | 57.30 | 62.73 | 68.16 | 73.59 | 79.01 | 84.43 | 89.85 | 95.27 | 100.69 | 111.53 | v | |
| 1240.4 | 1249.8 | 1259.2 | 1268.6 | 1278.0 | 1287.4 | 1335.0 | 1383.3 | 1432.4 | 1482.4 | 1533.2 | 1584.9 | 1637.6 | 1691.1 | 1745.6 | 1857.3 | h | **11** |
| 1.9066 | 1.9173 | 1.9279 | 1.9382 | 1.9484 | 1.9583 | 2.0055 | 2.0490 | 2.0897 | 2.1278 | 2.1639 | 2.1981 | 2.2308 | 2.2621 | 2.2922 | 2.3493 | s | (197.75) |
| 42.51 | 43.51 | 44.51 | 45.52 | 46.52 | 47.52 | 52.51 | 57.50 | 62.48 | 67.45 | 72.42 | 77.40 | 82.36 | 87.33 | 92.30 | 102.24 | v | |
| 1240.3 | 1249.7 | 1259.0 | 1268.5 | 1277.9 | 1287.3 | 1335.0 | 1383.3 | 1432.4 | 1482.4 | 1533.2 | 1584.9 | 1637.5 | 1691.0 | 1745.5 | 1857.3 | h | **12** |
| 1.8969 | 1.9076 | 1.9182 | 1.9285 | 1.9387 | 1.9486 | 1.9958 | 2.0394 | 2.0800 | 2.1182 | 2.1543 | 2.1885 | 2.2212 | 2.2525 | 2.2827 | 2.3397 | s | (201.96) |
| 39.22 | 40.15 | 41.08 | 42.00 | 42.93 | 43.85 | 48.47 | 53.07 | 57.66 | 62.26 | 66.85 | 71.44 | 76.03 | 80.61 | 85.20 | 94.37 | v | |
| 1240.2 | 1249.5 | 1258.9 | 1268.4 | 1277.8 | 1287.3 | 1334.9 | 1383.2 | 1432.4 | 1482.3 | 1533.2 | 1584.9 | 1637.5 | 1691.0 | 1745.5 | 1857.3 | h | **13** |
| 1.8879 | 1.8987 | 1.9093 | 1.9196 | 1.9298 | 1.9398 | 1.9870 | 2.0305 | 2.0712 | 2.1094 | 2.1454 | 2.1797 | 2.2124 | 2.2437 | 2.2738 | 2.3309 | s | (205.88) |
| 36.41 | 37.27 | 38.13 | 38.99 | 39.85 | 40.71 | 45.00 | 49.27 | 53.54 | 57.81 | 62.07 | 66.33 | 70.59 | 74.85 | 79.11 | 87.63 | v | |
| 1240.0 | 1249.4 | 1258.8 | 1268.3 | 1277.7 | 1287.2 | 1334.8 | 1383.2 | 1432.3 | 1482.3 | 1533.1 | 1584.9 | 1637.5 | 1691.0 | 1745.5 | 1857.3 | h | **14** |
| 1.8796 | 1.8904 | 1.9010 | 1.9114 | 1.9216 | 1.9315 | 1.9788 | 2.0224 | 2.0630 | 2.1012 | 2.1372 | 2.1715 | 2.2042 | 2.2355 | 2.2657 | 2.3227 | s | (209.56) |
| 34.68 | 35.50 | 36.32 | 37.14 | 37.96 | 38.78 | 42.86 | 46.94 | 51.00 | 55.07 | 59.13 | 63.19 | 67.25 | 71.31 | 75.37 | 83.48 | v | |
| 1239.9 | 1249.3 | 1258.8 | 1268.2 | 1277.6 | 1287.1 | 1334.8 | 1383.2 | 1432.3 | 1482.3 | 1533.1 | 1584.8 | 1637.5 | 1691.0 | 1745.5 | 1857.3 | h | **14.696** |
| 1.8743 | 1.8850 | 1.8956 | 1.9060 | 1.9162 | 1.9261 | 1.9734 | 2.0170 | 2.0576 | 2.0958 | 2.1319 | 2.1662 | 2.1989 | 2.2302 | 2.2603 | 2.3174 | s | (212.00) |
| 33.97 | 34.78 | 35.58 | 36.38 | 37.19 | 37.99 | 41.99 | 45.98 | 49.97 | 53.95 | 57.93 | 61.91 | 65.89 | 69.86 | 73.84 | 81.79 | v | |
| 1239.9 | 1249.3 | 1258.7 | 1268.2 | 1277.6 | 1287.1 | 1334.8 | 1383.1 | 1432.3 | 1482.3 | 1533.1 | 1584.8 | 1637.5 | 1691.0 | 1745.5 | 1857.3 | h | **15** |
| 1.8719 | 1.8827 | 1.8933 | 1.9037 | 1.9139 | 1.9238 | 1.9711 | 2.0147 | 2.0554 | 2.0936 | 2.1296 | 2.1639 | 2.1966 | 2.2279 | 2.2580 | 2.3151 | s | (213.03) |
| 31.84 | 32.59 | 33.35 | 34.10 | 34.86 | 35.61 | 39.36 | 43.10 | 46.84 | 50.58 | 54.31 | 58.04 | 61.77 | 65.49 | 69.22 | 76.67 | v | |
| 1239.8 | 1249.2 | 1258.6 | 1268.0 | 1277.5 | 1287.0 | 1334.7 | 1383.1 | 1432.3 | 1482.2 | 1533.1 | 1584.8 | 1637.4 | 1691.0 | 1745.5 | 1857.3 | h | **16** |
| 1.8647 | 1.8755 | 1.8861 | 1.8965 | 1.9067 | 1.9167 | 1.9639 | 2.0076 | 2.0482 | 2.0864 | 2.1225 | 2.1568 | 2.1895 | 2.2208 | 2.2509 | 2.3080 | s | (216.32) |

## Table 3. Superheated Vapor

| Abs. Press. Lb./Sq. In. (Sat. Temp.) | | Sat. Liquid | Sat. Vapor | 220° | 230° | 240° | 250° | 260° | 270° | 280° | 290° | 300° | 320° | 340° | 360° | 380° | 400° |
|---|---|---|---|---|---|---|---|---|---|---|---|---|---|---|---|---|---|
| | | | | Temperature—Degrees Fahrenheit | | | | | | | | | | | | | |
| | v | 0.02 | 23.39 | 23.41 | 23.78 | 24.16 | 24.53 | 24.90 | 25.26 | 25.63 | 26.00 | 26.36 | 27.08 | 27.80 | 28.52 | 29.24 | 29.95 |
| 17 | h | 187.6 | 1153.1 | 1153.4 | 1158.4 | 1163.4 | 1168.3 | 1173.1 | 1177.9 | 1182.7 | 1187.5 | 1192.3 | 1201.8 | 1211.3 | 1220.7 | 1230.2 | 1239.6 |
| (219.44) | s | 0.3231 | 1.7449 | 1.7453 | 1.7526 | 1.7597 | 1.7666 | 1.7734 | 1.7801 | 1.7866 | 1.7931 | 1.7994 | 1.8117 | 1.8238 | 1.8354 | 1.8468 | 1.8579 |
| | v | 0.02 | 22.17 | | 22.44 | 22.79 | 23.14 | 23.49 | 23.84 | 24.19 | 24.53 | 24.88 | 25.56 | 26.25 | 26.93 | 27.60 | 28.28 |
| 18 | h | 190.6 | 1154.2 | | 1158.0 | 1163.0 | 1167.9 | 1172.8 | 1177.7 | 1182.5 | 1187.3 | 1192.1 | 1201.6 | 1211.1 | 1220.6 | 1230.0 | 1239.5 |
| (222.41) | s | 0.3275 | 1.7403 | | 1.7458 | 1.7530 | 1.7599 | 1.7668 | 1.7735 | 1.7800 | 1.7865 | 1.7928 | 1.8052 | 1.8173 | 1.8290 | 1.8404 | 1.8515 |
| | v | 0.02 | 21.08 | | 21.24 | 21.57 | 21.91 | 22.24 | 22.57 | 22.90 | 23.23 | 23.56 | 24.20 | 24.85 | 25.50 | 26.14 | 26.78 |
| 19 | h | 193.4 | 1155.3 | | 1157.7 | 1162.7 | 1167.6 | 1172.5 | 1177.4 | 1182.2 | 1187.1 | 1191.9 | 1201.4 | 1210.9 | 1220.4 | 1229.9 | 1239.4 |
| (225.24) | s | 0.3317 | 1.7360 | | 1.7394 | 1.7466 | 1.7536 | 1.7605 | 1.7672 | 1.7738 | 1.7803 | 1.7867 | 1.7991 | 1.8111 | 1.8229 | 1.8343 | 1.8454 |
| | v | 0.02 | 20.09 | | 20.15 | 20.48 | 20.79 | 21.11 | 21.43 | 21.74 | 22.05 | 22.36 | 22.98 | 23.60 | 24.21 | 24.82 | 25.43 |
| 20 | h | 196.2 | 1156.3 | | 1157.3 | 1162.3 | 1167.3 | 1172.2 | 1177.1 | 1182.0 | 1186.8 | 1191.6 | 1201.2 | 1210.8 | 1220.3 | 1229.7 | 1239.2 |
| (227.96) | s | 0.3356 | 1.7319 | | 1.7333 | 1.7405 | 1.7476 | 1.7545 | 1.7613 | 1.7679 | 1.7744 | 1.7808 | 1.7932 | 1.8053 | 1.8170 | 1.8285 | 1.8396 |
| | v | 0.017 | 19.192 | | | 19.482 | 19.786 | 20.09 | 20.39 | 20.69 | 20.99 | 21.29 | 21.88 | 22.46 | 23.05 | 23.63 | 24.21 |
| 21 | h | 198.8 | 1157.2 | | | 1162.0 | 1167.0 | 1171.9 | 1176.8 | 1181.7 | 1186.6 | 1191.4 | 1201.0 | 1210.6 | 1220.1 | 1229.6 | 1239.1 |
| (230.57) | s | 0.3395 | 1.7280 | | | 1.7348 | 1.7419 | 1.7488 | 1.7556 | 1.7622 | 1.7687 | 1.7751 | 1.7876 | 1.7997 | 1.8115 | 1.8230 | 1.8341 |
| | v | 0.017 | 18.375 | | | 18.579 | 18.870 | 19.160 | 19.449 | 19.736 | 20.02 | 20.31 | 20.87 | 21.43 | 21.99 | 22.55 | 23.11 |
| 22 | h | 201.3 | 1158.1 | | | 1161.6 | 1166.6 | 1171.6 | 1176.5 | 1181.4 | 1186.3 | 1191.2 | 1200.8 | 1210.4 | 1219.9 | 1229.5 | 1239.0 |
| (233.07) | s | 0.3431 | 1.7242 | | | 1.7292 | 1.7363 | 1.7433 | 1.7501 | 1.7568 | 1.7633 | 1.7698 | 1.7823 | 1.7944 | 1.8062 | 1.8177 | 1.8289 |
| | v | 0.017 | 17.627 | | | 17.754 | 18.034 | 18.312 | 18.589 | 18.864 | 19.138 | 19.411 | 19.953 | 20.49 | 21.03 | 21.56 | 22.09 |
| 23 | h | 203.8 | 1159.0 | | | 1161.2 | 1166.3 | 1171.3 | 1176.3 | 1181.2 | 1186.1 | 1190.9 | 1200.6 | 1210.2 | 1219.8 | 1229.3 | 1238.8 |
| (235.49) | s | 0.3466 | 1.7206 | | | 1.7239 | 1.7311 | 1.7381 | 1.7449 | 1.7516 | 1.7582 | 1.7646 | 1.7772 | 1.7894 | 1.8012 | 1.8127 | 1.8238 |
| | v | 0.017 | 16.938 | | | 16.997 | 17.267 | 17.535 | 17.801 | 18.065 | 18.329 | 18.590 | 19.111 | 19.628 | 20.14 | 20.65 | 21.17 |
| 24 | h | 206.1 | 1159.8 | | | 1160.9 | 1166.0 | 1171.0 | 1176.0 | 1180.9 | 1185.8 | 1190.7 | 1200.4 | 1210.0 | 1219.6 | 1229.2 | 1238.7 |
| (237.82) | s | 0.3500 | 1.7172 | | | 1.7188 | 1.7260 | 1.7330 | 1.7399 | 1.7466 | 1.7532 | 1.7597 | 1.7723 | 1.7845 | 1.7963 | 1.8078 | 1.8190 |
| | v | 0.017 | 16.303 | | | | 16.561 | 16.819 | 17.076 | 17.330 | 17.584 | 17.836 | 18.337 | 18.834 | 19.329 | 19.821 | 20.31 |
| 25 | h | 208.4 | 1160.6 | | | | 1165.6 | 1170.7 | 1175.7 | 1180.6 | 1185.6 | 1190.5 | 1200.2 | 1209.8 | 1219.4 | 1229.0 | 1238.5 |
| (240.07) | s | 0.3533 | 1.7139 | | | | 1.7212 | 1.7282 | 1.7351 | 1.7418 | 1.7485 | 1.7550 | 1.7676 | 1.7798 | 1.7917 | 1.8032 | 1.8144 |
| | v | 0.017 | 15.715 | | | | 15.910 | 16.159 | 16.406 | 16.652 | 16.896 | 17.139 | 17.622 | 18.101 | 18.577 | 19.051 | 19.524 |
| 26 | h | 210.6 | 1161.3 | | | | 1165.3 | 1170.4 | 1175.4 | 1180.4 | 1185.3 | 1190.2 | 1200.0 | 1209.7 | 1219.3 | 1228.9 | 1238.4 |
| (242.25) | s | 0.3564 | 1.7108 | | | | 1.7164 | 1.7235 | 1.7305 | 1.7372 | 1.7439 | 1.7504 | 1.7631 | 1.7753 | 1.7872 | 1.7987 | 1.8100 |
| | v | 0.017 | 15.170 | | | | 15.307 | 15.548 | 15.786 | 16.024 | 16.259 | 16.494 | 16.960 | 17.422 | 17.882 | 18.339 | 18.794 |
| 27 | h | 212.8 | 1162.0 | | | | 1165.0 | 1170.1 | 1175.1 | 1180.1 | 1185.1 | 1190.0 | 1199.8 | 1209.5 | 1219.1 | 1228.7 | 1238.3 |
| (244.36) | s | 0.3594 | 1.7078 | | | | 1.7119 | 1.7190 | 1.7260 | 1.7328 | 1.7395 | 1.7460 | 1.7587 | 1.7710 | 1.7829 | 1.7944 | 1.8057 |
| | v | 0.017 | 14.663 | | | | 14.747 | 14.980 | 15.211 | 15.440 | 15 669 | 15.895 | 16.345 | 16.792 | 17.235 | 17.677 | 18.116 |
| 28 | h | 214.8 | 1162.7 | | | | 1164.6 | 1169.7 | 1174.8 | 1179.8 | 1184.8 | 1189.8 | 1199.6 | 1209.3 | 1218.9 | 1228.6 | 1238.1 |
| (246.41) | s | 0.3623 | 1.7048 | | | | 1.7075 | 1.7146 | 1.7217 | 1.7285 | 1.7352 | 1.7417 | 1.7545 | 1.7668 | 1.7787 | 1.7903 | 1.8016 |
| | v | 0.017 | 14.189 | | | | 14.225 | 14.451 | 14.675 | 14.897 | 15.118 | 15.337 | 15.773 | 16.205 | 16.634 | 17.061 | 17.485 |
| 29 | h | 216.9 | 1163.4 | | | | 1164.3 | 1169.4 | 1174.5 | 1179.6 | 1184.6 | 1189.5 | 1199.3 | 1209.1 | 1218.8 | 1228.4 | 1238.0 |
| (248.40) | s | 0.3652 | 1.7020 | | | | 1.7032 | 1.7104 | 1.7175 | 1.7243 | 1.7310 | 1.7376 | 1.7504 | 1.7627 | 1.7747 | 1.7863 | 1.7976 |
| | v | 0.017 | 13.746 | | | | | 13.957 | 14.174 | 14.390 | 14.604 | 14.816 | 15.238 | 15.657 | 16.072 | 16.485 | 16.897 |
| 30 | h | 218.8 | 1164.1 | | | | | 1169.1 | 1174.2 | 1179.3 | 1184.3 | 1189.3 | 1199.1 | 1208.9 | 1218.6 | 1228.3 | 1237.9 |
| (250.33) | s | 0.3680 | 1.6993 | | | | | 1.7063 | 1.7134 | 1.7203 | 1.7270 | 1.7336 | 1.7464 | 1.7588 | 1.7708 | 1.7824 | 1.7937 |
| | v | 0.017 | 13.330 | | | | | 13.495 | 13.706 | 13.915 | 14.123 | 14.329 | 14.738 | 15.144 | 15.547 | 15.947 | 16.346 |
| 31 | h | 220.7 | 1164.7 | | | | | 1168.8 | 1173.9 | 1179.0 | 1184.0 | 1189.0 | 1198.9 | 1208.7 | 1218.4 | 1228.1 | 1237.7 |
| (252.22) | s | 0.3707 | 1.6967 | | | | | 1.7024 | 1.7095 | 1.7164 | 1.7231 | 1.7298 | 1.7426 | 1.7550 | 1.7670 | 1.7787 | 1.7900 |
| | v | 0.017 | 12.940 | | | | | 13.062 | 13.267 | 13.470 | 13.672 | 13.872 | 14.270 | 14.664 | 15.055 | 15.443 | 15.829 |
| 32 | h | 222.6 | 1165.4 | | | | | 1168.5 | 1173.6 | 1178.7 | 1183.8 | 1188.8 | 1198.7 | 1208.5 | 1218.3 | 1227.9 | 1237.6 |
| (254.05) | s | 0.3733 | 1.6941 | | | | | 1.6985 | 1.7056 | 1.7126 | 1.7194 | 1.7260 | 1.7389 | 1.7513 | 1.7634 | 1.7750 | 1.7864 |
| | v | 0.017 | 12.572 | | | | | 12.656 | 12.855 | 13.052 | 13.249 | 13.443 | 13.830 | 14.212 | 14.592 | 14.969 | 15.344 |
| 33 | h | 224.4 | 1166.0 | | | | | 1168.1 | 1173.3 | 1178.4 | 1183.5 | 1188.6 | 1198.5 | 1208.3 | 1218.1 | 1227.8 | 1237.4 |
| (255.84) | s | 0.3758 | 1.6917 | | | | | 1.6947 | 1.7019 | 1.7089 | 1.7157 | 1.7224 | 1.7353 | 1.7477 | 1.7598 | 1.7715 | 1.7829 |
| | v | 0.017 | 12.226 | | | | | 12.273 | 12.467 | 12.659 | 12.850 | 13.040 | 13.415 | 13.788 | 14.157 | 14.523 | 14.888 |
| 34 | h | 226.2 | 1166.5 | | | | | 1167.8 | 1173.0 | 1178.2 | 1183.3 | 1188.3 | 1198.3 | 1208.2 | 1217.9 | 1227.6 | 1237.3 |
| (257.58) | s | 0.3783 | 1.6893 | | | | | 1.6911 | 1.6983 | 1.7053 | 1.7121 | 1.7188 | 1.7318 | 1.7443 | 1.7564 | 1.7681 | 1.7794 |

Temperature—Degrees Fahrenheit

| 420° | 440° | 460° | 480° | 500° | 550° | 600° | 650° | 700° | 800° | 900° | 1000° | 1100° | 1200° | 1400° | 1600° | | Abs. Press. Lb./Sq. In. (Sat. Temp.) |
|---|---|---|---|---|---|---|---|---|---|---|---|---|---|---|---|---|---|
| 30.67 | 31.38 | 32.09 | 32.80 | 33.51 | 35.28 | 37.04 | 38.80 | 40.56 | 44.08 | 47.60 | 51.11 | 54.62 | 58.13 | 65.15 | 72.16 | v | |
| 1249.1 | 1258.5 | 1267.9 | 1277.4 | 1286.9 | 1310.7 | 1334.6 | 1358.7 | 1383.0 | 1432.2 | 1482.2 | 1533.1 | 1584.8 | 1637.4 | 1745.5 | 1857.3 | h | **17** |
| 1.8687 | 1.8794 | 1.8897 | 1.8999 | 1.9099 | 1.9341 | 1.9572 | 1.9795 | 2.0009 | 2.0415 | 2.0797 | 2.1158 | 2.1501 | 2.1828 | 2.2442 | 2.3013 | s | (219.44) |
| 28.95 | 29.63 | 30.30 | 30.97 | 31.64 | 33.31 | 34.98 | 36.64 | 38.31 | 41.63 | 44.95 | 48.27 | 51.59 | 54.90 | 61.53 | 68.15 | v | |
| 1248.9 | 1258.4 | 1267.8 | 1277.3 | 1286.8 | 1310.6 | 1334.6 | 1358.7 | 1383.0 | 1432.2 | 1482.2 | 1533.0 | 1584.8 | 1637.4 | 1745.5 | 1857.2 | h | **18** |
| 1.8623 | 1.8730 | 1.8834 | 1.8936 | 1.9035 | 1.9277 | 1.9509 | 1.9731 | 1.9945 | 2.0352 | 2.0734 | 2.1095 | 2.1438 | 2.1765 | 2.2379 | 2.2950 | s | (222.41) |
| 27.42 | 28.06 | 28.70 | 29.33 | 29.97 | 31.55 | 33.13 | 34.71 | 36.29 | 39.44 | 42.58 | 45.73 | 48.87 | 52.01 | 58.29 | 64.57 | v | |
| 1248.8 | 1258.3 | 1267.7 | 1277.2 | 1286.7 | 1310.5 | 1334.5 | 1358.6 | 1382.9 | 1432.1 | 1482.1 | 1533.0 | 1584.7 | 1637.4 | 1745.4 | 1857.2 | h | **19** |
| 1.8563 | 1.8669 | 1.8773 | 1.8875 | 1.8975 | 1.9217 | 1.9449 | 1.9671 | 1.9885 | 2.0292 | 2.0674 | 2.1035 | 2.1378 | 2.1705 | 2.2320 | 2.2890 | s | (225.24) |
| 26.04 | 26.65 | 27.25 | 27.86 | 28.46 | 29.97 | 31.47 | 32.97 | 34.47 | 37.46 | 40.45 | 43.44 | 46.42 | 49.41 | 55.37 | 61.34 | v | |
| 1248.7 | 1258.2 | 1267.6 | 1277.1 | 1286.6 | 1310.5 | 1334.4 | 1358.6 | 1382.9 | 1432.1 | 1482.1 | 1533.0 | 1584.7 | 1637.4 | 1745.4 | 1857.2 | h | **20** |
| 1.8505 | 1.8612 | 1.8716 | 1.8818 | 1.8918 | 1.9160 | 1.9392 | 1.9614 | 1.9829 | 2.0235 | 2.0618 | 2.0978 | 2.1321 | 2.1648 | 2.2263 | 2.2834 | s | (227.96) |
| 24.79 | 25.37 | 25.95 | 26.52 | 27.10 | 28.54 | 29.97 | 31.40 | 32.82 | 35.68 | 38.52 | 41.37 | 44.21 | 47.05 | 52.74 | 58.42 | v | |
| 1248.6 | 1258.1 | 1267.5 | 1277.0 | 1286.5 | 1310.4 | 1334.4 | 1358.5 | 1382.8 | 1432.0 | 1482.1 | 1532.9 | 1584.7 | 1637.3 | 1745.4 | 1857.2 | h | **21** |
| 1.8450 | 1.8557 | 1.8661 | 1.8763 | 1.8863 | 1.9106 | 1.9338 | 1.9560 | 1.9774 | 2.0181 | 2.0564 | 2.0925 | 2.1267 | 2.1594 | 2.2209 | 2.2780 | s | (230.57) |
| 23.66 | 24.21 | 24.76 | 25.31 | 25.86 | 27.23 | 28.60 | 29.97 | 31.33 | 34.05 | 36.77 | 39.49 | 42.20 | 44.91 | 50.34 | 55.76 | v | |
| 1248.4 | 1257.9 | 1267.4 | 1276.9 | 1286.4 | 1310.3 | 1334.3 | 1358.5 | 1382.8 | 1432.0 | 1482.0 | 1532.9 | 1584.7 | 1637.3 | 1745.4 | 1857.2 | h | **22** |
| 1.8398 | 1.8505 | 1.8609 | 1.8711 | 1.8811 | 1.9054 | 1.9286 | 1.9509 | 1.9723 | 2.0130 | 2.0512 | 2.0873 | 2.1216 | 2.1543 | 2.2158 | 2.2729 | s | (233.07) |
| 22.62 | 23.15 | 23.68 | 24.21 | 24.73 | 26.04 | 27.35 | 28.66 | 29.96 | 32.57 | 35.17 | 37.77 | 40.36 | 42.96 | 48.15 | 53.33 | v | |
| 1248.3 | 1257.8 | 1267.3 | 1276.8 | 1286.4 | 1310.2 | 1334.2 | 1358.4 | 1382.7 | 1432.0 | 1482.0 | 1532.9 | 1584.6 | 1637.3 | 1745.4 | 1857.2 | h | **23** |
| 1.8348 | 1.8455 | 1.8559 | 1.8661 | 1.8762 | 1.9004 | 1.9236 | 1.9459 | 1.9674 | 2.0081 | 2.0463 | 2.0824 | 2.1167 | 2.1494 | 2.2109 | 2.2680 | s | (235.49) |
| 21.67 | 22.18 | 22.69 | 23.19 | 23.70 | 24.96 | 26.21 | 27.46 | 28.71 | 31.21 | 33.70 | 36.19 | 38.68 | 41.17 | 46.14 | 51.11 | v | |
| 1248.2 | 1257.7 | 1267.2 | 1276.7 | 1286.3 | 1310.2 | 1334.2 | 1358.4 | 1382.7 | 1431.9 | 1482.0 | 1532.9 | 1584.6 | 1637.3 | 1745.4 | 1857.2 | h | **24** |
| 1.8300 | 1.8407 | 1.8511 | 1.8614 | 1.8714 | 1.8957 | 1.9189 | 1.9412 | 1.9626 | 2.0034 | 2.0416 | 2.0777 | 2.1120 | 2.1447 | 2.2062 | 2.2633 | s | (237.82) |
| 20.80 | 21.29 | 21.77 | 22.26 | 22.74 | 23.95 | 25.16 | 26.36 | 27.56 | 29.96 | 32.35 | 34.74 | 37.13 | 39.52 | 44.30 | 49.07 | v | |
| 1248.1 | 1257.6 | 1267.1 | 1276.6 | 1286.2 | 1310.1 | 1334.1 | 1358.3 | 1382.6 | 1431.9 | 1481.9 | 1532.8 | 1584.6 | 1637.3 | 1745.3 | 1857.2 | h | **25** |
| 1.8254 | 1.8361 | 1.8465 | 1.8568 | 1.8668 | 1.8911 | 1.9144 | 1.9367 | 1.9581 | 1.9988 | 2.0371 | 2.0732 | 2.1075 | 2.1402 | 2.2017 | 2.2587 | s | (240.07) |
| 19.994 | 20.46 | 20.93 | 21.40 | 21.86 | 23.03 | 24.19 | 25.34 | 26.50 | 28.80 | 31.11 | 33.41 | 35.70 | 38.00 | 42.59 | 47.18 | v | |
| 1248.0 | 1257.5 | 1267.0 | 1276.5 | 1286.1 | 1310.0 | 1334.0 | 1358.2 | 1382.6 | 1431.8 | 1481.9 | 1532.8 | 1584.6 | 1637.2 | 1745.3 | 1857.1 | h | **26** |
| 1.8210 | 1.8317 | 1.8421 | 1.8524 | 1.8624 | 1.8868 | 1.9100 | 1.9323 | 1.9538 | 1.9945 | 2.0327 | 2.0689 | 2.1031 | 2.1359 | 2.1973 | 2.2544 | s | (242.25) |
| 19.248 | 19.700 | 20.15 | 20.60 | 21.05 | 22.17 | 23.29 | 24.40 | 25.51 | 27.73 | 29.95 | 32.17 | 34.38 | 36.59 | 41.01 | 45.43 | v | |
| 1247.8 | 1257.4 | 1266.9 | 1276.4 | 1286.0 | 1309.9 | 1334.0 | 1358.2 | 1382.5 | 1431.8 | 1481.9 | 1532.8 | 1584.6 | 1637.2 | 1745.3 | 1857.1 | h | **27** |
| 1.8167 | 1.8274 | 1.8379 | 1.8482 | 1.8582 | 1.8825 | 1.9058 | 1.9281 | 1.9496 | 1.9903 | 2.0285 | 2.0647 | 2.0990 | 2.1317 | 2.1932 | 2.2503 | s | (244.36) |
| 18.554 | 18.991 | 19.426 | 19.860 | 20.29 | 21.37 | 22.45 | 23.53 | 24.60 | 26.74 | 28.88 | 31.02 | 33.15 | 35.28 | 39.55 | 43.81 | v | |
| 1247.7 | 1257.2 | 1266.8 | 1276.3 | 1285.9 | 1309.9 | 1333.9 | 1358.1 | 1382.5 | 1431.8 | 1481.8 | 1532.8 | 1584.5 | 1637.2 | 1745.3 | 1857.1 | h | **28** |
| 1.8126 | 1.8233 | 1.8338 | 1.8441 | 1.8541 | 1.8785 | 1.9017 | 1.9241 | 1.9455 | 1.9863 | 2.0245 | 2.0607 | 2.0950 | 2.1277 | 2.1892 | 2.2462 | s | (246.41) |
| 17.909 | 18.331 | 18.751 | 19.171 | 19.590 | 20.63 | 21.67 | 22.71 | 23.75 | 25.82 | 27.88 | 29.95 | 32.01 | 34.07 | 38.18 | 42.30 | v | |
| 1247.6 | 1257.1 | 1266.7 | 1276.2 | 1285.8 | 1309.8 | 1333.9 | 1358.1 | 1382.4 | 1431.7 | 1481.8 | 1532.7 | 1584.5 | 1637.2 | 1745.3 | 1857.1 | h | **29** |
| 1.8086 | 1.8193 | 1.8298 | 1.8401 | 1.8502 | 1.8746 | 1.8978 | 1.9201 | 1.9416 | 1.9824 | 2.0206 | 2.0568 | 2.0911 | 2.1238 | 2.1853 | 2.2424 | s | (248.40) |
| 17.306 | 17.714 | 18.121 | 18.528 | 18.933 | 19.943 | 20.95 | 21.95 | 22.96 | 24.96 | 26.95 | 28.95 | 30.94 | 32.93 | 36.91 | 40.89 | v | |
| 1247.5 | 1257.0 | 1266.6 | 1276.2 | 1285.7 | 1309.7 | 1333.8 | 1358.0 | 1382.4 | 1431.7 | 1481.8 | 1532.7 | 1584.5 | 1637.2 | 1745.3 | 1857.1 | h | **30** |
| 1.8047 | 1.8155 | 1.8260 | 1.8363 | 1.8464 | 1.8708 | 1.8940 | 1.9164 | 1.9379 | 1.9786 | 2.0169 | 2.0530 | 2.0873 | 2.1201 | 2.1815 | 2.2386 | s | (250.33) |
| 16.743 | 17.138 | 17.532 | 17.926 | 18.318 | 19.296 | 20.27 | 21.24 | 22.21 | 24.15 | 26.08 | 28.01 | 29.94 | 31.87 | 35.72 | 39.57 | v | |
| 1247.3 | 1256.9 | 1266.5 | 1276.1 | 1285.6 | 1309.6 | 1333.7 | 1358.0 | 1382.3 | 1431.6 | 1481.7 | 1532.7 | 1584.5 | 1637.1 | 1745.2 | 1857.1 | h | **31** |
| 1.8010 | 1.8118 | 1.8223 | 1.8326 | 1.8427 | 1.8671 | 1.8904 | 1.9127 | 1.9342 | 1.9750 | 2.0133 | 2.0494 | 2.0837 | 2.1164 | 2.1779 | 2.2350 | s | (252.22) |
| 16.214 | 16.598 | 16.980 | 17.361 | 17.741 | 18.689 | 19.634 | 20.58 | 21.52 | 23.39 | 25.26 | 27.13 | 29.00 | 30.87 | 34.60 | 38.33 | v | |
| 1247.2 | 1256.8 | 1266.4 | 1276.0 | 1285.5 | 1309.6 | 1333.7 | 1357.9 | 1382.3 | 1431.6 | 1481.7 | 1532.6 | 1584.4 | 1637.1 | 1745.2 | 1857.1 | h | **32** |
| 1.7974 | 1.8082 | 1.8187 | 1.8290 | 1.8391 | 1.8635 | 1.8868 | 1.9092 | 1.9307 | 1.9715 | 2.0097 | 2.0459 | 2.0802 | 2.1129 | 2.1744 | 2.2315 | s | (254.05) |
| 15.718 | 16.090 | 16.461 | 16.831 | 17.200 | 18.120 | 19.036 | 19.950 | 20.86 | 22.68 | 24.50 | 26.31 | 28.12 | 29.93 | 33.55 | 37.17 | v | |
| 1247.1 | 1256.7 | 1266.3 | 1275.9 | 1285.4 | 1309.5 | 1333.6 | 1357.8 | 1382.2 | 1431.6 | 1481.7 | 1532.6 | 1584.4 | 1637.1 | 1745.2 | 1857.1 | h | **33** |
| 1.7939 | 1.8047 | 1.8153 | 1.8256 | 1.8357 | 1.8601 | 1.8834 | 1.9058 | 1.9273 | 1.9681 | 2.0063 | 2.0425 | 2.0768 | 2.1095 | 2.1710 | 2.2281 | s | (255.84) |
| 15.251 | 15.612 | 15.973 | 16.332 | 16.690 | 17.584 | 18.474 | 19.361 | 20.25 | 22.01 | 23.78 | 25.54 | 27.30 | 29.05 | 32.57 | 36.08 | v | |
| 1246.9 | 1256.6 | 1266.2 | 1275.8 | 1285.4 | 1309.4 | 1333.5 | 1357.8 | 1382.2 | 1431.5 | 1481.6 | 1532.6 | 1584.4 | 1637.1 | 1745.2 | 1857.0 | h | **34** |
| 1.7905 | 1.8013 | 1.8119 | 1.8222 | 1.8323 | 1.8567 | 1.8800 | 1.9024 | 1.9239 | 1.9647 | 2.0030 | 2.0392 | 2.0735 | 2.1062 | 2.1677 | 2.2248 | s | (257.58) |

## Table 3. Superheated Vapor

| Abs. Press. Lb./Sq. In. (Sat. Temp.) | | Sat. Liquid | Sat. Vapor | Temperature—Degrees Fahrenheit 260° | 270° | 280° | 290° | 300° | 320° | 340° | 360° | 380° | 400° | 420° | 440° | 460° | 480° |
|---|---|---|---|---|---|---|---|---|---|---|---|---|---|---|---|---|---|
| | v | 0.017 | 11.898 | 11.911 | 12.101 | 12.288 | 12.474 | 12.659 | 13.025 | 13.387 | 13.746 | 14.103 | 14.457 | 14.810 | 15.162 | 15.512 | 15.862 |
| **35** | h | 227.9 | 1167.1 | 1167.5 | 1172.7 | 1177.9 | 1183.0 | 1188.1 | 1198.1 | 1208.0 | 1217.8 | 1227.5 | 1237.2 | 1246.8 | 1256.4 | 1266.1 | 1275.7 |
| (259.28) | s | 0.3807 | 1.6870 | 1.6875 | 1.6948 | 1.7018 | 1.7087 | 1.7154 | 1.7284 | 1.7409 | 1.7530 | 1.7647 | 1.7761 | 1.7872 | 1.7980 | 1.8086 | 1.8189 |
| | v | 0.017 | 11.588 | | 11.755 | 11.938 | 12.119 | 12.299 | 12.656 | 13.009 | 13.358 | 13.705 | 14.051 | 14.394 | 14.736 | 15.077 | 15.417 |
| **36** | h | 229.6 | 1167.6 | | 1172.4 | 1177.6 | 1182.7 | 1187.8 | 1197.9 | 1207.8 | 1217.6 | 1227.3 | 1237.0 | 1246.7 | 1256.3 | 1266.0 | 1275.6 |
| (260.95) | s | 0.3831 | 1.6848 | | 1.6913 | 1.6984 | 1.7053 | 1.7120 | 1.7251 | 1.7376 | 1.7497 | 1.7615 | 1.7729 | 1.7840 | 1.7948 | 1.8054 | 1.8158 |
| | v | 0.017 | 11.294 | | 11.428 | 11.606 | 11.783 | 11.959 | 12.307 | 12.651 | 12.991 | 13.330 | 13.666 | 14.001 | 14.334 | 14.666 | 14.997 |
| **37** | h | 231.3 | 1168.2 | | 1172.1 | 1177.3 | 1182.5 | 1187.6 | 1197.6 | 1207.6 | 1217.4 | 1227.2 | 1236.9 | 1246.6 | 1256.2 | 1265.8 | 1275.5 |
| (262.57) | s | 0.3854 | 1.6826 | | 1.6880 | 1.6951 | 1.7020 | 1.7087 | 1.7218 | 1.7344 | 1.7466 | 1.7583 | 1.7698 | 1.7809 | 1.7917 | 1.8023 | 1.8127 |
| | v | 0.017 | 11.015 | | 11.118 | 11.292 | 11.465 | 11.637 | 11.976 | 12.312 | 12.644 | 12.974 | 13.302 | 13.628 | 13.953 | 14.276 | 14.599 |
| **38** | h | 232.9 | 1168.7 | | 1171.8 | 1177.0 | 1182.2 | 1187.3 | 1197.4 | 1207.4 | 1217.3 | 1227.0 | 1236.7 | 1246.4 | 1256.1 | 1265.7 | 1275.4 |
| (264.16) | s | 0.3876 | 1.6805 | | 1.6847 | 1.6918 | 1.6988 | 1.7056 | 1.7187 | 1.7313 | 1.7435 | 1.7552 | 1.7667 | 1.7778 | 1.7887 | 1.7993 | 1.8097 |
| | v | 0.017 | 10.750 | | 10.824 | 10.994 | 11.163 | 11.331 | 11.662 | 11.990 | 12.314 | 12.636 | 12.956 | 13.274 | 13.591 | 13.906 | 14.221 |
| **39** | h | 234.5 | 1169.2 | | 1171.5 | 1176.7 | 1181.9 | 1187.1 | 1197.2 | 1207.2 | 1217.1 | 1226.9 | 1236.6 | 1246.3 | 1256.0 | 1265.6 | 1275.3 |
| (265.72) | s | 0.3898 | 1.6784 | | 1.6815 | 1.6886 | 1.6956 | 1.7024 | 1.7156 | 1.7282 | 1.7404 | 1.7522 | 1.7637 | 1.7748 | 1.7857 | 1.7963 | 1.8067 |
| | v | 0.017 | 10.498 | | 10.544 | 10.711 | 10.876 | 11.040 | 11.364 | 11.684 | 12.001 | 12.315 | 12.628 | 12.938 | 13.247 | 13.555 | 13.862 |
| **40** | h | 236.0 | 1169.7 | | 1171.2 | 1176.5 | 1181.7 | 1186.8 | 1197.0 | 1207.0 | 1216.9 | 1226.7 | 1236.5 | 1246.2 | 1255.9 | 1265.5 | 1275.2 |
| (267.25) | s | 0.3919 | 1.6763 | | 1.6783 | 1.6855 | 1.6925 | 1.6994 | 1.7126 | 1.7252 | 1.7375 | 1.7493 | 1.7608 | 1.7719 | 1.7828 | 1.7934 | 1.8038 |
| | v | 0.017 | 10.258 | | 10.279 | 10.442 | 10.603 | 10.764 | 11.081 | 11.393 | 11.703 | 12.010 | 12.315 | 12.619 | 12.920 | 13.221 | 13.521 |
| **41** | h | 237.6 | 1170.2 | | 1170.9 | 1176.2 | 1181.4 | 1186.6 | 1196.8 | 1206.8 | 1216.7 | 1226.6 | 1236.3 | 1246.1 | 1255.8 | 1265.4 | 1275.1 |
| (268.74) | s | 0.3940 | 1.6743 | | 1.6753 | 1.6825 | 1.6895 | 1.6964 | 1.7096 | 1.7223 | 1.7346 | 1.7464 | 1.7579 | 1.7691 | 1.7800 | 1.7906 | 1.8010 |
| | v | 0.017 | 10.029 | | | 10.185 | 10.344 | 10.501 | 10.810 | 11.116 | 11.419 | 11.719 | 12.018 | 12.314 | 12.609 | 12.903 | 13.195 |
| **42** | h | 239.0 | 1170.7 | | | 1175.9 | 1181.1 | 1186.3 | 1196.6 | 1206.6 | 1216.6 | 1226.4 | 1236.2 | 1245.9 | 1255.6 | 1265.3 | 1275.0 |
| (270.21) | s | 0.3960 | 1.6724 | | | 1.6795 | 1.6866 | 1.6935 | 1.7067 | 1.7195 | 1.7318 | 1.7436 | 1.7551 | 1.7663 | 1.7772 | 1.7879 | 1.7983 |
| | v | 0.017 | 9.810 | | | 9.941 | 10.096 | 10.249 | 10.553 | 10.852 | 11.149 | 11.442 | 11.734 | 12.024 | 12.312 | 12.599 | 12.885 |
| **43** | h | 240.5 | 1171.1 | | | 1175.6 | 1180.9 | 1186.1 | 1196.3 | 1206.4 | 1216.4 | 1226.3 | 1236.1 | 1245.8 | 1255.5 | 1265.2 | 1274.9 |
| (271.64) | s | 0.3980 | 1.6706 | | | 1.6766 | 1.6837 | 1.6906 | 1.7039 | 1.7167 | 1.7290 | 1.7409 | 1.7524 | 1.7636 | 1.7746 | 1.7852 | 1.7956 |
| | v | 0.017 | 9.601 | | | 9.707 | 9.859 | 10.010 | 10.307 | 10.600 | 10.890 | 11.178 | 11.463 | 11.747 | 12.029 | 12.310 | 12.590 |
| **44** | h | 242.0 | 1171.6 | | | 1175.3 | 1180.6 | 1185.8 | 1196.1 | 1206.2 | 1216.2 | 1226.1 | 1235.9 | 1245.7 | 1255.4 | 1265.1 | 1274.8 |
| (273.05) | s | 0.4000 | 1.6687 | | | 1.6738 | 1.6809 | 1.6878 | 1.7012 | 1.7140 | 1.7263 | 1.7382 | 1.7498 | 1.7610 | 1.7719 | 1.7826 | 1.7930 |
| | v | 0.017 | 9.401 | | | 9.484 | 9.633 | 9.781 | 10.072 | 10.359 | 10.643 | 10.925 | 11.204 | 11.482 | 11.758 | 12.033 | 12.307 |
| **45** | h | 243.4 | 1172.0 | | | 1175.0 | 1180.3 | 1185.6 | 1195.9 | 1206.0 | 1216.0 | 1225.9 | 1235.8 | 1245.6 | 1255.3 | 1265.0 | 1274.7 |
| (274.44) | s | 0.4019 | 1.6669 | | | 1.6709 | 1.6781 | 1.6850 | 1.6985 | 1.7113 | 1.7237 | 1.7356 | 1.7472 | 1.7584 | 1.7693 | 1.7800 | 1.7904 |
| | v | 0.017 | 9.209 | | | 9.271 | 9.417 | 9.562 | 9.847 | 10.129 | 10.407 | 10.683 | 10.957 | 11.229 | 11.499 | 11.768 | 12.036 |
| **46** | h | 244.8 | 1172.4 | | | 1174.7 | 1180.0 | 1185.3 | 1195.7 | 1205.8 | 1215.9 | 1225.8 | 1235.6 | 1245.4 | 1255.2 | 1264.9 | 1274.6 |
| (275.80) | s | 0.4038 | 1.6652 | | | 1.6682 | 1.6754 | 1.6824 | 1.6958 | 1.7087 | 1.7211 | 1.7330 | 1.7446 | 1.7559 | 1.7668 | 1.7775 | 1.7879 |
| | v | 0.017 | 9.025 | | | 9.066 | 9.210 | 9.352 | 9.632 | 9.908 | 10.181 | 10.451 | 10.720 | 10.986 | 11.251 | 11.515 | 11.777 |
| **47** | h | 246.1 | 1172.9 | | | 1174.4 | 1179.8 | 1185.1 | 1195.5 | 1205.6 | 1215.7 | 1225.6 | 1235.5 | 1245.3 | 1255.1 | 1264.8 | 1274.5 |
| (277.13) | s | 0.4057 | 1.6634 | | | 1.6655 | 1.6727 | 1.6797 | 1.6932 | 1.7061 | 1.7185 | 1.7305 | 1.7421 | 1.7534 | 1.7644 | 1.7751 | 1.7855 |
| | v | 0.017 | 8.848 | | | 8.870 | 9.011 | 9.151 | 9.426 | 9.697 | 9.965 | 10.230 | 10.493 | 10.754 | 11.014 | 11.272 | 11.529 |
| **48** | h | 247.5 | 1173.3 | | | 1174.1 | 1179.5 | 1184.8 | 1195.2 | 1205.4 | 1215.5 | 1225.5 | 1235.4 | 1245.2 | 1254.9 | 1264.7 | 1274.4 |
| (278.45) | s | 0.4075 | 1.6617 | | | 1.6629 | 1.6701 | 1.6771 | 1.6907 | 1.7036 | 1.7161 | 1.7281 | 1.7397 | 1.7510 | 1.7620 | 1.7727 | 1.7831 |
| | v | 0.017 | 8.678 | | | 8.682 | 8.821 | 8.958 | 9.228 | 9.494 | 9.757 | 10.017 | 10.275 | 10.531 | 10.786 | 11.039 | 11.291 |
| **49** | h | 248.8 | 1173.7 | | | 1173.8 | 1179.2 | 1184.6 | 1195.0 | 1205.2 | 1215.3 | 1225.3 | 1235.2 | 1245.0 | 1254.8 | 1264.6 | 1274.3 |
| (279.74) | s | 0.4093 | 1.6601 | | | 1.6603 | 1.6675 | 1.6746 | 1.6882 | 1.7011 | 1.7136 | 1.7256 | 1.7373 | 1.7486 | 1.7596 | 1.7703 | 1.7807 |
| | v | 0.017 | 8.515 | | | | 8.638 | 8.773 | 9.038 | 9.299 | 9.557 | 9.812 | 10.065 | 10.317 | 10.567 | 10.815 | 11.062 |
| **50** | h | 250.1 | 1174.1 | | | | 1178.9 | 1184.3 | 1194.8 | 1205.0 | 1215.2 | 1225.2 | 1235.1 | 1244.9 | 1254.7 | 1264.5 | 1274.2 |
| (281.01) | s | 0.4110 | 1.6585 | | | | 1.6650 | 1.6721 | 1.6857 | 1.6987 | 1.7112 | 1.7233 | 1.7349 | 1.7462 | 1.7572 | 1.7680 | 1.7784 |
| | v | 0.017 | 8.359 | | | | 8.462 | 8.595 | 8.855 | 9.112 | 9.365 | 9.616 | 9.865 | 10.111 | 10.356 | 10.600 | 10.843 |
| **51** | h | 251.4 | 1174.4 | | | | 1178.6 | 1184.0 | 1194.6 | 1204.9 | 1215.0 | 1225.0 | 1234.9 | 1244.8 | 1254.6 | 1264.4 | 1274.1 |
| (282.26) | s | 0.4127 | 1.6569 | | | | 1.6625 | 1.6696 | 1.6833 | 1.6963 | 1.7089 | 1.7209 | 1.7326 | 1.7439 | 1.7549 | 1.7657 | 1.7762 |

Temperature—Degrees Fahrenheit

| 500° | 520° | 540° | 560° | 580° | 600° | 650° | 700° | 750° | 800° | 900° | 1000° | 1100° | 1200° | 1400° | 1600° | | Abs. Press. Lb./Sq. In. (Sat. Temp.) |
|---|---|---|---|---|---|---|---|---|---|---|---|---|---|---|---|---|---|
| 16.210 | 16.558 | 16.905 | 17.251 | 17.597 | 17.943 | 18.805 | 19.666 | 20.52 | 21.38 | 23.10 | 24.81 | 26.51 | 28.22 | 31.63 | 35.04 | v | |
| 1285.3 | 1294.9 | 1304.5 | 1314.1 | 1323.8 | 1333.5 | 1357.7 | 1382.1 | 1406.7 | 1431.5 | 1481.6 | 1532.6 | 1584.4 | 1637.1 | 1745.2 | 1857.0 | h | 35 |
| 1.8290 | 1.8390 | 1.8487 | 1.8582 | 1.8676 | 1.8768 | 1.8992 | 1.9207 | 1.9414 | 1.9615 | 1.9998 | 2.0360 | 2.0703 | 2.1030 | 2.1645 | 2.2216 | s | (259.28) |
| 15.756 | 16.094 | 16.432 | 16.769 | 17.106 | 17.442 | 18.281 | 19.117 | 19.953 | 20.79 | 22.45 | 24.12 | 25.78 | 27.44 | 30.76 | 34.07 | v | |
| 1285.2 | 1294.8 | 1304.4 | 1314.1 | 1323.7 | 1333.4 | 1357.7 | 1382.1 | 1406.7 | 1431.5 | 1481.6 | 1532.5 | 1584.3 | 1637.0 | 1745.2 | 1857.0 | h | 36 |
| 1.8259 | 1.8358 | 1.8455 | 1.8551 | 1.8644 | 1.8737 | 1.8961 | 1.9176 | 1.9383 | 1.9584 | 1.9967 | 2.0329 | 2.0672 | 2.0999 | 2.1614 | 2.2185 | s | (260.95) |
| 15.327 | 15.656 | 15.985 | 16.313 | 16.641 | 16.968 | 17.784 | 18.599 | 19.412 | 20.22 | 21.84 | 23.46 | 25.08 | 26.69 | 29.92 | 33.15 | v | |
| 1285.1 | 1294.7 | 1304.3 | 1314.0 | 1323.6 | 1333.3 | 1357.6 | 1382.0 | 1406.6 | 1431.4 | 1481.5 | 1532.5 | 1584.3 | 1637.0 | 1745.2 | 1857.0 | h | 37 |
| 1.8228 | 1.8327 | 1.8424 | 1.8520 | 1.8614 | 1.8706 | 1.8930 | 1.9145 | 1.9353 | 1.9554 | 1.9936 | 2.0298 | 2.0641 | 2.0969 | 2.1584 | 2.2155 | s | (262.57) |
| 14.920 | 15.241 | 15.561 | 15.881 | 16.200 | 16.519 | 17.314 | 18.107 | 18.899 | 19.689 | 21.27 | 22.84 | 24.42 | 25.99 | 29.14 | 32.28 | v | |
| 1285.0 | 1294.6 | 1304.3 | 1313.9 | 1323.6 | 1333.3 | 1357.6 | 1382.0 | 1406.6 | 1431.4 | 1481.5 | 1532.5 | 1584.3 | 1637.0 | 1745.1 | 1857.0 | h | 38 |
| 1.8198 | 1.8297 | 1.8394 | 1.8490 | 1.8584 | 1.8676 | 1.8900 | 1.9115 | 1.9323 | 1.9524 | 1.9907 | 2.0269 | 2.0612 | 2.0939 | 2.1554 | 2.2126 | s | (264.16) |
| 14.535 | 14.847 | 15.159 | 15.471 | 15.782 | 16.093 | 16.868 | 17.641 | 18.413 | 19.183 | 20.72 | 22.26 | 23.79 | 25.32 | 28.39 | 31.45 | v | |
| 1284.9 | 1294.5 | 1304.2 | 1313.8 | 1323.5 | 1333.2 | 1357.5 | 1381.9 | 1406.6 | 1431.3 | 1481.5 | 1532.4 | 1584.3 | 1637 0 | 1745.1 | 1857.0 | h | 39 |
| 1.8168 | 1.8268 | 1.8365 | 1.8461 | 1.8555 | 1.8647 | 1.8871 | 1.9087 | 1.9294 | 1.9495 | 1.9878 | 2.0240 | 2.0583 | 2.0911 | 2.1526 | 2.2097 | s | (265.72) |
| 14.168 | 14.473 | 14.778 | 15.082 | 15.385 | 15.688 | 16.444 | 17.198 | 17.951 | 18.702 | 20.20 | 21.70 | 23.20 | 24.69 | 27.68 | 30.66 | v | |
| 1284.8 | 1294.5 | 1304.1 | 1313.8 | 1323.4 | 1333.1 | 1357.4 | 1381.9 | 1406.5 | 1431.3 | 1481.4 | 1532.4 | 1584.3 | 1637.0 | 1745.1 | 1857.0 | h | 40 |
| 1.8140 | 1.8239 | 1.8337 | 1.8432 | 1.8526 | 1.8619 | 1.8843 | 1.9058 | 1.9266 | 1.9467 | 1.9850 | 2.0212 | 2.0555 | 2.0883 | 2.1498 | 2.2069 | s | (267.25) |
| 13.819 | 14.117 | 14.414 | 14.711 | 15.008 | 15.303 | 16.041 | 16.777 | 17.512 | 18.244 | 19.708 | 21.17 | 22.63 | 24.09 | 27.00 | 29.91 | v | |
| 1284.7 | 1294.4 | 1304.0 | 1313.7 | 1323.4 | 1333.1 | 1357.4 | 1381.8 | 1406.5 | 1431.3 | 1481.4 | 1532.4 | 1584.2 | 1636.9 | 1745.1 | 1857.0 | h | 41 |
| 1.8112 | 1.8211 | 1.8309 | 1.8405 | 1.8499 | 1.8591 | 1.8815 | 1.9031 | 1.9239 | 1.9439 | 1.9823 | 2.0184 | 2.0528 | 2.0855 | 2.1470 | 2.2042 | s | (268.74) |
| 13.487 | 13.778 | 14.069 | 14.358 | 14.648 | 14.937 | 15.657 | 16.376 | 17.093 | 17.809 | 19.238 | 20.66 | 22.09 | 23.51 | 26.36 | 29.20 | v | |
| 1284.6 | 1294.3 | 1303.9 | 1313.6 | 1323.3 | 1333.0 | 1357.3 | 1381.8 | 1406.4 | 1431.2 | 1481.4 | 1532.4 | 1584.2 | 1636.9 | 1745.1 | 1856.9 | h | 42 |
| 1.8084 | 1.8184 | 1.8282 | 1.8378 | 1.8471 | 1.8564 | 1.8788 | 1.9004 | 1.9212 | 1.9413 | 1.9796 | 2.0158 | 2.0501 | 2.0829 | 2.1444 | 2.2015 | s | (270.21) |
| 13.171 | 13.455 | 13.739 | 14.022 | 14.305 | 14.587 | 15.291 | 15.993 | 16.694 | 17.393 | 18.790 | 20.18 | 21.58 | 22.97 | 25.75 | 28.52 | v | |
| 1284.5 | 1294.2 | 1303.9 | 1313.5 | 1323.2 | 1332.9 | 1357.3 | 1381.7 | 1406.4 | 1431.2 | 1481.3 | 1532.3 | 1584.2 | 1636.9 | 1745.1 | 1856.9 | h | 43 |
| 1.8058 | 1.8158 | 1.8255 | 1.8351 | 1.8445 | 1.8538 | 1.8762 | 1.8978 | 1.9186 | 1.9387 | 1.9770 | 2.0132 | 2.0475 | 2.0803 | 2.1418 | 2.1989 | s | (271.64) |
| 12.868 | 13.147 | 13.424 | 13.701 | 13.977 | 14.254 | 14.942 | 15.628 | 16.313 | 16.997 | 18.361 | 19.724 | 21.08 | 22.44 | 25.16 | 27.87 | v | |
| 1284.4 | 1294.1 | 1303.8 | 1313.5 | 1323.2 | 1332.9 | 1357.2 | 1381.7 | 1406.3 | 1431.1 | 1481.3 | 1532.3 | 1584.2 | 1636.9 | 1745.0 | 1856.9 | h | 44 |
| 1.8032 | 1.8132 | 1.8229 | 1.8325 | 1.8419 | 1.8512 | 1.8736 | 1.8952 | 1.9160 | 1.9361 | 1.9744 | 2.0106 | 2.0450 | 2.0777 | 2.1392 | 2.1964 | s | (273.05) |
| 12.580 | 12.852 | 13.123 | 13.394 | 13.665 | 13.935 | 14.608 | 15.279 | 15.949 | 16.618 | 17.952 | 19.285 | 20.62 | 21.94 | 24.60 | 27.25 | v | |
| 1284.4 | 1294.0 | 1303.7 | 1313.4 | 1323.1 | 1332.8 | 1357.1 | 1381.6 | 1406.3 | 1431.1 | 1481.3 | 1532.3 | 1584.1 | 1636.9 | 1745.0 | 1856.9 | h | 45 |
| 1.8006 | 1.8106 | 1.8204 | 1.8300 | 1.8394 | 1.8487 | 1.8711 | 1.8927 | 1.9135 | 1.9336 | 1.9719 | 2.0081 | 2.0425 | 2.0752 | 2.1368 | 2.1939 | s | (274.44) |
| 12.303 | 12.570 | 12.836 | 13.101 | 13.366 | 13.630 | 14.288 | 14.945 | 15.601 | 16.255 | 17.561 | 18.864 | 20.17 | 21.47 | 24.07 | 26.66 | v | |
| 1284.3 | 1293.9 | 1303.6 | 1313.3 | 1323.0 | 1332.7 | 1357.1 | 1381.6 | 1406.2 | 1431.0 | 1481.3 | 1532.3 | 1584.1 | 1636.8 | 1745.0 | 1856.9 | h | 46 |
| 1.7981 | 1.8081 | 1.8179 | 1.8275 | 1.8369 | 1.8462 | 1.8686 | 1.8902 | 1.9110 | 1.9312 | 1.9695 | 2.0057 | 2.0400 | 2.0728 | 2.1343 | 2.1915 | s | (275.80) |
| 12.039 | 12.300 | 12.560 | 12.820 | 13.079 | 13.338 | 13.983 | 14.626 | 15.268 | 15.908 | 17.186 | 18.462 | 19.736 | 21.01 | 23.55 | 26.09 | v | |
| 1284.2 | 1293.9 | 1303.5 | 1313.2 | 1323.0 | 1332.7 | 1357.0 | 1381.5 | 1406.2 | 1431.0 | 1481.2 | 1532.2 | 1584.1 | 1636.8 | 1745.0 | 1856.9 | h | 47 |
| 1.7957 | 1.8057 | 1.8155 | 1.8251 | 1.8345 | 1.8438 | 1.8662 | 1.8878 | 1.9086 | 1.9288 | 1.9671 | 2.0033 | 2.0377 | 2.0704 | 2.1320 | 2.1891 | s | (277.13) |
| 11.786 | 12.042 | 12.296 | 12.551 | 12.804 | 13.058 | 13.690 | 14.320 | 14.948 | 15.575 | 16.827 | 18.077 | 19.325 | 20.57 | 23.06 | 25.55 | v | |
| 1284.1 | 1293.8 | 1303.5 | 1313.2 | 1322.9 | 1332.6 | 1357.0 | 1381.5 | 1406.1 | 1431.0 | 1481.2 | 1532.2 | 1584.1 | 1636.8 | 1745.0 | 1856.9 | h | 48 |
| 1.7933 | 1.8033 | 1.8131 | 1.8227 | 1.8321 | 1.8414 | 1.8639 | 1.8855 | 1.9063 | 1.9264 | 1.9648 | 2.0010 | 2.0353 | 2.0681 | 2.1296 | 2.1868 | s | (278.45) |
| 11.542 | 11.793 | 12.043 | 12.292 | 12.541 | 12 789 | 13.409 | 14.026 | 14.642 | 15.256 | 16.483 | 17.707 | 18.930 | 20.15 | 22.59 | 25.03 | v | |
| 1284.0 | 1293.7 | 1303.4 | 1313.1 | 1322.8 | 1332.5 | 1356.9 | 1381.4 | 1406.1 | 1430.9 | 1481.2 | 1532.2 | 1584.0 | 1636.8 | 1745.0 | 1856.9 | h | 49 |
| 1.7910 | 1.8010 | 1.8108 | 1.8204 | 1.8298 | 1.8391 | 1.8616 | 1.8832 | 1.9040 | 1.9241 | 1.9625 | 1.9987 | 2.0330 | 2.0658 | 2.1273 | 2.1845 | s | (279.74) |
| 11.309 | 11.555 | 11.800 | 12.044 | 12.288 | 12.532 | 13.139 | 13.744 | 14.348 | 14.950 | 16.152 | 17.352 | 18.550 | 19.747 | 22.14 | 24.53 | v | |
| 1283.9 | 1293.6 | 1303.3 | 1313.0 | 1322.7 | 1332.5 | 1356.9 | 1381.4 | 1406.1 | 1430.9 | 1481.1 | 1532.1 | 1584.0 | 1636.8 | 1745.0 | 1856.8 | h | 50 |
| 1.7887 | 1.7987 | 1.8085 | 1.8181 | 1.8275 | 1.8368 | 1.8593 | 1.8809 | 1.9017 | 1.9219 | 1.9602 | 1.9964 | 2.0308 | 2.0636 | 2.1251 | 2.1822 | s | (281.01) |
| 11.085 | 11.326 | 11.566 | 11.806 | 12.045 | 12.284 | 12.879 | 13.473 | 14.065 | 14.656 | 15.835 | 17.011 | 18.186 | 19.359 | 21.70 | 24.05 | v | |
| 1283.8 | 1293.5 | 1303.2 | 1312.9 | 1322.7 | 1332.4 | 1356.8 | 1381.3 | 1406.0 | 1430.9 | 1481.1 | 1532.1 | 1584.0 | 1636.7 | 1744.9 | 1856.8 | h | 51 |
| 1.7864 | 1.7964 | 1.8062 | 1.8159 | 1.8253 | 1.8346 | 1.8571 | 1.8787 | 1.8995 | 1.9197 | 1.9580 | 1.9942 | 2.0286 | 2.0614 | 2.1229 | 2.1801 | s | (282.26) |

# Table 3. Superheated Vapor

| Abs. Press. Lb./Sq. In. (Sat. Temp.) | | Sat. Liquid | Sat. Vapor | 290° | 300° | 320° | 340° | 360° | 380° | 400° | 420° | 440° | 460° | 480° | 500° | 520° | 540 |
|---|---|---|---|---|---|---|---|---|---|---|---|---|---|---|---|---|---|
| 52 | v | 0.017 | 8.208 | 8.293 | 8.423 | 8.680 | 8.932 | 9.181 | 9.427 | 9.671 | 9.914 | 10.154 | 10.394 | 10.632 | 10.869 | 11.106 | 11.34 |
| | h | 252.6 | 1174.8 | 1178.4 | 1183.8 | 1194.3 | 1204.7 | 1214.8 | 1224.8 | 1234.8 | 1244.6 | 1254.5 | 1264.2 | 1274.0 | 1283.7 | 1293.4 | 1303. |
| (283.49) | s | 0.4144 | 1.6553 | 1.6601 | 1.6672 | 1.6809 | 1.6940 | 1.7066 | 1.7186 | 1.7303 | 1.7417 | 1.7527 | 1.7635 | 1.7739 | 1.7842 | 1.7942 | 1.804 |
| 53 | v | 0.017 | 8.062 | 8.131 | 8.259 | 8.511 | 8.759 | 9.004 | 9.246 | 9.485 | 9.723 | 9.960 | 10.195 | 10.429 | 10.662 | 10.894 | 11.12 |
| | h | 253.9 | 1175.2 | 1178.1 | 1183.5 | 1194.1 | 1204.5 | 1214.6 | 1224.7 | 1234.6 | 1244.5 | 1254.3 | 1264.1 | 1273.9 | 1283.6 | 1293.3 | 1303. |
| (284.70) | s | 0.4161 | 1.6538 | 1.6577 | 1.6648 | 1.6786 | 1.6917 | 1.7043 | 1.7164 | 1.7281 | 1.7395 | 1.7505 | 1.7613 | 1.7718 | 1.7820 | 1.7920 | 1.801 |
| 54 | v | 0.017 | 7.922 | 7.974 | 8.100 | 8.348 | 8.592 | 8.833 | 9.071 | 9.306 | 9.540 | 9.772 | 10.103 | 10.233 | 10.462 | 10.690 | 10.91 |
| | h | 255.1 | 1175.6 | 1177.8 | 1183.2 | 1193.9 | 1204.3 | 1214.4 | 1224.5 | 1234.5 | 1244.4 | 1254.2 | 1264.0 | 1273.8 | 1283.5 | 1293.3 | 1303. |
| (285.90) | s | 0.4177 | 1.6523 | 1.6553 | 1.6625 | 1.6763 | 1.6894 | 1.7021 | 1.7142 | 1.7259 | 1.7373 | 1.7483 | 1.7591 | 1.7696 | 1.7799 | 1.7899 | 1.799 |
| 55 | v | 0.017 | 7.787 | 7.823 | 7.947 | 8.192 | 8.432 | 8.668 | 8.902 | 9.134 | 9.364 | 9.592 | 9.819 | 10.044 | 10.269 | 10.493 | 10.71 |
| | h | 256.3 | 1175.9 | 1177.5 | 1183.0 | 1193.6 | 1204.0 | 1214.3 | 1224.4 | 1234.3 | 1244.3 | 1254.1 | 1263.9 | 1273.7 | 1283.4 | 1293.2 | 1302. |
| (287.07) | s | 0.4193 | 1.6509 | 1.6530 | 1.6602 | 1.6740 | 1.6872 | 1.6999 | 1.7120 | 1.7238 | 1.7351 | 1.7462 | 1.7570 | 1.7675 | 1.7778 | 1.7878 | 1.797 |
| 56 | v | 0.017 | 7.656 | 7.678 | 7.800 | 8.040 | 8.277 | 8.509 | 8.739 | 8.967 | 9.193 | 9.418 | 9.641 | 9.863 | 10.084 | 10.304 | 10.52 |
| | h | 257.5 | 1176.3 | 1177.2 | 1182.7 | 1193.4 | 1203.8 | 1214.1 | 1224.2 | 1234.2 | 1244.1 | 1254.0 | 1263.8 | 1273.6 | 1283.3 | 1293.1 | 1302. |
| (288.23) | s | 0.4209 | 1.6494 | 1.6507 | 1.6579 | 1.6718 | 1.6850 | 1.6977 | 1.7099 | 1.7217 | 1.7330 | 1.7441 | 1.7549 | 1.7654 | 1.7757 | 1.7858 | 1.795 |
| 57 | v | 0.017 | 7.529 | 7.537 | 7.658 | 7.895 | 8.127 | 8.356 | 8.583 | 8.807 | 9.029 | 9.250 | 9.469 | 9.687 | 9.904 | 10.121 | 10.33 |
| | h | 258.7 | 1176.6 | 1176.9 | 1182.5 | 1193.2 | 1203.6 | 1213.9 | 1224.0 | 1234.1 | 1244.0 | 1253.9 | 1263.7 | 1273.5 | 1283.2 | 1293.0 | 1302. |
| (289.37) | s | 0.4225 | 1.6480 | 1.6484 | 1.6557 | 1.6696 | 1.6829 | 1.6956 | 1.7078 | 1.7196 | 1.7310 | 1.7421 | 1.7529 | 1.7634 | 1.7737 | 1.7838 | 1.793 |
| 58 | v | 0.017 | 7.407 | | 7.520 | 7.754 | 7.983 | 8.208 | 8.431 | 8.652 | 8.870 | 9.087 | 9.303 | 9.518 | 9.732 | 9.945 | 10.157 |
| | h | 259.8 | 1176.9 | | 1182.2 | 1192.9 | 1203.4 | 1213.7 | 1223.9 | 1233.9 | 1243.9 | 1253.8 | 1263.6 | 1273.4 | 1283.1 | 1292.9 | 1302.7 |
| (290.50) | s | 0.4240 | 1.6466 | | 1.6535 | 1.6675 | 1.6808 | 1.6935 | 1.7057 | 1.7175 | 1.7289 | 1.7400 | 1.7509 | 1.7614 | 1.7717 | 1.7818 | 1.7916 |
| 59 | v | 0.017 | 7.289 | | 7.387 | 7.618 | 7.843 | 8.065 | 8.285 | 8.502 | 8.717 | 8.931 | 9.143 | 9.354 | 9.564 | 9.774 | 9.983 |
| | h | 261.0 | 1177.3 | | 1181.9 | 1192.7 | 1203.2 | 1213.5 | 1223.7 | 1233.8 | 1243.7 | 1253.6 | 1263.5 | 1273.3 | 1283.1 | 1292.8 | 1302.6 |
| (291.61) | s | 0.4255 | 1.6452 | | 1.6513 | 1.6554 | 1.6787 | 1.6914 | 1.7037 | 1.7155 | 1.7270 | 1.7381 | 1.7489 | 1.7595 | 1.7698 | 1.7798 | 1.7897 |
| 60 | v | 0.017 | 7.175 | | 7.259 | 7.486 | 7.708 | 7.927 | 8.143 | 8.357 | 8.569 | 8.779 | 8.988 | 9.196 | 9.403 | 9.609 | 9.814 |
| | h | 262.1 | 1177.6 | | 1181.6 | 1192.5 | 1203.0 | 1213.4 | 1223.6 | 1233.6 | 1243.6 | 1253.5 | 1263.4 | 1273.2 | 1283.0 | 1292.7 | 1302.5 |
| (292.71) | s | 0.4270 | 1.6438 | | 1.6492 | 1.6633 | 1.6766 | 1.6894 | 1.7017 | 1.7135 | 1.7250 | 1.7361 | 1.7470 | 1.7575 | 1.7678 | 1.7779 | 1.7878 |
| 61 | v | 0.017 | 7.064 | | 7.135 | 7.359 | 7.578 | 7.794 | 8.006 | 8.217 | 8.426 | 8.633 | 8.838 | 9.043 | 9.247 | 9.449 | 9.652 |
| | h | 263.2 | 1177.9 | | 1181.4 | 1192.2 | 1202.8 | 1213.2 | 1223.4 | 1233.5 | 1243.5 | 1253.4 | 1263.3 | 1273.1 | 1282.9 | 1292.7 | 1302.4 |
| (293.79) | s | 0.4285 | 1.6425 | | 1.6471 | 1.6612 | 1.6746 | 1.6874 | 1.6997 | 1.7116 | 1.7231 | 1.7342 | 1.7450 | 1.7556 | 1.7659 | 1.7760 | 1.7859 |
| 62 | v | 0.017 | 6.957 | | 7.015 | 7.236 | 7.452 | 7.664 | 7.874 | 8.081 | 8.287 | 8.491 | 8.693 | 8.895 | 9.095 | 9.295 | 9.494 |
| | h | 264.3 | 1178.2 | | 1181.1 | 1192.0 | 1202.6 | 1213.0 | 1223.2 | 1233.3 | 1243.3 | 1253.3 | 1263.1 | 1273.0 | 1282.8 | 1292.6 | 1302.3 |
| (294.85) | s | 0.4300 | 1.6412 | | 1.6450 | 1.6592 | 1.6726 | 1.6854 | 1.6978 | 1.7097 | 1.7212 | 1.7323 | 1.7432 | 1.7538 | 1.7641 | 1.7741 | 1.7840 |
| 63 | v | 0.017 | 6.853 | | 6.899 | 7.116 | 7.329 | 7.539 | 7.746 | 7.950 | 8.151 | 8.353 | 8.553 | 8.751 | 8.949 | 9.146 | 9.342 |
| | h | 265.4 | 1178.5 | | 1180.8 | 1191.8 | 1202.4 | 1212.8 | 1223.1 | 1233.2 | 1243.2 | 1253.2 | 1263.0 | 1272.9 | 1282.7 | 1292.5 | 1302.3 |
| (295.90) | s | 0.4314 | 1.6399 | | 1.6429 | 1.6572 | 1.6706 | 1.6835 | 1.6959 | 1.7078 | 1.7193 | 1.7305 | 1.7413 | 1.7519 | 1.7622 | 1.7723 | 1.7822 |
| 64 | v | 0.017 | 6.752 | | 6.786 | 7.001 | 7.211 | 7.418 | 7.621 | 7.823 | 8.022 | 8.220 | 8.417 | 8.613 | 8.807 | 9.001 | 9.194 |
| | h | 266.4 | 1178.8 | | 1180.5 | 1191.5 | 1202.2 | 1212.6 | 1222.9 | 1233.0 | 1243.1 | 1253.0 | 1262.9 | 1272.8 | 1282.6 | 1292.4 | 1302.2 |
| (296.94) | s | 0.4328 | 1.6387 | | 1.6409 | 1.6552 | 1.6687 | 1.6816 | 1.6940 | 1.7059 | 1.7174 | 1.7286 | 1.7395 | 1.7501 | 1.7604 | 1.7705 | 1.7804 |
| 65 | v | 0.017 | 6.655 | | 6.676 | 6.889 | 7.096 | 7.300 | 7.501 | 7.700 | 7.896 | 8.091 | 8.285 | 8.478 | 8.670 | 8.861 | 9.051 |
| | h | 267.5 | 1179.1 | | 1180.3 | 1191.3 | 1202.0 | 1212.5 | 1222.7 | 1232.9 | 1242.9 | 1252.9 | 1262.8 | 1272.7 | 1282.5 | 1292.3 | 1302.1 |
| (297.97) | s | 0.4342 | 1.6374 | | 1.6389 | 1.6532 | 1.6668 | 1.6797 | 1.6921 | 1.7040 | 1.7156 | 1.7268 | 1.7377 | 1.7483 | 1.7586 | 1.7687 | 1.7786 |
| 66 | v | 0.017 | 6.560 | | 6.571 | 6.780 | 6.985 | 7.186 | 7.384 | 7.580 | 7.774 | 7.966 | 8.157 | 8.347 | 8.536 | 8.724 | 8.912 |
| | h | 268.5 | 1179.4 | | 1180.0 | 1191.1 | 1201.8 | 1212.3 | 1222.6 | 1232.7 | 1242.8 | 1252.8 | 1262.7 | 1272.6 | 1282.4 | 1292.2 | 1302.0 |
| (298.99) | s | 0.4356 | 1.6362 | | 1.6369 | 1.6513 | 1.6649 | 1.6779 | 1.6903 | 1.7022 | 1.7138 | 1.7250 | 1.7359 | 1.7465 | 1.7569 | 1.7670 | 1.7769 |
| 67 | v | 0.017 | 6.468 | | 6.468 | 6.675 | 6.877 | 7.075 | 7.271 | 7.464 | 7.655 | 7.845 | 8.034 | 8.221 | 8.407 | 8.592 | 8.777 |
| | h | 269.6 | 1179.7 | | 1179.7 | 1190.8 | 1201.6 | 1212.1 | 1222.4 | 1232.6 | 1242.7 | 1252.7 | 1262.6 | 1272.5 | 1282.3 | 1292.1 | 1301.9 |
| (299.99) | s | 0.4369 | 1.6350 | | 1.6350 | 1.6494 | 1.6630 | 1.6760 | 1.6885 | 1.7004 | 1.7120 | 1.7232 | 1.7342 | 1.7448 | 1.7552 | 1.7653 | 1.7752 |
| 68 | v | 0.017 | 6.378 | | | 6.572 | 6.772 | 6.968 | 7.161 | 7.352 | 7.540 | 7.727 | 7.913 | 8.098 | 8.281 | 8.464 | 8.646 |
| | h | 270.6 | 1180.0 | | | 1190.6 | 1201.4 | 1211.9 | 1222.2 | 1232.4 | 1242.5 | 1252.5 | 1262.5 | 1272.4 | 1282.2 | 1292.0 | 1301.9 |
| (300.98) | s | 0.4383 | 1.6338 | | | 1.6475 | 1.6612 | 1.6742 | 1.6867 | 1.6987 | 1.7103 | 1.7215 | 1.7324 | 1.7431 | 1.7534 | 1.7636 | 1.7735 |
| 69 | v | 0.017 | 6.291 | | | 6.473 | 6.670 | 6.864 | 7.054 | 7.242 | 7.428 | 7.613 | 7.796 | 7.978 | 8.160 | 8.340 | 8.519 |
| | h | 271.6 | 1180.3 | | | 1190.3 | 1201.2 | 1211.7 | 1222.1 | 1232.3 | 1242.4 | 1252.4 | 1262.4 | 1272.3 | 1282.1 | 1291.9 | 1301.8 |
| (301.96) | s | 0.4396 | 1.6326 | | | 1.6457 | 1.6594 | 1.6724 | 1.6849 | 1.6969 | 1.7085 | 1.7198 | 1.7307 | 1.7414 | 1.7518 | 1.7619 | 1.7718 |

Temperature—Degrees Fahrenheit

| 560° | 580° | 600° | 650° | 700° | 750° | 800° | 850° | 900° | 950° | 1000° | 1050° | 1100° | 1200° | 1400° | 1600° | | Abs. Press. Lb./Sq. In. (Sat. Temp.) |
|---|---|---|---|---|---|---|---|---|---|---|---|---|---|---|---|---|---|
| 11.577 | 11.812 | 12.046 | 12.630 | 13.212 | 13.793 | 14.373 | 14.951 | 15.529 | 16.107 | 16.685 | 17.260 | 17.836 | 18.987 | 21.29 | 23.58 | v | |
| 1312.9 | 1322.6 | 1332.3 | 1356.7 | 1381.3 | 1406.0 | 1430.8 | 1455.8 | 1481.1 | 1506.5 | 1532.1 | 1557.9 | 1584.0 | 1636.7 | 1744.9 | 1856.8 | h | **52** |
| 1.8137 | 1.8231 | 1.8324 | 1.8549 | 1.8765 | 1.8974 | 1.9175 | 1.9370 | 1.9559 | 1.9742 | 1.9921 | 2.0095 | 2.0264 | 2.0592 | 2.1208 | 2.1779 | s | (283.49) |
| 11.357 | 11.587 | 11.817 | 12.390 | 12.962 | 13.532 | 14.100 | 14.668 | 15.235 | 15.802 | 16.368 | 16.933 | 17.498 | 18.628 | 20.88 | 23.14 | v | |
| 1312.8 | 1322.5 | 1332.3 | 1356.7 | 1381.2 | 1405.9 | 1430.8 | 1455.8 | 1481.0 | 1506.4 | 1532.1 | 1557.9 | 1583.9 | 1636.7 | 1744.9 | 1856.8 | h | **53** |
| 1.8115 | 1.8210 | 1.8302 | 1.8528 | 1.8744 | 1.8952 | 1.9154 | 1.9349 | 1.9538 | 1.9721 | 1.9900 | 2.0074 | 2.0243 | 2.0571 | 2.1187 | 2.1758 | s | (284.70) |
| 11.144 | 11.371 | 11.596 | 12.159 | 12.720 | 13.280 | 13.838 | 14.395 | 14.952 | 15.508 | 16.064 | 16.619 | 17.174 | 18.282 | 20.50 | 22.71 | v | |
| 1312.7 | 1322.5 | 1332.2 | 1356.6 | 1381.2 | 1405.9 | 1430.7 | 1455.8 | 1481.0 | 1506.4 | 1532.0 | 1557.9 | 1583.9 | 1636.7 | 1744.9 | 1856.8 | h | **54** |
| 1.8094 | 1.8189 | 1.8281 | 1.8507 | 1.8723 | 1.8932 | 1.9133 | 1.9328 | 1.9517 | 1.9700 | 1.9879 | 2.0053 | 2.0223 | 2.0551 | 2.1166 | 2.1737 | s | (285.90) |
| 10.940 | 11.162 | 11.384 | 11.937 | 12.488 | 13.037 | 13.586 | 14.133 | 14.679 | 15.226 | 15.771 | 16.316 | 16.861 | 17.950 | 20.12 | 22.30 | v | |
| 1312.6 | 1322.4 | 1332.1 | 1356.6 | 1381.1 | 1405.8 | 1430.7 | 1455.7 | 1481.0 | 1506.4 | 1532.0 | 1557.9 | 1583.9 | 1636.7 | 1744.9 | 1856.8 | h | **55** |
| 1.8073 | 1.8168 | 1.8261 | 1.8486 | 1.8703 | 1.8911 | 1.9112 | 1.9307 | 1.9496 | 1.9680 | 1.9859 | 2.0033 | 2.0202 | 2.0530 | 2.1146 | 2.1717 | s | (287.07) |
| 10.742 | 10.961 | 11.179 | 11.722 | 12.263 | 12.803 | 13.342 | 13.879 | 14.416 | 14.953 | 15.489 | 16.024 | 16.559 | 17.629 | 19.764 | 21.90 | v | |
| 1312.6 | 1322.3 | 1332.1 | 1356.5 | 1381.1 | 1405.8 | 1430.7 | 1455.7 | 1480.9 | 1506.3 | 1532.0 | 1557.8 | 1583.9 | 1636.6 | 1744.9 | 1856.8 | h | **56** |
| 1.8053 | 1.8147 | 1.8240 | 1.8466 | 1.8682 | 1.8891 | 1.9092 | 1.9287 | 1.9476 | 1.9660 | 1.9839 | 2.0013 | 2.0182 | 2.0510 | 2.1126 | 2.1697 | s | (288.23) |
| 10.552 | 10.767 | 10.981 | 11.515 | 12.047 | 12.578 | 13.107 | 13.635 | 14.163 | 14.690 | 15.216 | 15.742 | 16.268 | 17.319 | 19.417 | 21.51 | v | |
| 1312.5 | 1322.2 | 1332.0 | 1356.5 | 1381.0 | 1405.7 | 1430.6 | 1455.7 | 1480.9 | 1506.3 | 1532.0 | 1557.8 | 1583.9 | 1636.6 | 1744.8 | 1856.8 | h | **57** |
| 1.8033 | 1.8127 | 1.8220 | 1.8446 | 1.8663 | 1.8871 | 1.9073 | 1.9268 | 1.9457 | 1.9640 | 1.9819 | 1.9993 | 2.0163 | 2.0491 | 2.1106 | 2.1678 | s | (289.37) |
| 10.368 | 10.580 | 10.790 | 11.315 | 11.838 | 12.360 | 12.880 | 13.399 | 13.918 | 14.436 | 14.953 | 15.470 | 15.987 | 17.020 | 19.082 | 21.14 | v | |
| 1312.4 | 1322.2 | 1331.9 | 1356.4 | 1381.0 | 1405.7 | 1430.6 | 1455.6 | 1480.9 | 1506.3 | 1531.9 | 1557.8 | 1583.8 | 1636.6 | 1744.8 | 1856.7 | h | **58** |
| 1.8013 | 1.8108 | 1.8201 | 1.8427 | 1.8643 | 1.8852 | 1.9053 | 1.9248 | 1.9437 | 1.9621 | 1.9800 | 1.9974 | 2.0143 | 2.0472 | 2.1087 | 2.1658 | s | (290.50) |
| 10.191 | 10.399 | 10.606 | 11.122 | 11.636 | 12.149 | 12.660 | 13.171 | 13.681 | 14.190 | 14.699 | 15.207 | 15.716 | 16.731 | 18.759 | 20.78 | v | |
| 1312.3 | 1322.1 | 1331.9 | 1356.3 | 1380.9 | 1405.7 | 1430.5 | 1455.6 | 1480.8 | 1506.3 | 1531.9 | 1557.7 | 1583.8 | 1636.6 | 1744.8 | 1856.7 | h | **59** |
| 1.7994 | 1.8088 | 1.8181 | 1.8407 | 1.8624 | 1.8833 | 1.9034 | 1.9229 | 1.9418 | 1.9602 | 1.9781 | 1.9955 | 2.0125 | 2.0453 | 2.1068 | 2.1640 | s | (291.61) |
| 10.019 | 10.223 | 10.427 | 10.935 | 11.441 | 11.945 | 12.449 | 12.951 | 13.452 | 13.953 | 14.454 | 14.954 | 15.453 | 16.451 | 18.446 | 20.44 | v | |
| 1312.3 | 1322.0 | 1331.8 | 1356.3 | 1380.9 | 1405.6 | 1430.5 | 1455.6 | 1480.8 | 1506.2 | 1531.9 | 1557.7 | 1583.8 | 1636.6 | 1744.8 | 1856.7 | h | **60** |
| 1.7975 | 1.8069 | 1.8162 | 1.8388 | 1.8605 | 1.8814 | 1.9015 | 1.9210 | 1.9400 | 1.9583 | 1.9762 | 1.9936 | 2.0106 | 2.0434 | 2.1049 | 2.1621 | s | (292.71) |
| 9.853 | 10.054 | 10.255 | 10.754 | 11.252 | 11.748 | 12.243 | 12.738 | 13.231 | 13.724 | 14.216 | 14.708 | 15.199 | 16.181 | 18.143 | 20.10 | v | |
| 1312.2 | 1322.0 | 1331.7 | 1356.2 | 1380.8 | 1405.6 | 1430.4 | 1455.5 | 1480.8 | 1506.2 | 1531.8 | 1557.7 | 1583.8 | 1636.5 | 1744.8 | 1856.7 | h | **61** |
| 1.7956 | 1.8051 | 1.8144 | 1.8370 | 1.8586 | 1.8795 | 1.8997 | 1.9192 | 1.9381 | 1.9565 | 1.9744 | 1.9918 | 2.0088 | 2.0416 | 2.1031 | 2.1603 | s | (293.79) |
| 9.692 | 9.890 | 10.088 | 10.580 | 11.070 | 11.558 | 12.045 | 12.531 | 13.017 | 13.502 | 13.986 | 14.470 | 14.954 | 15.920 | 17.850 | 19.778 | v | |
| 1312.1 | 1321.9 | 1331.7 | 1356.2 | 1380.8 | 1405.5 | 1430.4 | 1455.5 | 1480.7 | 1506.2 | 1531.8 | 1557.7 | 1583.7 | 1636.5 | 1744.8 | 1856.7 | h | **62** |
| 1.7937 | 1.8032 | 1.8125 | 1.8351 | 1.8568 | 1.8777 | 1.8979 | 1.9174 | 1.9363 | 1.9547 | 1.9726 | 1.9900 | 2.0070 | 2.0398 | 2.1013 | 2.1585 | s | (294.85) |
| 9.537 | 9.732 | 9.926 | 10.410 | 10.893 | 11.373 | 11.853 | 12.332 | 12.809 | 13.287 | 13.764 | 14.240 | 14.716 | 15.667 | 17.566 | 19.464 | v | |
| 1312.0 | 1321.8 | 1331.6 | 1356.1 | 1380.7 | 1405.5 | 1430.4 | 1455.4 | 1480.7 | 1506.1 | 1531.8 | 1557.6 | 1583.7 | 1636.5 | 1744.8 | 1856.7 | h | **63** |
| 1.7919 | 1.8014 | 1.8107 | 1.8333 | 1.8550 | 1.8759 | 1.8961 | 1.9156 | 1.9345 | 1.9529 | 1.9708 | 1.9882 | 2.0052 | 2.0380 | 2.0996 | 2.1567 | s | (295.90) |
| 9.386 | 9.578 | 9.770 | 10.246 | 10.721 | 11.195 | 11.667 | 12.138 | 12.608 | 13.078 | 13.548 | 14.017 | 14.485 | 15.422 | 17.292 | 19.159 | v | |
| 1312.0 | 1321.8 | 1331.5 | 1356.1 | 1380.7 | 1405.4 | 1430.3 | 1455.4 | 1480.7 | 1506.1 | 1531.8 | 1557.6 | 1583.7 | 1636.5 | 1744.7 | 1856.7 | h | **64** |
| 1.7901 | 1.7996 | 1.8090 | 1 8316 | 1.8533 | 1.8742 | 1.8943 | 1.9138 | 1.9328 | 1.9512 | 1.9690 | 1.9864 | 2.0034 | 2.0362 | 2.0978 | 2.1550 | s | (296.94) |
| 9.240 | 9.429 | 9.618 | 10.087 | 10.555 | 11.021 | 11.486 | 11.951 | 12.414 | 12.877 | 13.339 | 13.801 | 14.262 | 15.184 | 17.025 | 18.864 | v | |
| 1311.9 | 1321.7 | 1331.5 | 1356.0 | 1380.6 | 1405.4 | 1430.3 | 1455.4 | 1480.6 | 1506.1 | 1531.7 | 1557.6 | 1583.7 | 1636.4 | 1744.7 | 1856.7 | h | **65** |
| 1.7884 | 1.7979 | 1.8072 | 1.8298 | 1.8515 | 1.8724 | 1.8926 | 1.9121 | 1.9310 | 1.9494 | 1.9673 | 1.9847 | 2.0017 | 2.0345 | 2.0961 | 2.1533 | s | (297.97) |
| 9.099 | 9.285 | 9.471 | 9.933 | 10.394 | 10.853 | 11.311 | 11.769 | 12.225 | 12.681 | 13.136 | 13.591 | 14.045 | 14.953 | 16.767 | 18.578 | v | |
| 1311.8 | 1321.6 | 1331.4 | 1355.9 | 1380.6 | 1405.3 | 1430.2 | 1455.3 | 1480.6 | 1506.1 | 1531.7 | 1557.6 | 1583.6 | 1636.4 | 1744.7 | 1856.6 | h | **66** |
| 1.7866 | 1.7961 | 1.8055 | 1.8281 | 1.8498 | 1.8707 | 1.8909 | 1.9104 | 1.9294 | 1.9477 | 1.9656 | 1.9830 | 2.0000 | 2.0328 | 2.0944 | 2.1516 | s | (298.99) |
| 8.961 | 9.145 | 9.328 | 9.784 | 10.238 | 10.690 | 11.142 | 11.592 | 12.042 | 12.491 | 12.939 | 13.388 | 13.835 | 14.730 | 16.517 | 18.301 | v | |
| 1311.7 | 1321.5 | 1331.3 | 1355.9 | 1380.5 | 1405.3 | 1430.2 | 1455.3 | 1480.6 | 1506.0 | 1531.7 | 1557.5 | 1583.6 | 1636.4 | 1744.7 | 1856.6 | h | **67** |
| 1.7849 | 1.7944 | 1.8038 | 1.8264 | 1.8481 | 1.8690 | 1.8892 | 1.9087 | 1.9277 | 1.9461 | 1.9639 | 1.9814 | 1.9984 | 2.0312 | 2.0927 | 2.1499 | s | (299.99) |
| 8.828 | 9.009 | 9.189 | 9.639 | 10.086 | 10.532 | 10.977 | 11.421 | 11.864 | 12.307 | 12.749 | 13.190 | 13.631 | 14.513 | 16.273 | 18.032 | v | |
| 1311.7 | 1321.5 | 1331.3 | 1355.8 | 1380.5 | 1405.2 | 1430.2 | 1455.3 | 1480.5 | 1506.0 | 1531.6 | 1557.5 | 1583.6 | 1636.4 | 1744.7 | 1856.6 | h | **68** |
| 1.7832 | 1.7927 | 1.8021 | 1.8247 | 1.8464 | 1.8674 | 1.8876 | 1.9071 | 1.9260 | 1.9444 | 1.9623 | 1.9797 | 1.9967 | 2.0295 | 2.0911 | 2.1483 | s | (300.98) |
| 8.698 | 8.877 | 9.055 | 9.498 | 9.939 | 10.379 | 10.817 | 11.255 | 11.691 | 12.128 | 12.563 | 12.999 | 13.433 | 14.302 | 16.037 | 17.770 | v | |
| 1311.6 | 1321.4 | 1331.2 | 1355.8 | 1380.4 | 1405.2 | 1430.1 | 1455.2 | 1480.5 | 1506.0 | 1531.6 | 1557.5 | 1583.6 | 1636.4 | 1744.7 | 1856.6 | h | **69** |
| 1.7816 | 1.7911 | 1.8004 | 1.8231 | 1.8448 | 1.8657 | 1.8859 | 1.9055 | 1.9244 | 1.9428 | 1.9607 | 1.9781 | 1.9951 | 2.0279 | 2.0895 | 2.1467 | s | (301.96) |

## Table 3. Superheated Vapor

| Abs. Press. Lb./Sq. In. (Sat. Temp.) | | Sat. Liquid | Sat. Vapor | Temperature—Degrees Fahrenheit 310° | 320° | 330° | 340° | 350° | 360° | 370° | 380° | 390° | 400° | 420° | 440° | 460° | 480° |
|---|---|---|---|---|---|---|---|---|---|---|---|---|---|---|---|---|---|
| | v | 0.017 | 6.206 | 6.277 | 6.376 | 6.474 | 6.571 | 6.667 | 6.762 | 6.857 | 6.950 | 7.044 | 7.136 | 7.320 | 7.502 | 7.683 | 7.863 |
| **70** | h | 272.6 | 1180.6 | 1184.5 | 1190.1 | 1195.6 | 1201.0 | 1206.3 | 1211.5 | 1216.7 | 1221.9 | 1227.0 | 1232.1 | 1242.3 | 1252.3 | 1262.2 | 1272.2 |
| (302.92) | s | 0.4409 | 1.6315 | 1.6367 | 1.6438 | 1.6508 | 1.6576 | 1.6642 | 1.6707 | 1.6770 | 1.6832 | 1.6892 | 1.6952 | 1.7068 | 1.7181 | 1.7291 | 1.7397 |
| | v | 0.017 | 6.124 | 6.184 | 6.283 | 6.379 | 6.475 | 6.570 | 6.664 | 6.757 | 6.850 | 6.942 | 7.033 | 7.214 | 7.394 | 7.573 | 7.750 |
| **71** | h | 273.6 | 1180.8 | 1184.3 | 1189.9 | 1195.3 | 1200.7 | 1206.1 | 1211.4 | 1216.6 | 1221.8 | 1226.9 | 1232.0 | 1242.1 | 1252.2 | 1262.1 | 1272.1 |
| (303.88) | s | 0.4422 | 1.6303 | 1.6348 | 1.6420 | 1.6490 | 1.6558 | 1.6624 | 1.6689 | 1.6753 | 1.6815 | 1.6875 | 1.6935 | 1.7052 | 1.7164 | 1.7274 | 1.7381 |
| | v | 0.018 | 6.044 | 6.094 | 6.191 | 6.287 | 6.382 | 6.475 | 6.568 | 6.660 | 6.752 | 6.843 | 6.933 | 7.112 | 7.289 | 7.465 | 7.640 |
| **72** | h | 274.6 | 1181.1 | 1184.0 | 1189.6 | 1195.1 | 1200.5 | 1205.9 | 1211.2 | 1216.4 | 1221.6 | 1226.7 | 1231.9 | 1242.0 | 1252.0 | 1262.0 | 1271.9 |
| (304.83) | s | 0.4435 | 1.6292 | 1.6330 | 1.6402 | 1.6472 | 1.6541 | 1.6607 | 1.6672 | 1.6735 | 1.6797 | 1.6859 | 1.6918 | 1.7035 | 1.7148 | 1.7258 | 1.7364 |
| | v | 0.018 | 5.966 | 6.007 | 6.103 | 6.197 | 6.291 | 6.383 | 6.475 | 6.566 | 6.656 | 6.746 | 6.835 | 7.012 | 7.187 | 7.361 | 7.534 |
| **73** | h | 275.5 | 1181.3 | 1183.8 | 1189.4 | 1194.9 | 1200.3 | 1205.7 | 1211.0 | 1216.2 | 1221.4 | 1226.6 | 1231.7 | 1241.9 | 1251.9 | 1261.9 | 1271.8 |
| (305.76) | s | 0.4447 | 1.6281 | 1.6312 | 1.6385 | 1.6455 | 1.6523 | 1.6590 | 1.6655 | 1.6719 | 1.6781 | 1.6842 | 1.6902 | 1.7019 | 1.7132 | 1.7242 | 1.7348 |
| | v | 0.018 | 5.890 | 5.921 | 6.016 | 6.110 | 6.202 | 6.294 | 6.384 | 6.474 | 6.563 | 6.652 | 6.740 | 6.915 | 7.088 | 7.259 | 7.430 |
| **74** | h | 276.5 | 1181.6 | 1183.5 | 1189.1 | 1194.7 | 1200.1 | 1205.5 | 1210.8 | 1216.0 | 1221.3 | 1226.4 | 1231.6 | 1241.7 | 1251.8 | 1261.8 | 1271.7 |
| (306.68) | s | 0.4460 | 1.6270 | 1.6294 | 1.6367 | 1.6438 | 1.6506 | 1.6573 | 1.6638 | 1.6702 | 1.6764 | 1.6826 | 1.6886 | 1.7003 | 1.7116 | 1.7226 | 1.7333 |
| | v | 0.018 | 5.816 | 5.838 | 5.932 | 6.025 | 6.116 | 6.207 | 6.296 | 6.385 | 6.473 | 6.561 | 6.648 | 6.820 | 6.991 | 7.161 | 7.329 |
| **75** | h | 277.4 | 1181.9 | 1183.2 | 1188.9 | 1194.4 | 1199.9 | 1205.3 | 1210.6 | 1215.9 | 1221.1 | 1226.3 | 1231.4 | 1241.6 | 1251.7 | 1261.7 | 1271.6 |
| (307.60) | s | 0.4472 | 1.6259 | 1.6277 | 1.6350 | 1.6421 | 1.6489 | 1.6556 | 1.6622 | 1.6686 | 1.6748 | 1.6809 | 1.6869 | 1.6987 | 1.7100 | 1.7210 | 1.7317 |
| | v | 0.018 | 5.743 | 5.757 | 5.850 | 5.942 | 6.032 | 6.122 | 6.210 | 6.298 | 6.385 | 6.472 | 6.558 | 6.728 | 6.897 | 7.065 | 7.231 |
| **76** | h | 278.4 | 1182.1 | 1183.0 | 1188.6 | 1194.2 | 1199.7 | 1205.1 | 1210.4 | 1215.7 | 1220.9 | 1226.1 | 1231.3 | 1241.5 | 1251.6 | 1261.6 | 1271.5 |
| (308.50) | s | 0.4484 | 1.6248 | 1.6260 | 1.6333 | 1.6404 | 1.6473 | 1.6540 | 1.6605 | 1.6669 | 1.6732 | 1.6793 | 1.6853 | 1.6971 | 1.7084 | 1.7194 | 1.7302 |
| | v | 0.018 | 5.673 | 5.679 | 5.770 | 5.861 | 5.951 | 6.039 | 6.127 | 6.214 | 6.300 | 6.385 | 6.470 | 6.638 | 6.805 | 6.971 | 7.135 |
| **77** | h | 279.3 | 1182.4 | 1182.7 | 1188.4 | 1194.0 | 1199.5 | 1204.9 | 1210.2 | 1215.5 | 1220.8 | 1225.9 | 1231.1 | 1241.3 | 1251.4 | 1261.5 | 1271.4 |
| (309.40) | s | 0.4496 | 1.6238 | 1.6242 | 1.6316 | 1.6387 | 1.6456 | 1.6523 | 1.6589 | 1.6653 | 1.6716 | 1.6777 | 1.6838 | 1.6955 | 1.7069 | 1.7179 | 1.7286 |
| | v | 0.018 | 5.604 | | 5.693 | 5.782 | 5.871 | 5.958 | 6.045 | 6.131 | 6.216 | 6.301 | 6.385 | 6.551 | 6.716 | 6.880 | 7.042 |
| **78** | h | 280.2 | 1182.6 | | 1188.1 | 1193.7 | 1199.2 | 1204.7 | 1210.0 | 1215.3 | 1220.6 | 1225.8 | 1231.0 | 1241.2 | 1251.3 | 1261.3 | 1271.3 |
| (310.29) | s | 0.4508 | 1.6228 | | 1.6299 | 1.6370 | 1.6440 | 1.6507 | 1.6573 | 1.6637 | 1.6700 | 1.6762 | 1.6822 | 1.6940 | 1.7054 | 1.7164 | 1.7271 |
| | v | 0.018 | 5.537 | | 5.617 | 5.706 | 5.793 | 5.880 | 5.966 | 6.051 | 6.135 | 6.218 | 6.301 | 6.466 | 6.629 | 6.790 | 6.951 |
| **79** | h | 281.1 | 1182.8 | | 1187.9 | 1193.5 | 1199.0 | 1204.5 | 1209.8 | 1215.2 | 1220.4 | 1225.6 | 1230.8 | 1241.1 | 1251.2 | 1261.2 | 1271.2 |
| (311.16) | s | 0.4520 | 1.6217 | | 1.6282 | 1.6354 | 1.6423 | 1.6491 | 1.6557 | 1.6622 | 1.6685 | 1.6746 | 1.6807 | 1.6924 | 1.7038 | 1.7149 | 1.7256 |
| | v | 0.018 | 5.472 | | 5.543 | 5.631 | 5.718 | 5.803 | 5.888 | 5.972 | 6.055 | 6.138 | 6.220 | 6.383 | 6.544 | 6.704 | 6.862 |
| **80** | h | 282.0 | 1183.1 | | 1187.6 | 1193.3 | 1198.8 | 1204.3 | 1209.7 | 1215.0 | 1220.3 | 1225.5 | 1230.7 | 1240.9 | 1251.1 | 1261.1 | 1271.1 |
| (312.03) | s | 0.4531 | 1.6207 | | 1.6266 | 1.6338 | 1.6407 | 1.6475 | 1.6541 | 1.6606 | 1.6669 | 1.6731 | 1.6791 | 1.6909 | 1.7023 | 1.7134 | 1.7242 |
| | v | 0.018 | 5.408 | | 5.471 | 5.558 | 5.644 | 5.728 | 5.812 | 5.896 | 5.978 | 6.060 | 6.141 | 6.302 | 6.461 | 6.619 | 6.776 |
| **81** | h | 282.9 | 1183.3 | | 1187.4 | 1193.0 | 1198.6 | 1204.1 | 1209.5 | 1214.8 | 1220.1 | 1225.3 | 1230.5 | 1240.8 | 1250.9 | 1261.0 | 1271.0 |
| (312.89) | s | 0.4543 | 1.6197 | | 1.6250 | 1.6322 | 1.6392 | 1.6460 | 1.6526 | 1.6591 | 1.6654 | 1.6716 | 1.6776 | 1.6895 | 1.7009 | 1.7119 | 1.7227 |
| | v | 0.018 | 5.346 | | 5.401 | 5.487 | 5.572 | 5.656 | 5.739 | 5.821 | 5.903 | 5.984 | 6.064 | 6.223 | 6.380 | 6.537 | 6.691 |
| **82** | h | 283.8 | 1183.5 | | 1187.1 | 1192.8 | 1198.4 | 1203.9 | 1209.3 | 1214.6 | 1219.9 | 1225.2 | 1230.3 | 1240.6 | 1250.8 | 1260.9 | 1270.9 |
| (313.74) | s | 0.4554 | 1.6187 | | 1.6233 | 1.6306 | 1.6376 | 1.6444 | 1.6510 | 1.6575 | 1.6639 | 1.6701 | 1.6761 | 1.6880 | 1.6994 | 1.7105 | 1.7213 |
| | v | 0.018 | 5.285 | | 5.332 | 5.417 | 5.502 | 5.585 | 5.667 | 5.748 | 5.829 | 5.909 | 5.988 | 6.146 | 6.302 | 6.456 | 6.609 |
| **83** | h | 284.7 | 1183.8 | | 1186.9 | 1192.6 | 1198.2 | 1203.7 | 1209.1 | 1214.4 | 1219.7 | 1225.0 | 1230.2 | 1240.5 | 1250.7 | 1260.8 | 1270.8 |
| (314.59) | s | 0.4565 | 1.6177 | | 1.6217 | 1.6290 | 1.6360 | 1.6428 | 1.6495 | 1.6560 | 1.6624 | 1.6686 | 1.6747 | 1.6865 | 1.6980 | 1.7091 | 1.7199 |
| | v | 0.018 | 5.226 | | 5.265 | 5.349 | 5.433 | 5.515 | 5.597 | 5.677 | 5.757 | 5.836 | 5.915 | 6.071 | 6.225 | 6.377 | 6.529 |
| **84** | h | 285.5 | 1184.0 | | 1186.6 | 1192.3 | 1197.9 | 1203.5 | 1208.9 | 1214.3 | 1219.6 | 1224.8 | 1230.0 | 1240.4 | 1250.6 | 1260.7 | 1270.7 |
| (315.42) | s | 0.4576 | 1.6168 | | 1.6202 | 1.6274 | 1.6345 | 1.6413 | 1.6480 | 1.6545 | 1.6609 | 1.6671 | 1.6732 | 1.6851 | 1.6965 | 1.7076 | 1.7184 |
| | v | 0.018 | 5.168 | | 5.200 | 5.283 | 5.366 | 5.447 | 5.528 | 5.608 | 5.687 | 5.765 | 5.843 | 5.997 | 6.150 | 6.301 | 6.450 |
| **85** | h | 286.4 | 1184.2 | | 1186.4 | 1192.1 | 1197.7 | 1203.2 | 1208.7 | 1214.1 | 1219.4 | 1224.7 | 1229.9 | 1240.2 | 1250.4 | 1260.6 | 1270.6 |
| (316.25) | s | 0.4587 | 1.6158 | | 1.6186 | 1.6259 | 1.6330 | 1.6398 | 1.6465 | 1.6530 | 1.6594 | 1.6657 | 1.6718 | 1.6837 | 1.6951 | 1.7062 | 1.7171 |
| | v | 0.018 | 5.111 | | 5.136 | 5.219 | 5.300 | 5.381 | 5.461 | 5.540 | 5.618 | 5.696 | 5.773 | 5.925 | 6.076 | 6.225 | 6.374 |
| **86** | h | 287.2 | 1184.4 | | 1186.1 | 1191.9 | 1197.5 | 1203.0 | 1208.5 | 1213.9 | 1219.2 | 1224.5 | 1229.7 | 1240.1 | 1250.3 | 1260.4 | 1270.5 |
| (317.07) | s | 0.4598 | 1.6149 | | 1.6170 | 1.6243 | 1.6315 | 1.6383 | 1.6450 | 1.6516 | 1.6580 | 1 6642 | 1.6703 | 1.6823 | 1.6937 | 1.7049 | 1.7157 |

| 500° | 520° | 540° | 560° | 580° | 600° | 650° | 700° | 750° | 800° | 850° | 900° | 1000° | 1200° | 1400° | 1600° | | Abs. Press. Lb./Sq. In. (Sat. Temp.) |
|---|---|---|---|---|---|---|---|---|---|---|---|---|---|---|---|---|---|
| | | | | | | | | | | | | | | | | | Temperature—Degrees Fahrenheit |
| 8.041 | 8.219 | 8.396 | 8.573 | 8.749 | 8.924 | 9.361 | 9.796 | 10.230 | 10.662 | 11.093 | 11.524 | 12.383 | 14.097 | 15.808 | 17.516 | v | |
| 1282.0 | 1291.9 | 1301.7 | 1311.5 | 1321.3 | 1331.1 | 1355.7 | 1380.4 | 1405.2 | 1430.1 | 1455.2 | 1480.5 | 1531.6 | 1636.3 | 1744.6 | 1856.6 | h | 70 |
| 1.7501 | 1.7603 | 1.7702 | 1.7799 | 1.7894 | 1.7988 | 1.8215 | 1.8432 | 1.8641 | 1.8843 | 1.9039 | 1.9228 | 1.9591 | 2.0263 | 2.0879 | 2.1451 | s | (302.92) |
| 7.926 | 8.101 | 8.276 | 8.450 | 8.624 | 8.797 | 9.228 | 9.657 | 10.085 | 10.511 | 10.936 | 11.361 | 12.208 | 13.898 | 15.585 | 17.269 | v | |
| 1281.9 | 1291.8 | 1301.6 | [illegible] | 1321.2 | 1331.1 | 1355.7 | 1380.3 | 1405.1 | 1430.0 | 1455.2 | 1480.4 | 1531.6 | 1636.3 | 1744.6 | 1856.6 | h | 71 |
| 1.7485 | 1.7586 | 1.7686 | 1.7783 | 1.7878 | 1.7972 | 1.8199 | 1.8416 | 1.8625 | 1.8827 | 1.9023 | 1.9212 | 1.9575 | 2.0247 | 2.0863 | 2.1435 | s | (303.88) |
| 7.814 | 7.987 | 8.160 | 8.331 | 8.503 | 8.674 | 9.098 | 9.522 | 9.944 | 10.364 | 10.783 | 11.202 | 12.038 | 13.705 | 15.368 | 17.029 | v | |
| 1281.8 | 1291.7 | 1301.5 | 1311.4 | 1321.2 | 1331.0 | 1355.6 | 1380.3 | 1405.1 | 1430.0 | 1455.1 | 1480.4 | 1531.5 | 1636.3 | 1744.6 | 1856.6 | h | 72 |
| 1.7469 | 1.7570 | 1.7670 | 1.7767 | 1.7862 | 1.7956 | 1.8183 | 1.8400 | 1.8610 | 1.8812 | 1.9007 | 1.9197 | 1.9559 | 2.0232 | 2.0848 | 2.1420 | s | (304.83) |
| 7.705 | 7.876 | 8.046 | 8.216 | 8.385 | 8.553 | 8.973 | 9.390 | 9.806 | 10.221 | 10.635 | 11.048 | 11.873 | 13.517 | 15.158 | 16.796 | v | |
| 1281.7 | 1291.6 | 1301.4 | 1311.3 | 1321.1 | 1330.9 | 1355.5 | 1380.2 | 1405.0 | 1430.0 | 1455.1 | 1480.4 | 1531.5 | 1636.3 | 1744.6 | 1856.6 | h | 73 |
| 1.7453 | 1.7554 | 1.7654 | 1.7751 | 1.7847 | 1.7940 | 1.8167 | 1.8385 | 1.8594 | 1.8796 | 1.8992 | 1.9181 | 1.9544 | 2.0217 | 2.0832 | 2.1404 | s | (305.76) |
| 7.599 | 7.768 | 7.936 | 8.103 | 8.270 | 8.436 | 8.850 | 9.262 | 9.673 | 10.082 | 10.490 | 10.898 | 11.712 | 13.334 | 14.952 | 16.569 | v | |
| 1281.6 | 1291.5 | 1301.4 | 1311.2 | 1321.0 | 1330.9 | 1355.5 | 1380.2 | 1405.0 | 1429.9 | 1455.0 | 1480.3 | 1531.5 | 1636.3 | 1744.6 | 1856.6 | h | 74 |
| 1.7437 | 1.7539 | 1.7638 | 1.7736 | 1.7831 | 1.7925 | 1.8152 | 1.8369 | 1.8579 | 1.8781 | 1.8977 | 1.9166 | 1.9529 | 2.0202 | 2.0817 | 2.1389 | s | (306.68) |
| 7.496 | 7.663 | 7.829 | 7.994 | 8.159 | 8.323 | 8.731 | 9.138 | 9.543 | 9.947 | 10.350 | 10.752 | 11.555 | 13.156 | 14.753 | 16.348 | v | |
| 1281.5 | 1291.4 | 1301.3 | 1311.1 | 1321.0 | 1330.8 | 1355.4 | 1380.1 | 1404.9 | 1429.9 | 1455.0 | 1480.3 | 1531.5 | 1636.2 | 1744.6 | 1856.5 | h | 75 |
| 1.7421 | 1.7523 | 1.7623 | 1.7720 | 1.7816 | 1.7910 | 1.8137 | 1.8354 | 1.8564 | 1.8766 | 1.8962 | 1.9151 | 1.9514 | 2.0187 | 2.0802 | 2.1374 | s | (307.60) |
| 7.396 | 7.560 | 7.724 | 7.887 | 8.050 | 8.212 | 8.615 | 9.017 | 9.417 | 9.815 | 10.213 | 10.610 | 11.403 | 12.982 | 14.558 | 16.132 | v | |
| 1281.4 | 1291.3 | 1301.2 | 1311.0 | 1320.9 | 1330.7 | 1355.4 | 1380.1 | 1404.9 | 1429.8 | 1455.0 | 1480.3 | 1531.4 | 1636.2 | 1744.5 | 1856.5 | h | 76 |
| 1.7406 | 1.7508 | 1.7608 | 1.7705 | 1.7801 | 1.7895 | 1.8122 | 1.8339 | 1.8549 | 1.8751 | 1.8947 | 1.9136 | 1.9499 | 2.0172 | 2.0788 | 2.1360 | s | (308.50) |
| 7.298 | 7.460 | 7.622 | 7.783 | 7.944 | 8.104 | 8.502 | 8.899 | 9.293 | 9.687 | 10.080 | 10.472 | 11.254 | 12.813 | 14.369 | 15.923 | v | |
| 1281.4 | 1291.2 | 1301.1 | 1311.0 | 1320.8 | 1330.7 | 1355.3 | 1380.0 | 1404.8 | 1429.8 | 1454.9 | 1480.2 | 1531.4 | 1636.2 | 1744.5 | 1856.5 | h | 77 |
| 1.7391 | 1.7493 | 1.7593 | 1.7690 | 1.7786 | 1.7880 | 1.8107 | 1.8325 | 1.8534 | 1.8736 | 1.8932 | 1.9122 | 1.9485 | 2.0157 | 2.0773 | 2.1345 | s | (309.40) |
| 7.203 | 7.363 | 7.523 | 7.682 | 7.841 | 7.999 | 8.392 | 8.784 | 9.173 | 9.562 | 9.950 | 10.337 | 11.109 | 12.649 | 14.185 | 15.718 | v | |
| 1281.3 | 1291.2 | 1301.0 | 1310.9 | 1320.7 | 1330.6 | 1355.3 | 1380.0 | 1404.8 | 1429.8 | 1454.9 | 1480.2 | 1531.4 | 1636.2 | 1744.5 | 1856.5 | h | 78 |
| 1.7376 | 1.7478 | 1.7578 | 1.7675 | 1.7771 | 1.7865 | 1.8092 | 1.8310 | 1.8520 | 1.8722 | 1.8918 | 1.9107 | 1.9470 | 2.0143 | 2.0759 | 2.1331 | s | (310.29) |
| 7.110 | 7.269 | 7.427 | 7.584 | 7.740 | 7.896 | 8.285 | 8.672 | 9.056 | 9.440 | 9.823 | 10.205 | 10.968 | 12.488 | 14.005 | 15.519 | v | |
| 1281.2 | 1291.1 | 1301.0 | 1310.8 | 1320.7 | 1330.5 | 1355.2 | 1379.9 | 1404.8 | 1429.7 | 1454.9 | 1480.2 | 1531.3 | 1636.2 | 1744.5 | 1856.5 | h | 79 |
| 1.7361 | 1.7463 | 1.7563 | 1.7661 | 1.7757 | 1.7850 | 1.8078 | 1.8296 | 1.8505 | 1.8708 | 1.8903 | 1.9093 | 1.9456 | 2.0129 | 2.0745 | 2.1317 | s | (311.16) |
| 7.020 | 7.176 | 7.332 | 7.488 | 7.642 | 7.797 | 8.180 | 8.562 | 8.942 | 9.322 | 9.700 | 10.077 | 10.830 | 12.332 | 13.830 | 15.325 | v | |
| 1281.1 | 1291.0 | 1300.9 | 1310.7 | 1320.6 | 1330.5 | 1355.1 | 1379.9 | 1404.7 | 1429.7 | 1454.8 | 1480.1 | 1531.3 | 1636.2 | 1744.5 | 1856.5 | h | 80 |
| 1.7346 | 1.7449 | 1.7549 | 1.7646 | 1.7742 | 1.7836 | 1.8063 | 1.8281 | 1.8491 | 1.8694 | 1.8889 | 1.9079 | 1.9442 | 2.0115 | 2.0731 | 2.1303 | s | (312.03) |
| 6.931 | 7.086 | 7.240 | 7.394 | 7.547 | 7.699 | 8.078 | 8.456 | 8.831 | 9.206 | 9.579 | 9.952 | 10.696 | 12.179 | 13.659 | 15.136 | v | |
| 1281.0 | 1290.9 | 1300.8 | 1310.7 | 1320.5 | 1330.4 | 1355.1 | 1379.8 | 1404.7 | 1429.6 | 1454.8 | 1480.1 | 1531.3 | 1636.1 | 1744.5 | 1856.5 | h | 81 |
| 1.7332 | 1.7434 | 1.7534 | 1.7632 | 1.7728 | 1.7822 | 1.8049 | 1.8267 | 1.8477 | 1.8680 | 1.8875 | 1.9065 | 1.9428 | 2.0101 | 2.0717 | 2.1289 | s | (312.89) |
| 6.845 | 6.998 | 7.151 | 7.302 | 7.453 | 7.604 | 7.979 | 8.352 | 8.723 | 9.093 | 9.462 | 9.830 | 10.565 | 12.029 | 13.492 | 14.951 | v | |
| 1280.9 | 1290.8 | 1300.7 | 1310.6 | 1320.5 | 1330.3 | 1355.0 | 1379.8 | 1404.6 | 1429.6 | 1454.8 | 1480.1 | 1531.3 | 1636.1 | 1744.4 | 1856.5 | h | 82 |
| 1.7318 | 1.7420 | 1.7520 | 1.7618 | 1.7714 | 1.7808 | 1.8036 | 1.8254 | 1.8464 | 1.8666 | 1.8862 | 1.9051 | 1.9415 | 2.0088 | 2.0704 | 2.1276 | s | (313.74) |
| 6.761 | 6.913 | 7.063 | 7.213 | 7.362 | 7.511 | 7.881 | 8.250 | 8.617 | 8.983 | 9.347 | 9.711 | 10.438 | 11.885 | 13.329 | 14.771 | v | |
| 1280.8 | 1290.7 | 1300.6 | 1310.5 | 1320.4 | 1330.3 | 1355.0 | 1379.7 | 1404.6 | 1429.6 | 1454.7 | 1480.0 | 1531.2 | 1636.1 | 1744.4 | 1856.4 | h | 83 |
| 1.7304 | 1.7406 | 1.7506 | 1.7604 | 1.7700 | 1.7794 | 1.8022 | 1.8240 | 1.8450 | 1.8652 | 1.8848 | 1.9038 | 1.9401 | 2.0074 | 2.0690 | 2.1262 | s | (314.59) |
| 6.679 | 6.829 | 6.978 | 7.126 | 7.274 | 7.421 | 7.787 | 8.151 | 8.513 | 8.875 | 9.235 | 9.595 | 10.313 | 11.743 | 13.170 | 14.595 | v | |
| 1280.7 | 1290.6 | 1300.5 | 1310.4 | 1320.3 | 1330.2 | 1354.9 | 1379.7 | 1404.5 | 1429.5 | 1454.7 | 1480.0 | 1531.2 | 1636.1 | 1744.4 | 1856.4 | h | 84 |
| 1.7290 | 1.7392 | 1.7492 | 1.7590 | 1.7686 | 1.7780 | 1.8008 | 1.8226 | 1.8436 | 1.8639 | 1.8835 | 1.9024 | 1.9388 | 2.0061 | 2.0677 | 2.1249 | s | (315.42) |
| 6.599 | 6.747 | 6.894 | 7.041 | 7.187 | 7.332 | 7.694 | 8.054 | 8.413 | 8.770 | 9.126 | 9.482 | 10.191 | 11.605 | 13.015 | 14.423 | v | |
| 1280.6 | 1290.5 | 1300.5 | 1310.4 | 1320.2 | 1330.1 | 1354.8 | 1379.6 | 1404.5 | 1429.5 | 1454.6 | 1480.0 | 1531.2 | 1636.0 | 1744.4 | 1856.4 | h | 85 |
| 1.7276 | 1.7378 | 1.7479 | 1.7577 | 1.7673 | 1.7767 | 1.7995 | 1.8213 | 1.8423 | 1.8626 | 1.8821 | 1.9011 | 1.9375 | 2.0048 | 2.0664 | 2.1236 | s | (316.25) |
| 6.521 | 6.667 | 6.813 | 6.958 | 7.102 | 7.246 | 7.604 | 7.960 | 8.314 | 8.667 | 9.019 | 9.371 | 10.072 | 11.470 | 12.863 | 14.255 | v | |
| 1280.5 | 1290.4 | 1300.4 | 1310.3 | 1320.2 | 1330.1 | 1354.8 | 1379.6 | 1404.4 | 1429.4 | 1454.6 | 1479.9 | 1531.2 | 1636.0 | 1744.4 | 1856.4 | h | 86 |
| 1.7262 | 1.7365 | 1.7465 | 1.7563 | 1.7659 | 1.7753 | 1.7981 | 1.8200 | 1.8410 | 1.8612 | 1.8808 | 1.8998 | 1.9362 | 2.0035 | 2.0651 | 2.1223 | s | (317.07) |

## Table 3. Superheated Vapor

| Abs. Press. Lb./Sq. In. (Sat. Temp.) | | Sat. Liquid | Sat. Vapor | Temperature—Degrees Fahrenheit 320° | 330° | 340° | 350° | 360° | 370° | 380° | 390° | 400° | 420° | 440° | 460° | 480° | 500° |
|---|---|---|---|---|---|---|---|---|---|---|---|---|---|---|---|---|---|
| | v | 0.018 | 5.055 | 5.073 | 5.155 | 5.236 | 5.316 | 5.395 | 5.474 | 5.551 | 5.628 | 5.704 | 5.855 | 6.004 | 6.152 | 6.299 | 6.444 |
| 87 | h | 288.1 | 1184.6 | 1185.9 | 1191.6 | 1197.3 | 1202.8 | 1208.3 | 1213.7 | 1219.1 | 1224.3 | 1229.6 | 1240.0 | 1250.2 | 1260.3 | 1270.4 | 1280.4 |
| (317.88) | s | 0.4609 | 1.6139 | 1.6155 | 1.6228 | 1.6300 | 1.6369 | 1.6436 | 1.6501 | 1.6565 | 1.6628 | 1.6689 | 1.6809 | 1.6924 | 1.7035 | 1.7143 | 1.7249 |
| | v | 0.018 | 5.001 | 5.012 | 5.094 | 5.174 | 5.253 | 5.331 | 5.409 | 5.486 | 5.562 | 5.637 | 5.787 | 5.934 | 6.081 | 6.226 | 6.370 |
| 88 | h | 288.9 | 1184.8 | 1185.6 | 1191.4 | 1197.1 | 1202.6 | 1208.1 | 1213.5 | 1218.9 | 1224.2 | 1229.4 | 1239.8 | 1250.1 | 1260.2 | 1270.3 | 1280.3 |
| (318.68) | s | 0.4620 | 1.6130 | 1.6140 | 1.6214 | 1.6285 | 1.6354 | 1.6421 | 1.6487 | 1.6551 | 1.6614 | 1.6675 | 1.6795 | 1.6910 | 1.7021 | 1.7130 | 1.7235 |
| | v | 0.018 | 4.948 | 4.952 | 5.033 | 5.113 | 5.191 | 5.269 | 5.346 | 5.422 | 5.497 | 5.572 | 5.720 | 5.866 | 6.011 | 6.154 | 6.297 |
| 89 | h | 289.7 | 1185.1 | 1185.4 | 1191.1 | 1196.8 | 1202.4 | 1207.9 | 1213.4 | 1218.7 | 1224.0 | 1229.3 | 1239.7 | 1249.9 | 1260.1 | 1270.2 | 1280.2 |
| (319.48) | s | 0.4630 | 1.6121 | 1.6125 | 1.6199 | 1.6270 | 1.6339 | 1.6407 | 1.6473 | 1.6537 | 1.6600 | 1.6661 | 1.6781 | 1.6896 | 1.7008 | 1.7117 | 1.7222 |
| | v | 0.018 | 4.896 | | 4.974 | 5.053 | 5.131 | 5.208 | 5.284 | 5.359 | 5.434 | 5.508 | 5.654 | 5.799 | 5.942 | 6.084 | 6.225 |
| 90 | h | 290.6 | 1185.3 | | 1190.9 | 1196.6 | 1202.2 | 1207.7 | 1213.2 | 1218.6 | 1223.9 | 1229.1 | 1239.5 | 1249.8 | 1260.0 | 1270.1 | 1280.1 |
| (320.27) | s | 0.4641 | 1.6112 | | 1.6184 | 1.6256 | 1.6325 | 1.6393 | 1.6459 | 1.6523 | 1.6586 | 1.6648 | 1.6767 | 1.6883 | 1.6995 | 1.7103 | 1.7209 |
| | v | 0.018 | 4.845 | | 4.916 | 4.995 | 5.072 | 5.148 | 5.223 | 5.298 | 5.372 | 5.445 | 5.590 | 5.733 | 5.875 | 6.016 | 6.155 |
| 91 | h | 291.4 | 1185.5 | | 1190.7 | 1196.4 | 1202.0 | 1207.5 | 1213.0 | 1218.4 | 1223.7 | 1229.0 | 1239.4 | 1249.7 | 1259.9 | 1270.0 | 1280.0 |
| (321.06) | s | 0.4651 | 1.6103 | | 1.6169 | 1.6241 | 1.6311 | 1.6379 | 1.6445 | 1.6509 | 1.6572 | 1.6634 | 1.6754 | 1.6870 | 1.6982 | 1.7090 | 1.7196 |
| | v | 0.018 | 4.796 | | 4.860 | 4.937 | 5.014 | 5.089 | 5.164 | 5.238 | 5.311 | 5.384 | 5.527 | 5.669 | 5.810 | 5.949 | 6.087 |
| 92 | h | 292.2 | 1185.7 | | 1190.4 | 1196.2 | 1201.8 | 1207.3 | 1212.8 | 1218.2 | 1223.5 | 1228.8 | 1239.3 | 1249.6 | 1259.8 | 1269.9 | 1279.9 |
| (321.83) | s | 0.4661 | 1.6094 | | 1.6155 | 1.6227 | 1.6297 | 1.6365 | 1.6431 | 1.6496 | 1.6559 | 1.6621 | 1.6741 | 1.6857 | 1.6969 | 1.7077 | 1.7183 |
| | v | 0.018 | 4.747 | | 4.804 | 4.881 | 4.957 | 5.032 | 5.106 | 5.179 | 5.252 | 5.324 | 5.466 | 5.606 | 5.745 | 5.883 | 6.020 |
| 93 | h | 293.0 | 1185.9 | | 1190.2 | 1195.9 | 1201.6 | 1207.1 | 1212.6 | 1218.0 | 1223.4 | 1228.7 | 1239.1 | 1249.4 | 1259.6 | 1269.8 | 1279.8 |
| (322.60) | s | 0.4672 | 1.6085 | | 1.6140 | 1.6213 | 1.6283 | 1.6351 | 1.6417 | 1.6482 | 1.6545 | 1.6607 | 1.6728 | 1.6844 | 1.6956 | 1.7065 | 1.7171 |
| | v | 0.018 | 4.699 | | 4.750 | 4.827 | 4.902 | 4.976 | 5.049 | 5.122 | 5.194 | 5.265 | 5.406 | 5.545 | 5.683 | 5.819 | 5.955 |
| 94 | h | 293.8 | 1186.1 | | 1189.9 | 1195.7 | 1201.4 | 1206.9 | 1212.4 | 1217.8 | 1223.2 | 1228.5 | 1239.0 | 1249.3 | 1259.5 | 1269.7 | 1279.7 |
| (323.36) | s | 0.4682 | 1.6076 | | 1.6126 | 1.6199 | 1.6269 | 1.6337 | 1.6404 | 1.6469 | 1.6532 | 1.6594 | 1.6715 | 1.6831 | 1.6943 | 1.7052 | 1.7158 |
| | v | 0.018 | 4.652 | | 4.697 | 4.773 | 4.847 | 4.921 | 4.994 | 5.065 | 5.137 | 5.207 | 5.347 | 5.485 | 5.621 | 5.756 | 5.891 |
| 95 | h | 294.6 | 1186.2 | | 1189.7 | 1195.5 | 1201.2 | 1206.7 | 1212.2 | 1217.7 | 1223.0 | 1228.3 | 1238.8 | 1249.2 | 1259.4 | 1269.5 | 1279.6 |
| (324.12) | s | 0.4692 | 1.6068 | | 1.6112 | 1.6185 | 1.6255 | 1.6324 | 1.6390 | 1.6455 | 1.6519 | 1.6581 | 1.6702 | 1.6818 | 1.6930 | 1.7039 | 1.7146 |
| | v | 0.018 | 4.606 | | 4.645 | 4.720 | 4.794 | 4.867 | 4.939 | 5.010 | 5.081 | 5.151 | 5.289 | 5.426 | 5.561 | 5.695 | 5.828 |
| 96 | h | 295.3 | 1186.4 | | 1189.4 | 1195.3 | 1201.0 | 1206.5 | 1212.1 | 1217.5 | 1222.9 | 1228.2 | 1238.7 | 1249.1 | 1259.3 | 1269.4 | 1279.5 |
| (324.87) | s | 0.4702 | 1.6060 | | 1.6098 | 1.6171 | 1.6241 | 1.6310 | 1.6377 | 1.6442 | 1.6506 | 1.6568 | 1.6689 | 1.6805 | 1.6918 | 1.7027 | 1.7133 |
| | v | 0.018 | 4.561 | | 4.595 | 4.669 | 4.742 | 4.814 | 4.886 | 4.957 | 5.027 | 5.096 | 5.233 | 5.368 | 5.502 | 5.635 | 5.766 |
| 97 | h | 296.1 | 1186.6 | | 1189.2 | 1195.0 | 1200.7 | 1206.3 | 1211.9 | 1217.3 | 1222.7 | 1228.0 | 1238.6 | 1248.9 | 1259.2 | 1269.3 | 1279.4 |
| (325.61) | s | 0.4711 | 1.6051 | | 1.6084 | 1.6157 | 1.6228 | 1.6297 | 1.6364 | 1.6429 | 1.6493 | 1.6555 | 1.6676 | 1.6793 | 1.6906 | 1.7015 | 1.7121 |
| | v | 0.018 | 4.517 | | 4.545 | 4.619 | 4.691 | 4.763 | 4.834 | 4.904 | 4.973 | 5.042 | 5.178 | 5.312 | 5.445 | 5.576 | 5.706 |
| 98 | h | 296.9 | 1186.8 | | 1189.0 | 1194.8 | 1200.5 | 1206.1 | 1211.7 | 1217.2 | 1222.6 | 1227.9 | 1238.4 | 1248.8 | 1259.1 | 1269.2 | 1279.3 |
| (326.35) | s | 0.4721 | 1.6043 | | 1.6070 | 1.6143 | 1.6215 | 1.6284 | 1.6351 | 1.6416 | 1.6480 | 1.6543 | 1.6664 | 1.6781 | 1.6893 | 1.7003 | 1.7109 |
| | v | 0.018 | 4.474 | | 4.496 | 4.569 | 4.641 | 4.712 | 4.783 | 4.852 | 4.921 | 4.989 | 5.124 | 5.257 | 5.388 | 5.518 | 5.647 |
| 99 | h | 297.6 | 1187.0 | | 1188.7 | 1194.6 | 1200.3 | 1205.9 | 1211.5 | 1217.0 | 1222.4 | 1227.7 | 1238.3 | 1248.7 | 1258.9 | 1269.1 | 1279.2 |
| (327.08) | s | 0.4731 | 1.6035 | | 1.6056 | 1.6130 | 1.6201 | 1.6271 | 1.6338 | 1.6403 | 1.6467 | 1.6530 | 1.6651 | 1.6768 | 1.6881 | 1.6991 | 1.7097 |
| | v | 0.018 | 4.432 | | 4.448 | 4.521 | 4.592 | 4.663 | 4.732 | 4.801 | 4.870 | 4.937 | 5.071 | 5.202 | 5.333 | 5.462 | 5.589 |
| 100 | h | 298.4 | 1187.2 | | 1188.5 | 1194.3 | 1200.1 | 1205.7 | 1211.3 | 1216.8 | 1222.2 | 1227.6 | 1238.1 | 1248.6 | 1258.8 | 1269.0 | 1279.1 |
| (327.81) | s | 0.4740 | 1.6026 | | 1.6043 | 1.6117 | 1.6188 | 1.6258 | 1.6325 | 1.6391 | 1.6455 | 1.6518 | 1.6639 | 1.6756 | 1.6869 | 1.6979 | 1.7085 |
| | v | 0.018 | 4.350 | | 4.355 | 4.427 | 4.497 | 4.567 | 4.635 | 4.703 | 4.770 | 4.836 | 4.968 | 5.097 | 5.225 | 5.352 | 5.477 |
| 102 | h | 299.9 | 1187.5 | | 1188.0 | 1193.9 | 1199.7 | 1205.3 | 1210.9 | 1216.4 | 1221.9 | 1227.3 | 1237.9 | 1248.3 | 1258.6 | 1268.8 | 1278.9 |
| (329.25) | s | 0.4759 | 1.6010 | | 1.6016 | 1.6090 | 1.6162 | 1.6232 | 1.6300 | 1.6366 | 1.6430 | 1.6493 | 1.6615 | 1.6732 | 1.6845 | 1.6955 | 1.7062 |
| | v | 0.018 | 4.271 | | | 4.336 | 4.406 | 4.474 | 4.541 | 4.608 | 4.674 | 4.740 | 4.869 | 4.996 | 5.121 | 5.246 | 5.369 |
| 104 | h | 301.4 | 1187.9 | | | 1193.4 | 1199.2 | 1204.9 | 1210.6 | 1216.1 | 1221.6 | 1226.9 | 1237.6 | 1248.0 | 1258.4 | 1268.6 | 1278.7 |
| (330.66) | s | 0.4778 | 1.5994 | | | 1.6064 | 1.6136 | 1.6207 | 1.6275 | 1.6341 | 1.6405 | 1.6468 | 1.6591 | 1.6708 | 1.6822 | 1.6932 | 1.7039 |
| | v | 0.018 | 4.194 | | | 4.249 | 4.318 | 4.385 | 4.451 | 4.517 | 4.582 | 4.646 | 4.773 | 4.898 | 5.022 | 5.144 | 5.266 |
| 106 | h | 302.8 | 1188.2 | | | 1193.0 | 1198.8 | 1204.5 | 1210.2 | 1215.7 | 1221.2 | 1226.6 | 1237.3 | 1247.8 | 1258.1 | 1268.4 | 1278.6 |
| (332.05) | s | 0.4796 | 1.5978 | | | 1.6038 | 1.6111 | 1.6182 | 1.6250 | 1.6316 | 1.6381 | 1.6445 | 1.6567 | 1.6685 | 1.6799 | 1.6909 | 1.7016 |
| | v | 0.018 | 4.120 | | | 4.165 | 4.233 | 4.299 | 4.365 | 4.429 | 4.493 | 4.557 | 4.681 | 4.805 | 4.926 | 5.046 | 5.166 |
| 108 | h | 304.3 | 1188.6 | | | 1192.5 | 1198.4 | 1204.1 | 1209.8 | 1215.4 | 1220.9 | 1226.3 | 1237.0 | 1247.5 | 1257.9 | 1268.2 | 1278.4 |
| (333.42) | s | 0.4814 | 1.5963 | | | 1.6013 | 1.6086 | 1.6157 | 1.6226 | 1.6292 | 1.6357 | 1.6421 | 1.6544 | 1.6662 | 1.6776 | 1.6887 | 1.6994 |

Temperature—Degrees Fahrenheit

| Abs. Press. Lb./Sq. In. (Sat. Temp.) | | 520° | 540° | 560° | 580° | 600° | 650° | 700° | 750° | 800° | 850° | 900° | 950° | 1000° | 1200° | 1400° | 1600° |
|---|---|---|---|---|---|---|---|---|---|---|---|---|---|---|---|---|---|
| | v | 6.589 | 6.733 | 6.876 | 7.019 | 7.162 | 7.515 | 7.867 | 8.218 | 8.567 | 8.915 | 9.263 | 9.610 | 9.956 | 11.338 | 12.715 | 14.091 |
| 87 | h | 1290.4 | 1300.3 | 1310.2 | 1320.1 | 1330.0 | 1354.7 | 1379.5 | 1404.4 | 1429.4 | 1454.6 | 1479.9 | 1505.4 | 1531.1 | 1636.0 | 1744.4 | 1856.4 |
| (317.88) | s | 1.7351 | 1.7452 | 1.7550 | 1.7646 | 1.7740 | 1.7968 | 1.8187 | 1.8397 | 1.8600 | 1.8795 | 1.8985 | 1.9170 | 1.9349 | 2.0022 | 2.0638 | 2.1210 |
| | v | 6.513 | 6.655 | 6.797 | 6.938 | 7.079 | 7.429 | 7.777 | 8.124 | 8.469 | 8.813 | 9.157 | 9.500 | 9.842 | 11.208 | 12.571 | 13.931 |
| 88 | h | 1290.3 | 1300.2 | 1310.1 | 1320.0 | 1329.9 | 1354.7 | 1379.5 | 1404.4 | 1429.4 | 1454.5 | 1479.9 | 1505.4 | 1531.1 | 1636.0 | 1744.4 | 1856.4 |
| (318.68) | s | 1.7338 | 1.7439 | 1.7537 | 1.7633 | 1.7727 | 1.7955 | 1.8174 | 1.8384 | 1.8587 | 1.8783 | 1.8973 | 1.9157 | 1.9336 | 2.0009 | 2.0626 | 2.1198 |
| | v | 6.438 | 6.579 | 6.719 | 6.859 | 6.998 | 7.345 | 7.689 | 8.032 | 8.373 | 8.714 | 9.053 | 9.393 | 9.731 | 11.082 | 12.429 | 13.774 |
| 89 | h | 1290.2 | 1300.1 | 1310.0 | 1320.0 | 1329.9 | 1354.6 | 1379.4 | 1404.3 | 1429.3 | 1454.5 | 1479.8 | 1505.4 | 1531.1 | 1636.0 | 1744.3 | 1856.4 |
| (319.48) | s | 1.7325 | 1.7425 | 1.7524 | 1.7620 | 1.7714 | 1.7943 | 1.8161 | 1.8371 | 1.8574 | 1.8770 | 1.8960 | 1.9144 | 1.9324 | 1.9997 | 2.0613 | 2.1185 |
| | v | 6.365 | 6.505 | 6.644 | 6.782 | 6.920 | 7.262 | 7.603 | 7.942 | 8.279 | 8.616 | 8.952 | 9.288 | 9.623 | 10.959 | 12.291 | 13.621 |
| 90 | h | 1290.1 | 1300.0 | 1310.0 | 1319.9 | 1329.8 | 1354.6 | 1379.4 | 1404.3 | 1429.3 | 1454.5 | 1479.8 | 1505.3 | 1531.0 | 1635.9 | 1744.3 | 1856.4 |
| (320.27) | s | 1.7312 | 1.7413 | 1.7511 | 1.7607 | 1.7702 | 1.7930 | 1.8149 | 1.8359 | 1.8562 | 1.8758 | 1.8948 | 1.9132 | 1.9311 | 1.9984 | 2.0601 | 2.1173 |
| | v | 6.294 | 6.432 | 6.569 | 6.706 | 6.843 | 7.181 | 7.518 | 7.854 | 8.188 | 8.521 | 8.853 | 9.185 | 9.517 | 10.838 | 12.156 | 13.471 |
| 91 | h | 1290.0 | 1300.0 | 1309.9 | 1319.8 | 1329.7 | 1354.5 | 1379.3 | 1404.2 | 1429.2 | 1454.4 | 1479.8 | 1505.3 | 1531.0 | 1635.9 | 1744.3 | 1856.3 |
| (321.06) | s | 1.7299 | 1.7400 | 1.7498 | 1.7594 | 1.7689 | 1.7917 | 1.8136 | 1.8346 | 1.8549 | 1.8745 | 1.8935 | 1.9119 | 1.9299 | 1.9972 | 2.0588 | 2.1161 |
| | v | 6.224 | 6.361 | 6.497 | 6.632 | 6.767 | 7.102 | 7.436 | 7.767 | 8.098 | 8.428 | 8.757 | 9.085 | 9.413 | 10.720 | 12.023 | 13.325 |
| 92 | h | 1289.9 | 1299.9 | 1309.8 | 1319.7 | 1329.7 | 1354.4 | 1379.3 | 1404.2 | 1429.2 | 1454.4 | 1479.7 | 1505.3 | 1531.0 | 1635.9 | 1744.3 | 1856.3 |
| (321.83) | s | 1.7286 | 1.7387 | 1.7485 | 1.7582 | 1.7676 | 1.7905 | 1.8124 | 1.8334 | 1.8537 | 1.8733 | 1.8923 | 1.9107 | 1.9286 | 1.9960 | 2.0576 | 2.1148 |
| | v | 6.156 | 6.291 | 6.426 | 6.560 | 6.693 | 7.025 | 7.355 | 7.683 | 8.010 | 8.337 | 8.662 | 8.987 | 9.311 | 10.604 | 11.894 | 13.181 |
| 93 | h | 1289.8 | 1299.8 | 1309.7 | 1319.7 | 1329.6 | 1354.4 | 1379.2 | 1404.1 | 1429.2 | 1454.4 | 1479.7 | 1505.2 | 1531.0 | 1635.9 | 1744.3 | 1856.3 |
| (322.60) | s | 1.7274 | 1.7375 | 1.7473 | 1.7570 | 1.7664 | 1.7893 | 1.8112 | 1.8322 | 1.8525 | 1.8721 | 1.8911 | 1.9095 | 1.9274 | 1.9948 | 2.0564 | 2.1136 |
| | v | 6.089 | 6.223 | 6.356 | 6.489 | 6.621 | 6.949 | 7.276 | 7.601 | 7.925 | 8.247 | 8.569 | 8.891 | 9.212 | 10.491 | 11.767 | 13.041 |
| 94 | h | 1289.7 | 1299.7 | 1309.7 | 1319.6 | 1329.5 | 1354.3 | 1379.2 | 1404.1 | 1429.1 | 1454.3 | 1479.7 | 1505.2 | 1530.9 | 1635.9 | 1744.3 | 1856.3 |
| (323.36) | s | 1.7261 | 1.7362 | 1.7461 | 1.7557 | 1.7652 | 1.7880 | 1.8099 | 1.8310 | 1.8513 | 1.8709 | 1.8899 | 1.9083 | 1.9263 | 1.9936 | 2.0552 | 2.1125 |
| | v | 6.024 | 6.156 | 6.288 | 6.419 | 6.550 | 6.875 | 7.199 | 7.520 | 7.841 | 8.160 | 8.479 | 8.797 | 9.114 | 10.381 | 11.643 | 12.903 |
| 95 | h | 1289.6 | 1299.6 | 1309.6 | 1319.5 | 1329.4 | 1354.3 | 1379.1 | 1404.0 | 1429.1 | 1454.3 | 1479.6 | 1505.2 | 1530.9 | 1635.8 | 1744.2 | 1856.3 |
| (324.12) | s | 1.7249 | 1.7350 | 1.7448 | 1.7545 | 1.7640 | 1.7868 | 1.8087 | 1.8298 | 1.8501 | 1.8697 | 1.8887 | 1.9071 | 1.9251 | 1.9924 | 2.0541 | 2.1113 |
| | v | 5.960 | 6.091 | 6.222 | 6.352 | 6.481 | 6.803 | 7.123 | 7.441 | 7.758 | 8.074 | 8.390 | 8.705 | 9.019 | 10.272 | 11.522 | 12.769 |
| 96 | h | 1289.6 | 1299.6 | 1309.5 | 1319.4 | 1329.4 | 1354.2 | 1379.1 | 1404.0 | 1429.0 | 1454.3 | 1479.6 | 1505.2 | 1530.9 | 1635.8 | 1744.2 | 1856.3 |
| (324.87) | s | 1.7237 | 1.7338 | 1.7436 | 1.7533 | 1.7628 | 1.7856 | 1.8075 | 1.8286 | 1.8489 | 1.8685 | 1.8875 | 1.9060 | 1.9239 | 1.9913 | 2.0529 | 2.1101 |
| | v | 5.897 | 6.027 | 6.156 | 6.285 | 6.413 | 6.732 | 7.049 | 7.364 | 7.678 | 7.991 | 8.303 | 8.614 | 8.925 | 10.166 | 11.403 | 12.637 |
| 97 | h | 1289.5 | 1299.5 | 1309.4 | 1319.4 | 1329.3 | 1354.1 | 1379.0 | 1403.9 | 1429.0 | 1454.2 | 1479.6 | 1505.1 | 1530.9 | 1635.8 | 1744.2 | 1856.3 |
| (325.61) | s | 1.7225 | 1.7326 | 1.7424 | 1.7521 | 1.7616 | 1.7845 | 1.8064 | 1.8274 | 1.8477 | 1.8674 | 1.8864 | 1.9048 | 1.9228 | 1.9901 | 2.0518 | 2.1090 |
| | v | 5.836 | 5.964 | 6.092 | 6.220 | 6.347 | 6.662 | 6.976 | 7.288 | 7.599 | 7.909 | 8.218 | 8.526 | 8.834 | 10.062 | 11.286 | 12.508 |
| 98 | h | 1289.4 | 1299.4 | 1309.4 | 1319.3 | 1329.2 | 1354.1 | 1379.0 | 1403.9 | 1429.0 | 1454.2 | 1479.5 | 1505.1 | 1530.8 | 1635.8 | 1744.2 | 1856.3 |
| (326.35) | s | 1.7213 | 1.7314 | 1.7412 | 1.7509 | 1.7604 | 1.7833 | 1.8052 | 1.8263 | 1.8466 | 1.8662 | 1.8852 | 1.9037 | 1.9216 | 1.9890 | 2.0506 | 2.1078 |
| | v | 5.775 | 5.903 | 6.030 | 6.156 | 6.282 | 6.594 | 6.905 | 7.214 | 7.521 | 7.828 | 8.134 | 8.440 | 8.744 | 9.960 | 11.172 | 12.381 |
| 99 | h | 1289.3 | 1299.3 | 1309.3 | 1319.2 | 1329.2 | 1354.0 | 1378.9 | 1403.9 | 1428.9 | 1454.1 | 1479.5 | 1505.1 | 1530.8 | 1635.8 | 1744.2 | 1856.2 |
| (327.08) | s | 1.7201 | 1.7302 | 1.7401 | 1.7497 | 1.7592 | 1.7821 | 1.8041 | 1.8251 | 1.8454 | 1.8651 | 1.8841 | 1.9025 | 1.9205 | 1.9878 | 2.0495 | 2.1067 |
| | v | 5.717 | 5.843 | 5.968 | 6.093 | 6.218 | 6.527 | 6.835 | 7.141 | 7.446 | 7.749 | 8.052 | 8.355 | 8.656 | 9.860 | 11.060 | 12.258 |
| 100 | h | 1289.2 | 1299.2 | 1309.2 | 1319.2 | 1329.1 | 1354.0 | 1378.9 | 1403.8 | 1428.9 | 1454.1 | 1479.5 | 1505.0 | 1530.8 | 1635.7 | 1744.2 | 1856.2 |
| (327.81) | s | 1.7189 | 1.7290 | 1.7389 | 1.7486 | 1.7581 | 1.7810 | 1.8029 | 1.8240 | 1.8443 | 1.8639 | 1.8829 | 1.9014 | 1.9193 | 1.9867 | 2.0484 | 2.1056 |
| | v | 5.602 | 5.726 | 5.849 | 5.972 | 6.094 | 6.398 | 6.699 | 6.999 | 7.298 | 7.596 | 7.893 | 8.190 | 8.486 | 9.666 | 10.843 | 12.017 |
| 102 | h | 1289.0 | 1299.1 | 1309.0 | 1319.0 | 1329.0 | 1353.9 | 1378.7 | 1403.7 | 1428.8 | 1454.0 | 1479.4 | 1505.0 | 1530.7 | 1635.7 | 1744.1 | 1856.2 |
| (329.25) | s | 1.7166 | 1.7267 | 1.7366 | 1.7463 | 1.7558 | 1.7787 | 1.8007 | 1.8218 | 1.8421 | 1.8617 | 1.8807 | 1.8992 | 1.9171 | 1.9845 | 2.0462 | 2.1034 |
| | v | 5.492 | 5.614 | 5.735 | 5.855 | 5.975 | 6.273 | 6.569 | 6.864 | 7.157 | 7.449 | 7.741 | 8.032 | 8.322 | 9.480 | 10.634 | 11.786 |
| 104 | h | 1288.8 | 1298.9 | 1308.9 | 1318.9 | 1328.8 | 1353.7 | 1378.6 | 1403.6 | 1428.7 | 1454.0 | 1479.3 | 1504.9 | 1530.7 | 1635.7 | 1744.1 | 1856.2 |
| (330.66) | s | 1.7143 | 1.7244 | 1.7343 | 1.7440 | 1.7535 | 1.7765 | 1.7985 | 1.8196 | 1.8399 | 1.8595 | 1.8785 | 1.8970 | 1.9150 | 1.9824 | 2.0440 | 2.1013 |
| | v | 5.386 | 5.505 | 5.624 | 5.743 | 5.861 | 6.153 | 6.444 | 6.733 | 7.021 | 7.308 | 7.594 | 7.879 | 8.164 | 9.301 | 10.433 | 11.563 |
| 106 | h | 1288.7 | 1298.7 | 1308.7 | 1318.7 | 1328.7 | 1353.6 | 1378.5 | 1403.5 | 1428.6 | 1453.9 | 1479.3 | 1504.9 | 1530.6 | 1635.6 | 1744.1 | 1856.2 |
| (332.05) | s | 1.7121 | 1.7222 | 1.7321 | 1.7418 | 1.7514 | 1.7743 | 1.7963 | 1.8174 | 1.8377 | 1.8574 | 1.8764 | 1.8949 | 1.9128 | 1.9802 | 2.0419 | 2.0992 |
| | v | 5.284 | 5.401 | 5.518 | 5.634 | 5.750 | 6.037 | 6.323 | 6.607 | 6.890 | 7.171 | 7.452 | 7.733 | 8.012 | 9.128 | 10.239 | 11.349 |
| 108 | h | 1288.5 | 1298.5 | 1308.6 | 1318.6 | 1328.6 | 1353.5 | 1378.4 | 1403.4 | 1428.6 | 1453.8 | 1479.2 | 1504.8 | 1530.6 | 1635.6 | 1744.0 | 1856.1 |
| (333.42) | s | 1.7099 | 1.7200 | 1.7300 | 1.7397 | 1.7492 | 1.7722 | 1.7942 | 1.8153 | 1.8356 | 1.8553 | 1.8743 | 1.8928 | 1.9108 | 1.9782 | 2.0398 | 2.0971 |

## Table 3. Superheated Vapor

| Abs. Press. Lb./Sq. In. (Sat. Temp.) | | Sat. Liquid | Sat. Vapor | 340° | 350° | 360° | 370° | 380° | 390° | 400° | 420° | 440° | 460° | 480° | 500° | 520° | 540° |
|---|---|---|---|---|---|---|---|---|---|---|---|---|---|---|---|---|---|
| | | | | Temperature—Degrees Fahrenheit | | | | | | | | | | | | | |
| 110 | v | 0.018 | 4.049 | 4.085 | 4.151 | 4.216 | 4.281 | 4.345 | 4.408 | 4.470 | 4.593 | 4.714 | 4.834 | 4.952 | 5.069 | 5.186 | 5.30 |
| | h | 305.7 | 1183.9 | 1192.0 | 1197.9 | 1203.7 | 1209.4 | 1215.0 | 1220.5 | 1226.0 | 1236.7 | 1247.3 | 1257.7 | 1268.0 | 1278.2 | 1288.3 | 1298. |
| (334.77) | s | 0.4832 | 1.5948 | 1.5988 | 1.6062 | 1.6133 | 1.6202 | 1.6269 | 1.6334 | 1.6398 | 1.6521 | 1.6640 | 1.6754 | 1.6865 | 1.6972 | 1.7077 | 1.717 |
| 112 | v | 0.018 | 3.981 | 4.007 | 4.072 | 4.137 | 4.200 | 4.263 | 4.325 | 4.387 | 4.508 | 4.627 | 4.745 | 4.861 | 4.976 | 5.091 | 5.20 |
| | h | 307.1 | 1189.2 | 1191.5 | 1197.5 | 1203.3 | 1209.0 | 1214.7 | 1220.2 | 1225.7 | 1236.4 | 1247.0 | 1257.4 | 1267.7 | 1278.0 | 1288.1 | 1298. |
| (336.11) | s | 0.4849 | 1.5934 | 1.5963 | 1.6037 | 1.6109 | 1.6178 | 1.6245 | 1.6311 | 1.6375 | 1.6499 | 1.6618 | 1.6732 | 1.6843 | 1.6951 | 1.7056 | 1.715 |
| 114 | v | 0.018 | 3.914 | 3.931 | 3.996 | 4.060 | 4.122 | 4.184 | 4.246 | 4.306 | 4.425 | 4.543 | 4.659 | 4.773 | 4.887 | 4.999 | 5.111 |
| | h | 308.4 | 1189.5 | 1191.1 | 1197.0 | 1202.9 | 1208.6 | 1214.3 | 1219.9 | 1225.4 | 1236.2 | 1246.8 | 1257.2 | 1267.5 | 1277.8 | 1287.9 | 1298. |
| (337.42) | s | 0.4866 | 1.5919 | 1.5939 | 1.6013 | 1.6085 | 1.6155 | 1.6223 | 1.6289 | 1.6353 | 1.6477 | 1.6596 | 1.6711 | 1.6822 | 1.6930 | 1.7035 | 1.713 |
| 116 | v | 0.018 | 3.850 | 3.859 | 3.923 | 3.985 | 4.047 | 4.108 | 4.169 | 4.228 | 4.346 | 4.462 | 4.576 | 4.688 | 4.800 | 4.911 | 5.021 |
| | h | 309.8 | 1189.8 | 1190.6 | 1196.6 | 1202.5 | 1208.2 | 1213.9 | 1219.5 | 1225.0 | 1235.9 | 1246.5 | 1257.0 | 1267.3 | 1277.6 | 1287.7 | 1297.9 |
| (338.72) | s | 0.4883 | 1.5905 | 1.5915 | 1.5990 | 1.6062 | 1.6132 | 1.6200 | 1.6266 | 1.6331 | 1.6455 | 1.6575 | 1.6690 | 1.6801 | 1.6909 | 1.7014 | 1.711 |
| 118 | v | 0.018 | 3.788 | 3.788 | 3.852 | 3.914 | 3.975 | 4.035 | 4.094 | 4.153 | 4.269 | 4.383 | 4.495 | 4.606 | 4.716 | 4.826 | 4.934 |
| | h | 311.1 | 1190.1 | 1190.1 | 1196.2 | 1202.1 | 1207.9 | 1213.6 | 1219.2 | 1224.7 | 1235.6 | 1246.2 | 1256.7 | 1267.1 | 1277.4 | 1287.6 | 1297.7 |
| (339.99) | s | 0.4900 | 1.5891 | 1.5891 | 1.5967 | 1.6039 | 1.6109 | 1.6178 | 1.6244 | 1.6309 | 1.6434 | 1.6554 | 1.6669 | 1.6781 | 1.6889 | 1.6994 | 1.709 |
| 120 | v | 0.018 | 3.728 | | 3.783 | 3.844 | 3.904 | 3.964 | 4.023 | 4.081 | 4.195 | 4.307 | 4.418 | 4.527 | 4.636 | 4.743 | 4.849 |
| | h | 312.4 | 1190.4 | | 1195.7 | 1201.6 | 1207.5 | 1213.2 | 1218.8 | 1224.4 | 1235.3 | 1246.0 | 1256.5 | 1266.9 | 1277.2 | 1287.4 | 1297.5 |
| (341.25) | s | 0.4916 | 1.5878 | | 1.5944 | 1.6017 | 1.6087 | 1.6156 | 1.6222 | 1.6287 | 1.6413 | 1.6533 | 1.6649 | 1.6760 | 1.6869 | 1.6974 | 1.707 |
| 122 | v | 0.018 | 3.670 | | 3.716 | 3.777 | 3.836 | 3.895 | 3.953 | 4.010 | 4.123 | 4.234 | 4.343 | 4.451 | 4.557 | 4.663 | 4.768 |
| | h | 313.8 | 1190.7 | | 1195.2 | 1201.2 | 1207.1 | 1212.8 | 1218.5 | 1224.0 | 1235.0 | 1245.7 | 1256.3 | 1266.7 | 1277.0 | 1287.2 | 1297.4 |
| (342.50) | s | 0.4932 | 1.5865 | | 1.5921 | 1.5994 | 1.6065 | 1.6134 | 1.6201 | 1.6266 | 1.6392 | 1.6512 | 1.6628 | 1.6740 | 1.6849 | 1.6954 | 1.7057 |
| 124 | v | 0.018 | 3.614 | | 3.652 | 3.712 | 3.771 | 3.829 | 3.886 | 3.942 | 4.054 | 4.163 | 4.270 | 4.377 | 4.482 | 4.586 | 4.689 |
| | h | 315.0 | 1190.9 | | 1194.8 | 1200.8 | 1206.7 | 1212.5 | 1218.1 | 1223.7 | 1234.7 | 1245.4 | 1256.0 | 1266.4 | 1276.8 | 1287.0 | 1297.2 |
| (343.72) | s | 0.4948 | 1.5851 | | 1.5898 | 1.5972 | 1.6043 | 1.6113 | 1.6180 | 1.6245 | 1.6371 | 1.6492 | 1.6608 | 1.6721 | 1.6829 | 1.6935 | 1.7038 |
| 126 | v | 0.018 | 3.560 | | 3.590 | 3.649 | 3.707 | 3.764 | 3.821 | 3.876 | 3.986 | 4.094 | 4.200 | 4.305 | 4.408 | 4.511 | 4.613 |
| | h | 316.3 | 1191.2 | | 1194.3 | 1200.4 | 1206.3 | 1212.1 | 1217.8 | 1223.4 | 1234.4 | 1245.2 | 1255.8 | 1266.2 | 1276.6 | 1286.8 | 1297.0 |
| (344.94) | s | 0.4964 | 1.5838 | | 1.5876 | 1.5950 | 1.6022 | 1.6092 | 1.6159 | 1.6225 | 1.6351 | 1.6472 | 1.6589 | 1.6701 | 1.6810 | 1.6916 | 1.7019 |
| 128 | v | 0.018 | 3.507 | | 3.529 | 3.588 | 3.645 | 3.702 | 3.758 | 3.813 | 3.921 | 4.027 | 4.132 | 4.235 | 4.337 | 4.439 | 4.539 |
| | h | 317.6 | 1191.5 | | 1193.9 | 1199.9 | 1205.9 | 1211.7 | 1217.4 | 1223.1 | 1234.1 | 1244.9 | 1255.5 | 1266.0 | 1276.4 | 1286.6 | 1296.8 |
| (346.13) | s | 0.4980 | 1.5825 | | 1.5854 | 1.5929 | 1.6001 | 1.6071 | 1.6138 | 1.6204 | 1.6331 | 1.6453 | 1.6569 | 1.6682 | 1.6791 | 1.6897 | 1.7000 |
| 130 | v | 0.018 | 3.455 | | 3.471 | 3.529 | 3.585 | 3.641 | 3.696 | 3.751 | 3.858 | 3.963 | 4.066 | 4.168 | 4.269 | 4.369 | 4.468 |
| | h | 318.8 | 1191.7 | | 1193.4 | 1199.5 | 1205.5 | 1211.3 | 1217.1 | 1222.7 | 1233.8 | 1244.6 | 1255.3 | 1265.8 | 1276.2 | 1286.5 | 1296.7 |
| (347.32) | s | 0.4995 | 1.5812 | | 1.5833 | 1.5908 | 1.5980 | 1.6050 | 1.6118 | 1.6184 | 1.6312 | 1.6433 | 1.6550 | 1.6663 | 1.6773 | 1.6879 | 1.6982 |
| 132 | v | 0.018 | 3.405 | | 3.414 | 3.471 | 3.527 | 3.583 | 3.637 | 3.691 | 3.796 | 3.900 | 4.002 | 4.103 | 4.202 | 4.300 | 4.398 |
| | h | 320.0 | 1192.0 | | 1192.9 | 1199.1 | 1205.1 | 1211.0 | 1216.7 | 1222.4 | 1233.5 | 1244.4 | 1255.1 | 1265.6 | 1276.0 | 1286.3 | 1296.5 |
| (348.48) | s | 0.5010 | 1.5800 | | 1.5811 | 1.5886 | 1.5959 | 1.6030 | 1.6098 | 1.6164 | 1.6292 | 1.6414 | 1.6532 | 1.6645 | 1.6754 | 1.6860 | 1.6964 |
| 134 | v | 0.018 | 3.357 | | 3.359 | 3.416 | 3.471 | 3.526 | 3.579 | 3.633 | 3.737 | 3.839 | 3.940 | 4.039 | 4.137 | 4.234 | 4.331 |
| | h | 321.2 | 1192.2 | | 1192.5 | 1198.7 | 1204.7 | 1210.6 | 1216.4 | 1222.1 | 1233.2 | 1244.1 | 1254.8 | 1265.4 | 1275.8 | 1286.1 | 1296.3 |
| (349.64) | s | 0.5025 | 1.5787 | | 1.5790 | 1.5866 | 1.5939 | 1.6009 | 1.6078 | 1.6145 | 1.6273 | 1.6395 | 1.6513 | 1.6626 | 1.6736 | 1.6842 | 1.6946 |
| 136 | v | 0.018 | 3.310 | | | 3.362 | 3.416 | 3.470 | 3.523 | 3.576 | 3.679 | 3.780 | 3.880 | 3.978 | 4.074 | 4.170 | 4.265 |
| | h | 322.4 | 1192.5 | | | 1198.2 | 1204.3 | 1210.2 | 1216.0 | 1221.7 | 1232.9 | 1243.8 | 1254.6 | 1265.1 | 1275.6 | 1285.9 | 1296.2 |
| (350.78) | s | 0.5040 | 1.5775 | | | 1.5845 | 1.5919 | 1.5989 | 1.6058 | 1.6125 | 1.6254 | 1.6377 | 1.6495 | 1.6608 | 1.6718 | 1.6825 | 1.6928 |
| 138 | v | 0.018 | 3.264 | | | 3.309 | 3.363 | 3.417 | 3.469 | 3.521 | 3.623 | 3.723 | 3.821 | 3.918 | 4.013 | 4.108 | 4.202 |
| | h | 323.6 | 1192.7 | | | 1197.8 | 1203.9 | 1209.8 | 1215.7 | 1221.4 | 1232.6 | 1243.6 | 1254.3 | 1264.9 | 1275.4 | 1285.7 | 1296.0 |
| (351.91) | s | 0.5054 | 1.5763 | | | 1.5825 | 1.5899 | 1.5970 | 1.6039 | 1.6106 | 1.6235 | 1.6358 | 1.6476 | 1.6590 | 1.6700 | 1.6807 | 1.6911 |
| 140 | v | 0.018 | 3.220 | | | 3.258 | 3.312 | 3.365 | 3.417 | 3.468 | 3.569 | 3.667 | 3.764 | 3.860 | 3.954 | 4.047 | 4.140 |
| | h | 324.8 | 1193.0 | | | 1197.3 | 1203.5 | 1209.4 | 1215.3 | 1221.1 | 1232.3 | 1243.3 | 1254.1 | 1264.7 | 1275.2 | 1285.5 | 1295.8 |
| (353.02) | s | 0.5069 | 1.5751 | | | 1.5804 | 1.5879 | 1.5950 | 1.6020 | 1.6087 | 1.6217 | 1.6340 | 1.6458 | 1.6573 | 1.6683 | 1.6790 | 1.6894 |
| 142 | v | 0.018 | 3.177 | | | 3.208 | 3.262 | 3.314 | 3.365 | 3.416 | 3.516 | 3.613 | 3.709 | 3.803 | 3.896 | 3.989 | 4.080 |
| | h | 326.0 | 1193.2 | | | 1196.9 | 1203.0 | 1209.1 | 1215.0 | 1220.7 | 1232.0 | 1243.0 | 1253.8 | 1264.5 | 1275.0 | 1285.3 | 1295.6 |
| (354.12) | s | 0.5083 | 1.5740 | | | 1.5784 | 1.5859 | 1.5931 | 1.6001 | 1.6068 | 1.6198 | 1.6322 | 1.6441 | 1.6555 | 1.6666 | 1.6773 | 1.6877 |

| Temperature—Degrees Fahrenheit | | | | | | | | | | | | | | | | | |
|---|---|---|---|---|---|---|---|---|---|---|---|---|---|---|---|---|---|
| **560°** | **580°** | **600°** | **620°** | **640°** | **660°** | **680°** | **700°** | **750°** | **800°** | **850°** | **900°** | **1000°** | **1200°** | **1400°** | **1600°** | | **Abs. Press. Lb./Sq. In.** (Sat. Temp.) |
| 5.416 | 5.530 | 5.644 | 5.757 | 5.870 | 5.983 | 6.095 | 6.207 | 6.486 | 6.763 | 7.040 | 7.316 | 7.866 | 8.961 | 10.053 | 11.142 | **v** | |
| 1308.4 | 1318.4 | 1328.4 | 1338.4 | 1348.4 | 1358.4 | 1368.4 | 1378.3 | 1403.4 | 1428.5 | 1453.7 | 1479.2 | 1530.5 | 1635.5 | 1744.0 | 1856.1 | **h** | **110** |
| 1.7278 | 1.7376 | 1.7471 | 1.7564 | 1.7656 | 1.7746 | 1.7834 | 1.7921 | 1.8132 | 1.8336 | 1.8532 | 1.8723 | 1.9087 | 1.9761 | 2.0378 | 2.0951 | **s** | (334.77) |
| 5.317 | 5.429 | 5.541 | 5.653 | 5.763 | 5.874 | 5.985 | 6.095 | 6.369 | 6.642 | 6.913 | 7.184 | 7.725 | 8.801 | 9.873 | 10.943 | **v** | |
| 1308.3 | 1318.3 | 1328.3 | 1338.3 | 1348.3 | 1358.3 | 1368.2 | 1378.2 | 1403.3 | 1428.4 | 1453.7 | 1479.1 | 1530.4 | 1635.5 | 1744.0 | 1856.1 | **h** | **112** |
| 1.7257 | 1.7355 | 1.7450 | 1.7543 | 1.7635 | 1.7725 | 1.7813 | 1.7900 | 1.8112 | 1.8315 | 1.8512 | 1.8702 | 1.9067 | 1.9741 | 2.0358 | 2.0931 | **s** | (336.11) |
| 5.222 | 5.332 | 5.442 | 5.552 | 5.661 | 5.770 | 5.878 | 5.986 | 6.256 | 6.524 | 6.791 | 7.058 | 7.589 | 8.646 | 9.699 | 10.751 | **v** | |
| 1308.1 | 1318.1 | 1328.2 | 1338.2 | 1348.1 | 1358.1 | 1368.1 | 1378.1 | 1403.2 | 1428.3 | 1453.6 | 1479.0 | 1530.4 | 1635.4 | 1744.0 | 1856.1 | **h** | **114** |
| 1.7237 | 1.7334 | 1.7429 | 1.7523 | 1.7615 | 1.7705 | 1.7793 | 1.7880 | 1.8092 | 1.8295 | 1.8492 | 1.8683 | 1.9047 | 1.9721 | 2.0338 | 2.0911 | **s** | (337.42) |
| 5.130 | 5.239 | 5.347 | 5.455 | 5.562 | 5.669 | 5.776 | 5.882 | 6.147 | 6.410 | 6.673 | 6.935 | 7.457 | 8.496 | 9.532 | 10.565 | **v** | |
| 1307.9 | 1318.0 | 1328.0 | 1338.0 | 1348.0 | 1358.0 | 1368.0 | 1378.0 | 1403.1 | 1428.2 | 1453.5 | 1479.0 | 1530.3 | 1635.4 | 1743.9 | 1856.0 | **h** | **116** |
| 1.7216 | 1.7314 | 1.7409 | 1.7503 | 1.7595 | 1.7685 | 1.7773 | 1.7860 | 1.8072 | 1.8276 | 1.8472 | 1.8663 | 1.9028 | 1.9702 | 2.0319 | 2.0892 | **s** | (338.72) |
| 5.041 | 5.148 | 5.255 | 5.361 | 5.466 | 5.571 | 5.676 | 5.781 | 6.041 | 6.301 | 6.559 | 6.817 | 7.330 | 8.352 | 9.370 | 10.386 | **v** | |
| 1307.8 | 1317.8 | 1327.9 | 1337.9 | 1347.9 | 1357.9 | 1367.9 | 1377.9 | 1403.0 | 1428.2 | 1453.5 | 1478.9 | 1530.3 | 1635.4 | 1743.9 | 1856.0 | **h** | **118** |
| 1.7196 | 1.7294 | 1.7389 | 1.7483 | 1.7575 | 1.7665 | 1.7754 | 1.7841 | 1.8052 | 1.8256 | 1.8453 | 1.8644 | 1.9008 | 1.9683 | 2.0300 | 2.0873 | **s** | (339.99) |
| 4.955 | 5.061 | 5.165 | 5.270 | 5.374 | 5.477 | 5.580 | 5.683 | 5.940 | 6.195 | 6.449 | 6.702 | 7.207 | 8.212 | 9.214 | 10.213 | **v** | |
| 1307.6 | 1317.7 | 1327.7 | 1337.8 | 1347.8 | 1357.8 | 1367.8 | 1377.8 | 1402.9 | 1428.1 | 1453.4 | 1478.8 | 1530.2 | 1635.3 | 1743.9 | 1856.0 | **h** | **120** |
| 1.7177 | 1.7274 | 1.7370 | 1.7464 | 1.7556 | 1.7646 | 1.7734 | 1.7822 | 1.8033 | 1.8237 | 1.8434 | 1.8625 | 1.8990 | 1.9664 | 2.0281 | 2.0854 | **s** | (341.25) |
| 4.872 | 4.976 | 5.079 | 5.182 | 5.284 | 5.386 | 5.488 | 5.589 | 5.841 | 6.092 | 6.342 | 6.592 | 7.088 | 8.077 | 9.062 | 10.045 | **v** | |
| 1307.5 | 1317.6 | 1327.6 | 1337.6 | 1347.7 | 1357.7 | 1367.7 | 1377.7 | 1402.8 | 1428.0 | 1453.3 | 1478.8 | 1530.2 | 1635.3 | 1743.8 | 1856.0 | **h** | **122** |
| 1.7157 | 1.7255 | 1.7351 | 1.7445 | 1.7537 | 1.7627 | 1.7716 | 1.7803 | 1.8015 | 1.8219 | 1.8416 | 1.8606 | 1.8971 | 1.9646 | 2.0263 | 2.0836 | **s** | (342.50) |
| 4.792 | 4.894 | 4.996 | 5.097 | 5.197 | 5.298 | 5.398 | 5.498 | 5.746 | 5.993 | 6.239 | 6.484 | 6.973 | 7.947 | 8.916 | 9.883 | **v** | |
| 1307.3 | 1317.4 | 1327.5 | 1337.5 | 1347.5 | 1357.6 | 1367.6 | 1377.6 | 1402.7 | 1427.9 | 1453.2 | 1478.7 | 1530.1 | 1635.2 | 1743.8 | 1855.9 | **h** | **124** |
| 1.7138 | 1.7236 | 1.7332 | 1.7426 | 1.7518 | 1.7608 | 1.7697 | 1.7784 | 1.7996 | 1.8200 | 1.8397 | 1.8588 | 1.8953 | 1.9628 | 2.0245 | 2.0818 | **s** | (343.72) |
| 4.714 | 4.815 | 4.915 | 5.014 | 5.113 | 5.212 | 5.311 | 5.409 | 5.654 | 5.897 | 6.139 | 6.381 | 6.862 | 7.820 | 8.774 | 9.726 | **v** | |
| 1307.2 | 1317.3 | 1327.3 | 1337.4 | 1347.4 | 1357.5 | 1367.5 | 1377.5 | 1402.6 | 1427.8 | 1453.2 | 1478.6 | 1530.1 | 1635.2 | 1743.8 | 1855.9 | **h** | **126** |
| 1.7119 | 1.7217 | 1.7313 | 1.7407 | 1.7500 | 1.7590 | 1.7679 | 1.7766 | 1.7978 | 1.8182 | 1.8379 | 1.8570 | 1.8935 | 1.9610 | 2.0227 | 2.0800 | **s** | (344.94) |
| 4.639 | 4.738 | 4.836 | 4.934 | 5.032 | 5.130 | 5.227 | 5.323 | 5.564 | 5.804 | 6.042 | 6.280 | 6.754 | 7.697 | 8.637 | 9.573 | **v** | |
| 1307.0 | 1317.1 | 1327.2 | 1337.2 | 1347.3 | 1357.3 | 1367.4 | 1377.4 | 1402.5 | 1427.7 | 1453.1 | 1478.6 | 1530.0 | 1635.2 | 1743.7 | 1855.9 | **h** | **128** |
| 1.7101 | 1.7199 | 1.7295 | 1.7389 | 1.7481 | 1.7572 | 1.7661 | 1.7748 | 1.7960 | 1.8164 | 1.8361 | 1.8552 | 1.8917 | 1.9592 | 2.0210 | 2.0783 | **s** | (346.13) |
| 4.566 | 4.663 | 4.761 | 4.857 | 4.953 | 5.049 | 5.145 | 5.240 | 5.478 | 5.714 | 5.949 | 6.183 | 6.650 | 7.579 | 8.503 | 9.426 | **v** | |
| 1306.8 | 1317.0 | 1327.0 | 1337.1 | 1347.2 | 1357.2 | 1367.3 | 1377.3 | 1402.5 | 1427.7 | 1453.0 | 1478.5 | 1529.9 | 1635.1 | 1743.7 | 1855.9 | **h** | **130** |
| 1.7083 | 1.7181 | 1.7277 | 1.7371 | 1.7463 | 1.7554 | 1.7643 | 1.7730 | 1.7942 | 1.8147 | 1.8344 | 1.8535 | 1.8900 | 1.9575 | 2.0193 | 2.0765 | **s** | (347.32) |
| 4.495 | 4.591 | 4.687 | 4.782 | 4.877 | 4.972 | 5.066 | 5.160 | 5.394 | 5.626 | 5.858 | 6.089 | 6.548 | 7.463 | 8.374 | 9.283 | **v** | |
| 1306.7 | 1316.8 | 1326.9 | 1337.0 | 1347.1 | 1357.1 | 1367.2 | 1377.2 | 1402.4 | 1427.6 | 1452.9 | 1478.4 | 1529.9 | 1635.1 | 1743.7 | 1855.8 | **h** | **132** |
| 1.7065 | 1.7163 | 1.7259 | 1.7353 | 1.7446 | 1.7536 | 1.7625 | 1.7713 | 1.7925 | 1.8129 | 1.8327 | 1.8518 | 1.8883 | 1.9558 | 2.0176 | 2.0748 | **s** | (348.48) |
| 4.426 | 4.521 | 4.616 | 4.709 | 4.803 | 4.896 | 4.989 | 5.082 | 5.312 | 5.541 | 5.770 | 5.997 | 6.450 | 7.351 | 8.249 | 9.144 | **v** | |
| 1306.5 | 1316.7 | 1326.8 | 1336.9 | 1346.9 | 1357.0 | 1367.0 | 1377.1 | 1402.3 | 1427.5 | 1452.9 | 1478.4 | 1529.8 | 1635.0 | 1743.6 | 1855.8 | **h** | **134** |
| 1.7047 | 1.7145 | 1.7241 | 1.7336 | 1.7428 | 1.7519 | 1.7608 | 1.7695 | 1.7908 | 1.8112 | 1.8310 | 1.8501 | 1.8866 | 1.9541 | 2.0159 | 2.0732 | **s** | (349.64) |
| 4.359 | 4.453 | 4.546 | 4.639 | 4.731 | 4.823 | 4.915 | 5.006 | 5.233 | 5.459 | 5.684 | 5.908 | 6.355 | 7.243 | 8.127 | 9.010 | **v** | |
| 1306.4 | 1316.5 | 1326.6 | 1336.7 | 1346.8 | 1356.9 | 1366.9 | 1377.0 | 1402.2 | 1427.4 | 1452.8 | 1478.3 | 1529.8 | 1635.0 | 1743.6 | 1855.8 | **h** | **136** |
| 1.7029 | 1.7128 | 1.7224 | 1.7319 | 1.7411 | 1.7502 | 1.7591 | 1.7679 | 1.7891 | 1.8095 | 1.8293 | 1.8484 | 1.8850 | 1.9525 | 2.0143 | 2.0715 | **s** | (350.78) |
| 4.295 | 4.387 | 4.479 | 4.570 | 4.661 | 4.752 | 4.842 | 4.932 | 5.156 | 5.379 | 5.601 | 5.822 | 6.262 | 7.138 | 8.009 | 8.879 | **v** | |
| 1306.2 | 1316.4 | 1326.5 | 1336.6 | 1346.7 | 1356.8 | 1366.8 | 1376.9 | 1402.1 | 1427.3 | 1452.7 | 1478.2 | 1529.7 | 1635.0 | 1743.6 | 1855.8 | **h** | **138** |
| 1.7012 | 1.7111 | 1.7207 | 1.7302 | 1.7394 | 1.7485 | 1.7574 | 1.7662 | 1.7874 | 1.8079 | 1.8277 | 1.8468 | 1.8833 | 1.9509 | 2.0126 | 2.0699 | **s** | (351.91) |
| 4.232 | 4.323 | 4.413 | 4.503 | 4.593 | 4.683 | 4.772 | 4.861 | 5.082 | 5.301 | 5.520 | 5.738 | 6.172 | 7.035 | 7.895 | 8.752 | **v** | |
| 1306.0 | 1316.2 | 1326.4 | 1336.5 | 1346.6 | 1356.6 | 1366.7 | 1376.8 | 1402.0 | 1427.3 | 1452.6 | 1478.2 | 1529.7 | 1634.9 | 1743.5 | 1855.7 | **h** | **140** |
| 1.6995 | 1.7094 | 1.7190 | 1.7285 | 1.7378 | 1.7468 | 1.7558 | 1.7645 | 1.7858 | 1.8063 | 1.8260 | 1.8451 | 1.8817 | 1.9493 | 2.0110 | 2.0683 | **s** | (353.02) |
| 4.171 | 4.261 | 4.350 | 4.439 | 4.527 | 4.616 | 4.704 | 4.791 | 5.009 | 5.226 | 5.441 | 5.656 | 6.085 | 6.936 | 7.783 | 8.628 | **v** | |
| 1305.9 | 1316.1 | 1326.2 | 1336.3 | 1346.4 | 1356.5 | 1366.6 | 1376.7 | 1401.9 | 1427.2 | 1452.6 | 1478.1 | 1529.6 | 1634.9 | 1743.5 | 1855.7 | **h** | **142** |
| 1.6978 | 1.7077 | 1.7174 | 1.7268 | 1.7361 | 1.7452 | 1.7541 | 1.7629 | 1.7842 | 1.8047 | 1.8244 | 1.8435 | 1.8801 | 1.9477 | 2.0095 | 2.0667 | **s** | (354.12) |

## Table 3. Superheated Vapor

| Abs. Press. Lb./Sq. In. (Sat. Temp.) | | Sat. Liquid | Sat. Vapor | Temperature—Degrees Fahrenheit 360° | 370° | 380° | 390° | 400° | 420° | 440° | 460° | 480° | 500° | 520° | 540° | 560° | 580° |
|---|---|---|---|---|---|---|---|---|---|---|---|---|---|---|---|---|---|
| | v | 0.018 | 3.134 | 3.160 | 3.213 | 3.265 | 3.316 | 3.366 | 3.464 | 3.561 | 3.655 | 3.748 | 3.840 | 3.931 | 4.022 | 4.111 | 4.200 |
| **144** | h | 327.1 | 1193.4 | 1196.5 | 1202.6 | 1208.7 | 1214.6 | 1220.4 | 1231.7 | 1242.8 | 1253.6 | 1264.2 | 1274.8 | 1285.2 | 1295.5 | 1305.7 | 1315.9 |
| (355.21) | s | 0.5097 | 1.5728 | 1.5765 | 1.5840 | 1.5912 | 1.5982 | 1.6050 | 1.6180 | 1.6304 | 1.6423 | 1.6538 | 1.6649 | 1.6756 | 1.6860 | 1.6962 | 1.7061 |
| | v | 0.018 | 3.094 | 3.113 | 3.165 | 3.217 | 3.267 | 3.317 | 3.414 | 3.509 | 3.603 | 3.695 | 3.786 | 3.876 | 3.965 | 4.053 | 4.141 |
| **146** | h | 328.3 | 1193.6 | 1196.0 | 1202.2 | 1208.3 | 1214.2 | 1220.0 | 1231.4 | 1242.5 | 1253.3 | 1264.0 | 1274.5 | 1285.0 | 1295.3 | 1305.6 | 1315.8 |
| (356.29) | s | 0.5111 | 1.5716 | 1.5745 | 1.5820 | 1.5893 | 1.5963 | 1.6031 | 1.6162 | 1.6287 | 1.6406 | 1.6521 | 1.6632 | 1.6739 | 1.6844 | 1.6945 | 1.7044 |
| | v | 0.018 | 3.054 | 3.067 | 3.119 | 3.170 | 3.220 | 3.269 | 3.365 | 3.459 | 3.552 | 3.643 | 3.733 | 3.822 | 3.910 | 3.997 | 4.084 |
| **148** | h | 329.4 | 1193.9 | 1195.6 | 1201.8 | 1207.9 | 1213.9 | 1219.7 | 1231.1 | 1242.2 | 1253.1 | 1263.8 | 1274.3 | 1284.8 | 1295.1 | 1305.4 | 1315.6 |
| (357.36) | s | 0.5124 | 1.5705 | 1.5726 | 1.5801 | 1.5874 | 1.5945 | 1.6013 | 1.6144 | 1.6269 | 1.6389 | 1.6504 | 1.6615 | 1.6723 | 1.6827 | 1.6929 | 1.7028 |
| | v | 0.018 | 3.015 | 3.023 | 3.074 | 3.124 | 3.174 | 3.223 | 3.318 | 3.411 | 3.502 | 3.592 | 3.681 | 3.769 | 3.856 | 3.942 | 4.028 |
| **150** | h | 330.5 | 1194.1 | 1195.1 | 1201.4 | 1207.5 | 1213.5 | 1219.4 | 1230.8 | 1242.0 | 1252.9 | 1263.6 | 1274.1 | 1284.6 | 1295.0 | 1305.2 | 1315.5 |
| (358.42) | s | 0.5138 | 1.5694 | 1.5706 | 1.5782 | 1.5856 | 1.5927 | 1.5995 | 1.6127 | 1.6252 | 1.6372 | 1.6487 | 1.6599 | 1.6707 | 1.6811 | 1.6913 | 1.7012 |
| | v | 0.018 | 2.977 | 2.980 | 3.030 | 3.080 | 3.129 | 3.177 | 3.272 | 3.364 | 3.454 | 3.543 | 3.631 | 3.718 | 3.804 | 3.889 | 3.973 |
| **152** | h | 331.6 | 1194.3 | 1194.7 | 1200.9 | 1207.1 | 1213.1 | 1219.0 | 1230.5 | 1241.7 | 1252.6 | 1263.4 | 1273.9 | 1284.4 | 1294.8 | 1305.1 | 1315.3 |
| (359.46) | s | 0.5151 | 1.5683 | 1.5687 | 1.5764 | 1.5837 | 1.5908 | 1.5977 | 1.6110 | 1.6235 | 1.6355 | 1.6471 | 1.6582 | 1.6690 | 1.6795 | 1.6897 | 1.6997 |
| | v | 0.018 | 2.940 | | 2.988 | 3.037 | 3.085 | 3.133 | 3.227 | 3.318 | 3.407 | 3.495 | 3.582 | 3.668 | 3.753 | 3.837 | 3.920 |
| **154** | h | 332.7 | 1194.5 | | 1200.5 | 1206.7 | 1212.8 | 1218.7 | 1230.2 | 1241.4 | 1252.4 | 1263.1 | 1273.7 | 1284.2 | 1294.6 | 1304.9 | 1315.2 |
| (360.49) | s | 0.5165 | 1.5672 | | 1.5745 | 1.5819 | 1.5891 | 1.5960 | 1.6093 | 1.6219 | 1.6339 | 1.6455 | 1.6566 | 1.6675 | 1.6780 | 1.6882 | 1.6981 |
| | v | 0.018 | 2.904 | | 2.946 | 2.995 | 3.043 | 3.090 | 3.183 | 3.273 | 3.362 | 3.449 | 3.534 | 3.619 | 3.703 | 3.786 | 3.869 |
| **156** | h | 333.8 | 1194.7 | | 1200.1 | 1206.3 | 1212.4 | 1218.3 | 1229.9 | 1241.1 | 1252.1 | 1262.9 | 1273.5 | 1284.0 | 1294.4 | 1304.8 | 1315.0 |
| (361.52) | s | 0.5178 | 1.5661 | | 1.5727 | 1.5801 | 1.5873 | 1.5942 | 1.6076 | 1.6202 | 1.6323 | 1.6439 | 1.6551 | 1.6659 | 1.6764 | 1.6866 | 1.6966 |
| | v | 0.018 | 2.869 | | 2.905 | 2.954 | 3.002 | 3.048 | 3.140 | 3.229 | 3.317 | 3.403 | 3.488 | 3.572 | 3.655 | 3.737 | 3.818 |
| **158** | h | 334.9 | 1194.9 | | 1199.7 | 1205.9 | 1212.0 | 1218.0 | 1229.6 | 1240.9 | 1251.9 | 1262.7 | 1273.3 | 1283.8 | 1294.3 | 1304.6 | 1314.9 |
| (362.53) | s | 0.5191 | 1.5650 | | 1.5709 | 1.5783 | 1.5855 | 1.5925 | 1.6059 | 1.6185 | 1.6306 | 1.6423 | 1.6535 | 1.6643 | 1.6749 | 1.6851 | 1.6951 |
| | v | 0.018 | 2.834 | | 2.866 | 2.914 | 2.961 | 3.008 | 3.098 | 3.187 | 3.273 | 3.359 | 3.443 | 3.525 | 3.607 | 3.689 | 3.769 |
| **160** | h | 335.9 | 1195.1 | | 1199.2 | 1205.5 | 1211.6 | 1217.6 | 1229.3 | 1240.6 | 1251.6 | 1262.4 | 1273.1 | 1283.7 | 1294.1 | 1304.4 | 1314.7 |
| (363.53) | s | 0.5204 | 1.5640 | | 1.5690 | 1.5766 | 1.5838 | 1.5908 | 1.6042 | 1.6169 | 1.6291 | 1.6407 | 1.6519 | 1.6628 | 1.6733 | 1.6836 | 1.6936 |
| | v | 0.018 | 2.801 | | 2.827 | 2.875 | 2.922 | 2.968 | 3.058 | 3.145 | 3.231 | 3.315 | 3.398 | 3.480 | 3.561 | 3.642 | 3.722 |
| **162** | h | 337.0 | 1195.3 | | 1198.8 | 1205.1 | 1211.3 | 1217.3 | 1229.0 | 1240.3 | 1251.4 | 1262.2 | 1272.9 | 1283.5 | 1293.9 | 1304.3 | 1314.6 |
| (364.53) | s | 0.5216 | 1.5630 | | 1.5672 | 1.5748 | 1.5821 | 1.5891 | 1.6025 | 1.6153 | 1.6275 | 1.6392 | 1.6504 | 1.6613 | 1.6718 | 1.6821 | 1.6921 |
| | v | 0.018 | 2.768 | | 2.790 | 2.837 | 2.883 | 2.929 | 3.018 | 3.105 | 3.190 | 3.273 | 3.355 | 3.436 | 3.516 | 3.596 | 3.675 |
| **164** | h | 338.0 | 1195.5 | | 1198.4 | 1204.7 | 1210.9 | 1216.9 | 1228.6 | 1240.0 | 1251.1 | 1262.0 | 1272.7 | 1283.3 | 1293.7 | 1304.1 | 1314.4 |
| (365.51) | s | 0.5229 | 1.5620 | | 1.5655 | 1.5731 | 1.5804 | 1.5874 | 1.6009 | 1.6137 | 1.6259 | 1.6376 | 1.6489 | 1.6598 | 1.6704 | 1.6806 | 1.6906 |
| | v | 0.018 | 2.736 | | 2.753 | 2.800 | 2.846 | 2.891 | 2.979 | 3.065 | 3.149 | 3.232 | 3.313 | 3.393 | 3.473 | 3.551 | 3.629 |
| **166** | h | 339.0 | 1195.7 | | 1197.9 | 1204.3 | 1210.5 | 1216.6 | 1228.3 | 1239.7 | 1250.9 | 1261.8 | 1272.5 | 1283.1 | 1293.6 | 1303.9 | 1314.3 |
| (366.48) | s | 0.5241 | 1.5610 | | 1.5637 | 1.5713 | 1.5787 | 1.5858 | 1.5993 | 1.6121 | 1.6244 | 1.6361 | 1.6474 | 1.6583 | 1.6689 | 1.6792 | 1.6892 |
| | v | 0.018 | 2.705 | | 2.717 | 2.764 | 2.809 | 2.854 | 2.941 | 3.027 | 3.110 | 3.191 | 3.272 | 3.351 | 3.430 | 3.508 | 3.585 |
| **168** | h | 340.1 | 1195.8 | | 1197.5 | 1203.9 | 1210.1 | 1216.2 | 1228.0 | 1239.5 | 1250.6 | 1261.5 | 1272.3 | 1282.9 | 1293.4 | 1303.8 | 1314.1 |
| (367.45) | s | 0.5254 | 1.5600 | | 1.5620 | 1.5696 | 1.5770 | 1.5841 | 1.5977 | 1.6106 | 1.6228 | 1.6346 | 1.6459 | 1.6568 | 1.6674 | 1.6777 | 1.6878 |
| | v | 0.018 | 2.675 | | 2.682 | 2.728 | 2.773 | 2.818 | 2.904 | 2.989 | 3.071 | 3.152 | 3.232 | 3.310 | 3.388 | 3.465 | 3.541 |
| **170** | h | 341.1 | 1196.0 | | 1197.1 | 1203.5 | 1209.7 | 1215.8 | 1227.7 | 1239.2 | 1250.4 | 1261.3 | 1272.1 | 1282.7 | 1293.2 | 1303.6 | 1314.0 |
| (368.41) | s | 0.5266 | 1.5590 | | 1.5603 | 1.5679 | 1.5753 | 1.5825 | 1.5961 | 1.6090 | 1.6213 | 1.6331 | 1.6444 | 1.6554 | 1.6660 | 1.6763 | 1.6864 |
| | v | 0.018 | 2.645 | | 2.648 | 2.694 | 2.738 | 2.782 | 2.868 | 2.952 | 3.034 | 3.114 | 3.193 | 3.270 | 3.347 | 3.424 | 3.499 |
| **172** | h | 342.1 | 1196.2 | | 1196.6 | 1203.1 | 1209.4 | 1215.5 | 1227.4 | 1238.9 | 1250.1 | 1261.1 | 1271.9 | 1282.5 | 1293.0 | 1303.5 | 1313.8 |
| (369.35) | s | 0.5278 | 1.5580 | | 1.5585 | 1.5662 | 1.5737 | 1.5809 | 1.5945 | 1.6075 | 1.6198 | 1.6316 | 1.6430 | 1.6539 | 1.6646 | 1.6749 | 1.6850 |
| | v | 0.018 | 2.616 | | | 2.660 | 2.704 | 2.748 | 2.833 | 2.916 | 2.997 | 3.076 | 3.154 | 3.231 | 3.308 | 3.383 | 3.458 |
| **174** | h | 343.1 | 1196.4 | | | 1202.7 | 1209.0 | 1215.1 | 1227.1 | 1238.6 | 1249.9 | 1260.8 | 1271.6 | 1282.3 | 1292.9 | 1303.3 | 1313.7 |
| (370.29) | s | 0.5290 | 1.5570 | | | 1.5646 | 1.5720 | 1.5792 | 1.5930 | 1.6060 | 1.6183 | 1.6301 | 1.6415 | 1.6525 | 1.6632 | 1.6735 | 1.6836 |
| | v | 0.018 | 2.587 | | | 2.627 | 2.671 | 2.714 | 2.799 | 2.881 | 2.961 | 3.040 | 3.117 | 3.193 | 3.269 | 3.343 | 3.417 |
| **176** | h | 344.1 | 1196.5 | | | 1202.3 | 1208.6 | 1214.8 | 1226.8 | 1238.3 | 1249.6 | 1260.6 | 1271.4 | 1282.1 | 1292.7 | 1303.1 | 1313.5 |
| (371.22) | s | 0.5302 | 1.5561 | | | 1.5629 | 1.5704 | 1.5776 | 1.5914 | 1.6045 | 1.6168 | 1.6287 | 1.6401 | 1.6511 | 1.6618 | 1.6721 | 1.6822 |
| | v | 0.018 | 2.559 | | | 2.594 | 2.638 | 2.681 | 2.765 | 2.846 | 2.926 | 3.004 | 3.080 | 3.156 | 3.230 | 3.304 | 3.377 |
| **178** | h | 345.1 | 1196.7 | | | 1201.8 | 1208.2 | 1214.4 | 1226.4 | 1238.1 | 1249.3 | 1260.4 | 1271.2 | 1281.9 | 1292.5 | 1303.0 | 1313.4 |
| (372.14) | s | 0.5313 | 1.5551 | | | 1.5613 | 1.5688 | 1.5761 | 1.5899 | 1.6030 | 1.6154 | 1.6273 | 1.6387 | 1.6497 | 1.6604 | 1.6707 | 1.6808 |

**Temperature—Degrees Fahrenheit**

| 600° | 620° | 640° | 660° | 680° | 700° | 750° | 800° | 850° | 900° | 950° | 1000° | 1100° | 1200° | 1400° | 1600° | | Abs. Press. Lb./Sq. In. (Sat. Temp.) |
|---|---|---|---|---|---|---|---|---|---|---|---|---|---|---|---|---|---|
| 4.288 | 4.376 | 4.463 | 4.550 | 4.637 | 4.724 | 4.939 | 5.152 | 5.365 | 5.577 | 5.789 | 6.000 | 6.420 | 6.839 | 7.675 | 8.508 | v | |
| 1326.1 | 1336.2 | 1346.3 | 1356.4 | 1366.5 | 1376.6 | 1401.8 | 1427.1 | 1452.5 | 1478.0 | 1503.7 | 1529.6 | 1581.8 | 1634.8 | 1743.5 | 1855.7 | h | **144** |
| 1.7157 | 1.7252 | 1.7345 | 1.7436 | 1.7525 | 1.7613 | 1.7826 | 1.8031 | 1.8228 | 1.8420 | 1.8605 | 1.8785 | 1.9132 | 1.9461 | 2.0079 | 2.0652 | s | (355.21) |
| 4.228 | 4.315 | 4.401 | 4.487 | 4.573 | 4.658 | 4.870 | 5.081 | 5.291 | 5.500 | 5.709 | 5.917 | 6.332 | 6.745 | 7.569 | 8.392 | v | |
| 1325.9 | 1336.1 | 1346.2 | 1356.3 | 1366.4 | 1376.5 | 1401.7 | 1427.0 | 1452.4 | 1478.0 | 1503.6 | 1529.5 | 1581.8 | 1634.8 | 1743.5 | 1855.7 | h | **146** |
| 1.7141 | 1.7236 | 1.7329 | 1.7420 | 1.7509 | 1.7597 | 1.7810 | 1.8015 | 1.8213 | 1.8404 | 1.8590 | 1.8770 | 1.9116 | 1.9446 | 2.0064 | 2.0637 | s | (356.29) |
| 4.170 | 4.255 | 4.340 | 4.425 | 4.510 | 4.594 | 4.803 | 5.012 | 5.219 | 5.425 | 5.631 | 5.836 | 6.246 | 6.653 | 7.467 | 8.278 | v | |
| 1325.8 | 1335.9 | 1346.1 | 1356.2 | 1366.3 | 1376.4 | 1401.6 | 1426.9 | 1452.4 | 1477.9 | 1503.6 | 1529.4 | 1581.7 | 1634.8 | 1743.4 | 1855.6 | h | **148** |
| 1.7125 | 1.7220 | 1.7313 | 1.7404 | 1.7493 | 1.7581 | 1.7794 | 1.7999 | 1.8197 | 1.8389 | 1.8574 | 1.8755 | 1.9101 | 1.9431 | 2.0049 | 2.0622 | s | (357.36) |
| 4.113 | 4.197 | 4.281 | 4.365 | 4.449 | 4.532 | 4.738 | 4.944 | 5.148 | 5.352 | 5.555 | 5.758 | 6.162 | 6.564 | 7.367 | 8.168 | v | |
| 1325.7 | 1335.8 | 1346.0 | 1356.1 | 1366.2 | 1376.3 | 1401.5 | 1426.9 | 1452.3 | 1477.8 | 1503.5 | 1529.4 | 1581.7 | 1634.7 | 1743.4 | 1855.6 | h | **150** |
| 1.7109 | 1.7204 | 1.7297 | 1.7389 | 1.7478 | 1.7566 | 1.7779 | 1.7984 | 1.8182 | 1.8374 | 1.8559 | 1.8740 | 1.9086 | 1.9416 | 2.0034 | 2.0607 | s | (358.42) |
| 4.057 | 4.141 | 4.224 | 4.306 | 4.389 | 4.471 | 4.675 | 4.878 | 5.080 | 5.281 | 5.482 | 5.682 | 6.080 | 6.478 | 7.270 | 8.060 | v | |
| 1325.5 | 1335.7 | 1345.8 | 1356.0 | 1366.1 | 1376.2 | 1401.4 | 1426.8 | 1452.2 | 1477.8 | 1503.5 | 1529.3 | 1581.6 | 1634.7 | 1743.4 | 1855.6 | h | **152** |
| 1.7094 | 1.7189 | 1.7282 | 1.7373 | 1.7463 | 1.7551 | 1.7764 | 1.7969 | 1.8167 | 1.8359 | 1.8544 | 1.8725 | 1.9071 | 1.9401 | 2.0019 | 2.0592 | s | (359.46) |
| 4.003 | 4.086 | 4.168 | 4.249 | 4.331 | 4.412 | 4.614 | 4.814 | 5.013 | 5.212 | 5.410 | 5.607 | 6.001 | 6.393 | 7.175 | 7.955 | v | |
| 1325.4 | 1335.6 | 1345.7 | 1355.8 | 1365.9 | 1376.1 | 1401.4 | 1426.7 | 1452.1 | 1477.7 | 1503.4 | 1529.3 | 1581.6 | 1634.6 | 1743.3 | 1855.6 | h | **154** |
| 1.7078 | 1.7173 | 1.7267 | 1.7358 | 1.7448 | 1.7536 | 1.7749 | 1.7954 | 1.8152 | 1.8344 | 1.8530 | 1.8710 | 1.9056 | 1.9386 | 2.0004 | 2.0577 | s | (360.49) |
| 3.951 | 4.032 | 4.113 | 4.194 | 4.274 | 4.354 | 4.554 | 4.752 | 4.948 | 5.145 | 5.340 | 5.535 | 5.924 | 6.311 | 7.083 | 7.853 | v | |
| 1325.2 | 1335.4 | 1345.6 | 1355.7 | 1365.8 | 1376.0 | 1401.3 | 1426.6 | 1452.1 | 1477.6 | 1503.3 | 1529.2 | 1581.5 | 1634.6 | 1743.3 | 1855.5 | h | **156** |
| 1.7063 | 1.7158 | 1.7252 | 1.7343 | 1.7433 | 1.7521 | 1.7734 | 1.7940 | 1.8138 | 1.8329 | 1.8515 | 1.8695 | 1.9042 | 1.9372 | 1.9990 | 2.0563 | s | (361.52) |
| 3.899 | 3.980 | 4.060 | 4.140 | 4.219 | 4.298 | 4.495 | 4.691 | 4.885 | 5.079 | 5.272 | 5.464 | 5.848 | 6.231 | 6.993 | 7.753 | v | |
| 1325.1 | 1335.3 | 1345.5 | 1355.6 | 1365.7 | 1375.9 | 1401.2 | 1426.5 | 1452.0 | 1477.6 | 1503.3 | 1529.2 | 1581.5 | 1634.6 | 1743.3 | 1855.5 | h | **158** |
| 1.7048 | 1.7143 | 1.7237 | 1.7328 | 1.7418 | 1.7506 | 1.7720 | 1.7925 | 1.8123 | 1.8315 | 1.8501 | 1.8681 | 1.9028 | 1.9358 | 1.9976 | 2.0549 | s | (362.53) |
| 3.849 | 3.929 | 4.008 | 4.087 | 4.166 | 4.244 | 4.438 | 4.631 | 4.823 | 5.015 | 5.205 | 5.396 | 5.775 | 6.152 | 6.906 | 7.656 | v | |
| 1325.0 | 1335.2 | 1345.3 | 1355.5 | 1365.6 | 1375.7 | 1401.1 | 1426.4 | 1451.9 | 1477.5 | 1503.2 | 1529.1 | 1581.4 | 1634.5 | 1743.2 | 1855.5 | h | **160** |
| 1.7033 | 1.7129 | 1.7222 | 1.7313 | 1.7403 | 1.7491 | 1.7705 | 1.7911 | 1.8109 | 1.8301 | 1.8487 | 1.8667 | 1.9014 | 1.9344 | 1.9962 | 2.0535 | s | (363.53) |
| 3.801 | 3.879 | 3.957 | 4.035 | 4.113 | 4.190 | 4.382 | 4.573 | 4.763 | 4.952 | 5.141 | 5.328 | 5.703 | 6.076 | 6.820 | 7.562 | v | |
| 1324.8 | 1335.0 | 1345.2 | 1355.4 | 1365.5 | 1375.6 | 1401.0 | 1426.4 | 1451.8 | 1477.4 | 1503.2 | 1529.1 | 1581.4 | 1634.5 | 1743.2 | 1855.5 | h | **162** |
| 1.7019 | 1.7114 | 1.7207 | 1.7299 | 1.7389 | 1.7477 | 1.7691 | 1.7897 | 1.8095 | 1.8287 | 1.8473 | 1.8653 | 1.9000 | 1.9330 | 1.9948 | 2.0521 | s | (364.53) |
| 3.753 | 3.831 | 3.908 | 3.985 | 4.062 | 4.138 | 4.328 | 4.517 | 4.704 | 4.891 | 5.077 | 5.263 | 5.633 | 6.002 | 6.737 | 7.469 | v | |
| 1324.7 | 1334.9 | 1345.1 | 1355.3 | 1365.4 | 1375.5 | 1400.9 | 1426.3 | 1451.8 | 1477.4 | 1503.1 | 1529.0 | 1581.3 | 1634.4 | 1743.2 | 1855.4 | h | **164** |
| 1.7004 | 1.7100 | 1.7193 | 1.7285 | 1.7375 | 1.7463 | 1.7677 | 1.7883 | 1.8081 | 1.8273 | 1.8459 | 1.8639 | 1.8986 | 1.9316 | 1.9934 | 2.0508 | s | (365.51) |
| 3.707 | 3.784 | 3.860 | 3.936 | 4.012 | 4.088 | 4.275 | 4.462 | 4.647 | 4.832 | 5.016 | 5.199 | 5.565 | 5.929 | 6.655 | 7.379 | v | |
| 1324.5 | 1334.8 | 1345.0 | 1355.1 | 1365.3 | 1375.4 | 1400.8 | 1426.2 | 1451.7 | 1477.3 | 1503.0 | 1529.0 | 1581.3 | 1634.4 | 1743.1 | 1855.4 | h | **166** |
| 1.6990 | 1.7085 | 1.7179 | 1.7271 | 1.7361 | 1.7449 | 1.7663 | 1.7869 | 1.8067 | 1.8259 | 1.8445 | 1.8625 | 1.8972 | 1.9302 | 1.9921 | 2.0494 | s | (366.48) |
| 3.661 | 3.737 | 3.813 | 3.888 | 3.963 | 4.038 | 4.223 | 4.408 | 4.591 | 4.774 | 4.956 | 5.137 | 5.498 | 5.858 | 6.576 | 7.291 | v | |
| 1324.4 | 1334.6 | 1344.8 | 1355.0 | 1365.2 | 1375.3 | 1400.7 | 1426.1 | 1451.6 | 1477.2 | 1503.0 | 1528.9 | 1581.3 | 1634.4 | 1743.1 | 1855.4 | h | **168** |
| 1.6976 | 1.7071 | 1.7165 | 1.7257 | 1.7347 | 1.7435 | 1.7649 | 1.7855 | 1.8053 | 1.8245 | 1.8431 | 1.8612 | 1.8959 | 1.9289 | 1.9907 | 2.0481 | s | (367.45) |
| 3.617 | 3.692 | 3.767 | 3.842 | 3.916 | 3.990 | 4.173 | 4.355 | 4.536 | 4.717 | 4.897 | 5.076 | 5.433 | 5.789 | 6.498 | 7.205 | v | |
| 1324.2 | 1334.5 | 1344.7 | 1354.9 | 1365.1 | 1375.2 | 1400.6 | 1426.0 | 1451.5 | 1477.2 | 1502.9 | 1528.8 | 1581.2 | 1634.3 | 1743.1 | 1855.4 | h | **170** |
| 1.6962 | 1.7057 | 1.7151 | 1.7243 | 1.7333 | 1.7421 | 1.7636 | 1.7842 | 1.8040 | 1.8232 | 1.8418 | 1.8599 | 1.8946 | 1.9276 | 1.9894 | 2.0468 | s | (368.41) |
| 3.574 | 3.648 | 3.722 | 3.796 | 3.869 | 3.942 | 4.123 | 4.304 | 4.483 | 4.662 | 4.839 | 5.016 | 5.370 | 5.721 | 6.422 | 7.121 | v | |
| 1324.1 | 1334.4 | 1344.6 | 1354.8 | 1365.0 | 1375.1 | 1400.5 | 1426.0 | 1451.5 | 1477.1 | 1502.9 | 1528.8 | 1581.2 | 1634.3 | 1743.0 | 1855.3 | h | **172** |
| 1.6948 | 1.7043 | 1.7137 | 1.7229 | 1.7319 | 1.7408 | 1.7622 | 1.7828 | 1.8027 | 1.8219 | 1.8405 | 1.8586 | 1.8933 | 1.9263 | 1.9881 | 2.0455 | s | (369.35) |
| 3.532 | 3.605 | 3.678 | 3.751 | 3.824 | 3.896 | 4.075 | 4.254 | 4.431 | 4.607 | 4.783 | 4.958 | 5.307 | 5.655 | 6.348 | 7.039 | v | |
| 1324.0 | 1334.2 | 1344.5 | 1354.7 | 1364.8 | 1375.0 | 1400.4 | 1425.9 | 1451.4 | 1477.0 | 1502.8 | 1528.7 | 1581.1 | 1634.2 | 1743.0 | 1855.3 | h | **174** |
| 1.6934 | 1.7030 | 1.7124 | 1.7216 | 1.7306 | 1.7394 | 1.7609 | 1.7815 | 1.8014 | 1.8206 | 1.8392 | 1.8573 | 1.8920 | 1.9250 | 1.9868 | 2.0442 | s | (370.29) |
| 3.490 | 3.563 | 3.636 | 3.708 | 3.780 | 3.851 | 4.028 | 4.205 | 4.380 | 4.554 | 4.728 | 4.901 | 5.247 | 5.591 | 6.276 | 6.959 | v | |
| 1323.8 | 1334.1 | 1344.3 | 1354.5 | 1364.7 | 1374.9 | 1400.3 | 1425.8 | 1451.3 | 1477.0 | 1502.7 | 1528.7 | 1581.1 | 1634.2 | 1743.0 | 1855.3 | h | **176** |
| 1.6920 | 1.7016 | 1.7110 | 1.7202 | 1.7292 | 1.7381 | 1.7596 | 1.7802 | 1.8001 | 1.8193 | 1.8379 | 1.8560 | 1.8907 | 1.9237 | 1.9856 | 2.0429 | s | (371.22) |
| 3.450 | 3.522 | 3.594 | 3.665 | 3.736 | 3.807 | 3.982 | 4.157 | 4.330 | 4.503 | 4.675 | 4.846 | 5.187 | 5.528 | 6.205 | 6.881 | v | |
| 1323.7 | 1334.0 | 1344.2 | 1354.4 | 1364.6 | 1374.8 | 1400.2 | 1425.7 | 1451.3 | 1476.9 | 1502.7 | 1528.6 | 1581.0 | 1634.2 | 1743.0 | 1855.3 | h | **178** |
| 1.6907 | 1.7003 | 1.7097 | 1.7189 | 1.7279 | 1.7368 | 1.7583 | 1.7789 | 1.7988 | 1.8180 | 1.8366 | 1.8547 | 1.8894 | 1.9224 | 1.9843 | 2.0416 | s | (372.14) |

# Table 3. Superheated Vapor

| Abs. Press. Lb./Sq. In. (Sat. Temp.) | | Sat. Liquid | Sat. Vapor | 380° | 390° | 400° | 420° | 440° | 460° | 480° | 500° | 520° | 540° | 560° | 580° | 600° | 620° |
|---|---|---|---|---|---|---|---|---|---|---|---|---|---|---|---|---|---|
| | | | | Temperature—Degrees Fahrenheit | | | | | | | | | | | | | |
| | v | 0.018 | 2.532 | 2.563 | 2.606 | 2.649 | 2.732 | 2.813 | 2.891 | 2.969 | 3.044 | 3.119 | 3.193 | 3.266 | 3.339 | 3.411 | 3.482 |
| **180** | h | 346.0 | 1196.9 | 1201.4 | 1207.8 | 1214.0 | 1226.1 | 1237.8 | 1249.1 | 1260.2 | 1271.0 | 1281.7 | 1292.3 | 1302.8 | 1313.2 | 1323.5 | 1333.8 |
| (373.06) | s | 0.5325 | 1.5542 | 1.5596 | 1.5672 | 1.5745 | 1.5884 | 1.6015 | 1.6139 | 1.6258 | 1.6373 | 1.6483 | 1.6590 | 1.6694 | 1.6795 | 1.6894 | 1.6990 |
| | v | 0.018 | 2.505 | 2.532 | 2.575 | 2.617 | 2.700 | 2.780 | 2.858 | 2.934 | 3.009 | 3.084 | 3.157 | 3.229 | 3.301 | 3.372 | 3.443 |
| **182** | h | 347.0 | 1197.0 | 1201.0 | 1207.4 | 1213.7 | 1225.8 | 1237.5 | 1248.8 | 1259.9 | 1270.8 | 1281.5 | 1292.1 | 1302.6 | 1313.0 | 1323.4 | 1333.7 |
| (373.96) | s | 0.5336 | 1.5532 | 1.5580 | 1.5656 | 1.5729 | 1.5869 | 1.6000 | 1.6125 | 1.6244 | 1.6359 | 1.6470 | 1.6577 | 1.6681 | 1.6782 | 1.6880 | 1.6977 |
| | v | 0.018 | 2.479 | 2.501 | 2.544 | 2.586 | 2.668 | 2.748 | 2.825 | 2.901 | 2.975 | 3.049 | 3.121 | 3.193 | 3.264 | 3.334 | 3.404 |
| **184** | h | 348.0 | 1197.2 | 1200.6 | 1207.0 | 1213.3 | 1225.5 | 1237.2 | 1248.6 | 1259.7 | 1270.6 | 1281.4 | 1292.0 | 1302.5 | 1312.9 | 1323.3 | 1333.6 |
| (374.86) | s | 0.5348 | 1.5523 | 1.5564 | 1.5640 | 1.5714 | 1.5854 | 1.5986 | 1.6111 | 1.6230 | 1.6345 | 1.6456 | 1.6563 | 1.6667 | 1.6769 | 1.6867 | 1.6964 |
| | v | 0.018 | 2.454 | 2.472 | 2.514 | 2.556 | 2.637 | 2.716 | 2.793 | 2.868 | 2.942 | 3.015 | 3.086 | 3.157 | 3.228 | 3.297 | 3.367 |
| **186** | h | 348.9 | 1197.3 | 1200.2 | 1206.6 | 1212.9 | 1225.2 | 1236.9 | 1248.3 | 1259.5 | 1270.4 | 1281.2 | 1291.8 | 1302.3 | 1312.7 | 1323.1 | 1333.4 |
| (375.75) | s | 0.5359 | 1.5514 | 1.5548 | 1.5625 | 1.5699 | 1.5839 | 1.5971 | 1.6097 | 1.6217 | 1.6332 | 1.6443 | 1.6550 | 1.6654 | 1.6756 | 1.6854 | 1.6951 |
| | v | 0.018 | 2.429 | 2.443 | 2.485 | 2.527 | 2.607 | 2.685 | 2.761 | 2.836 | 2.909 | 2.981 | 3.052 | 3.123 | 3.192 | 3.261 | 3.330 |
| **188** | h | 349.9 | 1197.5 | 1199.7 | 1206.2 | 1212.6 | 1224.8 | 1236.6 | 1248.1 | 1259.2 | 1270.2 | 1281.0 | 1291.6 | 1302.1 | 1312.6 | 1323.0 | 1333.3 |
| (376.64) | s | 0.5370 | 1.5506 | 1.5532 | 1.5609 | 1.5683 | 1.5824 | 1.5957 | 1.6083 | 1.6203 | 1.6318 | 1.6429 | 1.6537 | 1.6641 | 1.6743 | 1.6842 | 1.6938 |
| | v | 0.018 | 2.404 | 2.414 | 2.456 | 2.498 | 2.577 | 2.655 | 2.730 | 2.804 | 2.877 | 2.948 | 3.019 | 3.089 | 3.157 | 3.226 | 3.294 |
| **190** | h | 350.8 | 1197.6 | 1199.3 | 1205.8 | 1212.2 | 1224.5 | 1236.3 | 1247.8 | 1259.0 | 1270.0 | 1280.8 | 1291.4 | 1302.0 | 1312.4 | 1322.8 | 1333.2 |
| (377.51) | s | 0.5381 | 1.5497 | 1.5517 | 1.5594 | 1.5668 | 1.5810 | 1.5943 | 1.6069 | 1.6189 | 1.6305 | 1.6416 | 1.6524 | 1.6628 | 1.6730 | 1.6829 | 1.6925 |
| | v | 0.018 | 2.380 | 2.387 | 2.428 | 2.469 | 2.548 | 2.625 | 2.700 | 2.773 | 2.845 | 2.916 | 2.986 | 3.055 | 3.123 | 3.191 | 3.258 |
| **192** | h | 351.7 | 1197.8 | 1198.9 | 1205.4 | 1211.8 | 1224.2 | 1236.0 | 1247.5 | 1258.8 | 1269.7 | 1280.6 | 1291.2 | 1301.8 | 1312.3 | 1322.7 | 1333.0 |
| (378.38) | s | 0.5392 | 1.5488 | 1.5501 | 1.5578 | 1.5653 | 1.5795 | 1.5929 | 1.6055 | 1.6176 | 1.6291 | 1.6403 | 1.6511 | 1.6615 | 1.6717 | 1.6816 | 1.6913 |
| | v | 0.018 | 2.356 | 2.359 | 2.401 | 2.441 | 2.520 | 2.596 | 2.671 | 2.743 | 2.815 | 2.885 | 2.954 | 3.022 | 3.090 | 3.157 | 3.224 |
| **194** | h | 352.6 | 1197.9 | 1198.4 | 1205.0 | 1211.5 | 1223.8 | 1235.8 | 1247.3 | 1258.5 | 1269.5 | 1280.4 | 1291.1 | 1301.6 | 1312.1 | 1322.5 | 1332.9 |
| (379.24) | s | 0.5403 | 1.5479 | 1.5485 | 1.5563 | 1.5638 | 1.5781 | 1.5915 | 1.6041 | 1.6162 | 1.6278 | 1.6390 | 1.6498 | 1.6603 | 1.6705 | 1.6804 | 1.6901 |
| | v | 0.018 | 2.333 | | 2.374 | 2.414 | 2.492 | 2.568 | 2.642 | 2.714 | 2.784 | 2.854 | 2.923 | 2.990 | 3.058 | 3.124 | 3.190 |
| **196** | h | 353.6 | 1198.1 | | 1204.6 | 1211.1 | 1223.5 | 1235.5 | 1247.0 | 1258.3 | 1269.3 | 1280.2 | 1290.9 | 1301.5 | 1312.0 | 1322.4 | 1332.8 |
| (380.10) | s | 0.5414 | 1.5470 | | 1.5548 | 1.5623 | 1.5766 | 1.5901 | 1.6028 | 1.6149 | 1.6265 | 1.6377 | 1.6485 | 1.6590 | 1.6692 | 1.6791 | 1.6888 |
| | v | 0.018 | 2.310 | | 2.347 | 2.387 | 2.465 | 2.540 | 2.613 | 2.685 | 2.755 | 2.824 | 2.892 | 2.959 | 3.026 | 3.091 | 3.157 |
| **198** | h | 354.5 | 1198.2 | | 1204.2 | 1210.7 | 1223.2 | 1235.2 | 1246.8 | 1258.1 | 1269.1 | 1280.0 | 1290.7 | 1301.3 | 1311.8 | 1322.3 | 1332.6 |
| (380.95) | s | 0.5425 | 1.5462 | | 1.5533 | 1.5609 | 1.5752 | 1.5887 | 1.6014 | 1.6136 | 1.6252 | 1.6365 | 1.6473 | 1.6578 | 1.6680 | 1.6779 | 1.6876 |
| | v | 0.018 | 2.288 | | 2.321 | 2.361 | 2.438 | 2.513 | 2.585 | 2.656 | 2.726 | 2.794 | 2.862 | 2.928 | 2.994 | 3.060 | 3.124 |
| **200** | h | 355.4 | 1198.4 | | 1203.8 | 1210.3 | 1222.9 | 1234.9 | 1246.5 | 1257.8 | 1268.9 | 1279.8 | 1290.5 | 1301.1 | 1311.7 | 1322.1 | 1332.5 |
| (381.79) | s | 0.5435 | 1.5453 | | 1.5518 | 1.5594 | 1.5738 | 1.5873 | 1.6001 | 1.6123 | 1.6240 | 1.6352 | 1.6460 | 1.6566 | 1.6668 | 1.6767 | 1.6864 |
| | v | 0.018 | 2.266 | | 2.296 | 2.335 | 2.412 | 2.486 | 2.558 | 2.628 | 2.697 | 2.765 | 2.832 | 2.898 | 2.963 | 3.028 | 3.092 |
| **202** | h | 356.2 | 1198.5 | | 1203.4 | 1210.0 | 1222.5 | 1234.6 | 1246.2 | 1257.6 | 1268.7 | 1279.6 | 1290.4 | 1301.0 | 1311.5 | 1322.0 | 1332.4 |
| (382.62) | s | 0.5446 | 1.5445 | | 1.5503 | 1.5580 | 1.5724 | 1.5860 | 1.5988 | 1.6110 | 1.6227 | 1.6339 | 1.6448 | 1.6553 | 1.6656 | 1.6755 | 1.6852 |
| | v | 0.018 | 2.245 | | 2.271 | 2.310 | 2.386 | 2.460 | 2.531 | 2.601 | 2.670 | 2.737 | 2.803 | 2.868 | 2.933 | 2.998 | 3.061 |
| **204** | h | 357.1 | 1198.6 | | 1203.0 | 1209.6 | 1222.2 | 1234.3 | 1246.0 | 1257.3 | 1268.5 | 1279.4 | 1290.2 | 1300.8 | 1311.3 | 1321.8 | 1332.2 |
| (383.45) | s | 0.5456 | 1.5436 | | 1.5488 | 1.5565 | 1.5710 | 1.5846 | 1.5975 | 1.6097 | 1.6214 | 1.6327 | 1.6436 | 1.6541 | 1.6644 | 1.6743 | 1.6840 |
| | v | 0.018 | 2.224 | | 2.247 | 2.286 | 2.361 | 2.434 | 2.505 | 2.574 | 2.642 | 2.709 | 2.775 | 2.839 | 2.904 | 2.967 | 3.031 |
| **206** | h | 358.0 | 1198.7 | | 1202.6 | 1209.2 | 1221.9 | 1234.0 | 1245.7 | 1257.1 | 1268.3 | 1279.2 | 1290.0 | 1300.6 | 1311.2 | 1321.7 | 1332.1 |
| (384.27) | s | 0.5467 | 1.5428 | | 1.5474 | 1.5551 | 1.5697 | 1.5833 | 1.5962 | 1.6084 | 1.6202 | 1.6315 | 1.6424 | 1.6529 | 1.6632 | 1.6732 | 1.6829 |
| | v | 0.018 | 2.203 | | 2.223 | 2.261 | 2.336 | 2.409 | 2.479 | 2.548 | 2.615 | 2.682 | 2.747 | 2.811 | 2.875 | 2.938 | 3.001 |
| **208** | h | 358.9 | 1198.9 | | 1202.2 | 1208.8 | 1221.5 | 1233.7 | 1245.4 | 1256.9 | 1268.0 | 1279.0 | 1289.8 | 1300.5 | 1311.0 | 1321.5 | 1331.9 |
| (385.09) | s | 0.5477 | 1.5420 | | 1.5459 | 1.5537 | 1.5683 | 1.5820 | 1.5949 | 1.6072 | 1.6189 | 1.6303 | 1.6412 | 1.6517 | 1.6620 | 1.6720 | 1.6817 |
| | v | 0.018 | 2.183 | | 2.199 | 2.238 | 2.312 | 2.384 | 2.454 | 2.522 | 2.589 | 2.655 | 2.720 | 2.783 | 2.847 | 2.909 | 2.971 |
| **210** | h | 359.8 | 1199.0 | | 1201.8 | 1208.4 | 1221.2 | 1233.4 | 1245.2 | 1256.6 | 1267.8 | 1278.8 | 1289.6 | 1300.3 | 1310.9 | 1321.4 | 1331.8 |
| (385.90) | s | 0.5487 | 1.5412 | | 1.5445 | 1.5523 | 1.5670 | 1.5807 | 1.5936 | 1.6059 | 1.6177 | 1.6291 | 1.6400 | 1.6506 | 1.6608 | 1.6709 | 1.6806 |
| | v | 0.018 | 2.163 | | 2.176 | 2.214 | 2.288 | 2.360 | 2.429 | 2.497 | 2.563 | 2.628 | 2.693 | 2.756 | 2.819 | 2.881 | 2.942 |
| **212** | h | 360.6 | 1199.1 | | 1201.4 | 1208.0 | 1220.9 | 1233.1 | 1244.9 | 1256.4 | 1267.6 | 1278.6 | 1289.4 | 1300.1 | 1310.7 | 1321.2 | 1331.7 |
| (386.70) | s | 0.5497 | 1.5404 | | 1.5430 | 1.5508 | 1.5656 | 1.5793 | 1.5923 | 1.6047 | 1.6165 | 1.6279 | 1.6388 | 1.6494 | 1.6597 | 1.6697 | 1.6794 |

Temperature—Degrees Fahrenheit

| Abs. Press. Lb./Sq. In. (Sat. Temp.) | | 640° | 660° | 680° | 700° | 750° | 800° | 850° | 900° | 950° | 1000° | 1050° | 1100° | 1200° | 1400° | 1600° |
|---|---|---|---|---|---|---|---|---|---|---|---|---|---|---|---|---|
| **180** | v | 3.553 | 3.623 | 3.694 | 3.764 | 3.937 | 4.110 | 4.281 | 4.452 | 4.622 | 4.792 | 4.961 | 5.129 | 5.466 | 6.136 | 6.804 |
| (373.06) | h | 1344.1 | 1354.3 | 1364.5 | 1374.7 | 1400.2 | 1425.6 | 1451.2 | 1476.8 | 1502.6 | 1528.6 | 1554.7 | 1581.0 | 1634.1 | 1742.9 | 1855.2 |
| | s | 1.7084 | 1.7176 | 1.7266 | 1.7355 | 1.7570 | 1.7776 | 1.7975 | 1.8167 | 1.8354 | 1.8534 | 1.8710 | 1.8882 | 1.9212 | 1.9831 | 2.0404 |
| **182** | v | 3.513 | 3.583 | 3.653 | 3.722 | 3.893 | 4.064 | 4.234 | 4.403 | 4.571 | 4.739 | 4.906 | 5.073 | 5.406 | 6.069 | 6.729 |
| (373.96) | h | 1344.0 | 1354.2 | 1364.4 | 1374.6 | 1400.1 | 1425.5 | 1451.1 | 1476.8 | 1502.6 | 1528.5 | 1554.6 | 1580.9 | 1634.1 | 1742.9 | 1855.2 |
| | s | 1.7071 | 1.7163 | 1.7253 | 1.7342 | 1.7557 | 1.7763 | 1.7962 | 1.8155 | 1.8341 | 1.8522 | 1.8698 | 1.8869 | 1.9200 | 1.9818 | 2.0392 |
| **184** | v | 3.474 | 3.543 | 3.612 | 3.680 | 3.850 | 4.019 | 4.187 | 4.354 | 4.521 | 4.687 | 4.852 | 5.017 | 5.346 | 6.002 | 6.656 |
| (374.86) | h | 1343.8 | 1354.1 | 1364.3 | 1374.5 | 1400.0 | 1425.5 | 1451.0 | 1476.7 | 1502.5 | 1528.5 | 1554.6 | 1580.9 | 1634.0 | 1742.9 | 1855.2 |
| | s | 1.7058 | 1.7150 | 1.7241 | 1.7329 | 1.7544 | 1.7751 | 1.7950 | 1.8142 | 1.8329 | 1.8510 | 1.8686 | 1.8857 | 1.9187 | 1.9806 | 2.0380 |
| **186** | v | 3.436 | 3.504 | 3.572 | 3.640 | 3.808 | 3.976 | 4.142 | 4.307 | 4.472 | 4.636 | 4.800 | 4.963 | 5.289 | 5.938 | 6.584 |
| (375.75) | h | 1343.7 | 1354.0 | 1364.2 | 1374.4 | 1399.9 | 1425.4 | 1451.0 | 1476.6 | 1502.4 | 1528.4 | 1554.5 | 1580.8 | 1634.0 | 1742.8 | 1855.2 |
| | s | 1.7045 | 1.7137 | 1.7228 | 1.7317 | 1.7532 | 1.7739 | 1.7938 | 1.8130 | 1.8317 | 1.8497 | 1.8673 | 1.8845 | 1.9175 | 1.9794 | 2.0368 |
| **188** | v | 3.398 | 3.466 | 3.533 | 3.600 | 3.767 | 3.933 | 4.097 | 4.261 | 4.424 | 4.586 | 4.748 | 4.910 | 5.232 | 5.874 | 6.514 |
| (376.64) | h | 1343.6 | 1353.8 | 1364.1 | 1374.3 | 1399.8 | 1425.3 | 1450.9 | 1476.6 | 1502.4 | 1528.3 | 1554.5 | 1580.8 | 1634.0 | 1742.8 | 1855.1 |
| | s | 1.7033 | 1.7125 | 1.7215 | 1.7304 | 1.7520 | 1.7726 | 1.7925 | 1.8118 | 1.8304 | 1.8485 | 1.8661 | 1.8833 | 1.9163 | 1.9782 | 2.0356 |
| **190** | v | 3.361 | 3.429 | 3.495 | 3.562 | 3.727 | 3.891 | 4.053 | 4.215 | 4.376 | 4.537 | 4.698 | 4.858 | 5.177 | 5.812 | 6.446 |
| (377.51) | h | 1343.5 | 1353.7 | 1363.9 | 1374.2 | 1399.7 | 1425.2 | 1450.8 | 1476.5 | 1502.3 | 1528.3 | 1554.4 | 1580.7 | 1633.9 | 1742.8 | 1855.1 |
| | s | 1.7020 | 1.7112 | 1.7203 | 1.7292 | 1.7507 | 1.7714 | 1.7913 | 1.8106 | 1.8292 | 1.8473 | 1.8649 | 1.8821 | 1.9151 | 1.9770 | 2.0344 |
| **192** | v | 3.325 | 3.392 | 3.458 | 3.524 | 3.687 | 3.849 | 4.010 | 4.171 | 4.330 | 4.490 | 4.649 | 4.807 | 5.123 | 5.752 | 6.378 |
| (378.38) | h | 1343.3 | 1353.6 | 1363.8 | 1374.1 | 1399.6 | 1425.1 | 1450.7 | 1476.4 | 1502.3 | 1528.2 | 1554.4 | 1580.7 | 1633.9 | 1742.7 | 1855.1 |
| | s | 1.7008 | 1.7100 | 1.7191 | 1.7280 | 1.7495 | 1.7702 | 1.7901 | 1.8094 | 1.8281 | 1.8462 | 1.8638 | 1.8809 | 1.9140 | 1.9759 | 2.0332 |
| **194** | v | 3.290 | 3.356 | 3.422 | 3.487 | 3.648 | 3.809 | 3.968 | 4.127 | 4.285 | 4.443 | 4.600 | 4.757 | 5.069 | 5.692 | 6.312 |
| (379.24) | h | 1343.2 | 1353.5 | 1363.7 | 1374.0 | 1399.5 | 1425.1 | 1450.7 | 1476.4 | 1502.2 | 1528.2 | 1554.3 | 1580.6 | 1633.8 | 1742.7 | 1855.1 |
| | s | 1.6995 | 1.7088 | 1.7179 | 1.7268 | 1.7483 | 1.7690 | 1.7889 | 1.8082 | 1.8269 | 1.8450 | 1.8626 | 1.8797 | 1.9128 | 1.9747 | 2.0321 |
| **196** | v | 3.255 | 3.321 | 3.386 | 3.450 | 3.610 | 3.770 | 3.927 | 4.085 | 4.241 | 4.397 | 4.553 | 4.708 | 5.017 | 5.634 | 6.248 |
| (380.10) | h | 1343.1 | 1353.4 | 1363.6 | 1373.9 | 1399.4 | 1425.0 | 1450.6 | 1476.3 | 1502.1 | 1528.1 | 1554.3 | 1580.6 | 1633.8 | 1742.7 | 1855.0 |
| | s | 1.6983 | 1.7076 | 1.7167 | 1.7256 | 1.7471 | 1.7678 | 1.7878 | 1.8071 | 1.8257 | 1.8438 | 1.8614 | 1.8786 | 1.9117 | 1.9736 | 2.0310 |
| **198** | v | 3.222 | 3.286 | 3.351 | 3.415 | 3.573 | 3.731 | 3.887 | 4.043 | 4.198 | 4.352 | 4.506 | 4.660 | 4.967 | 5.577 | 6.185 |
| (380.95) | h | 1342.9 | 1353.2 | 1363.5 | 1373.7 | 1399.3 | 1424.9 | 1450.5 | 1476.2 | 1502.1 | 1528.1 | 1554.2 | 1580.5 | 1633.8 | 1742.6 | 1855.0 |
| | s | 1.6971 | 1.7064 | 1.7155 | 1.7244 | 1.7460 | 1.7667 | 1.7866 | 1.8059 | 1.8246 | 1.8427 | 1.8603 | 1.8775 | 1.9105 | 1.9724 | 2.0298 |
| **200** | v | 3.189 | 3.253 | 3.316 | 3.380 | 3.537 | 3.693 | 3.848 | 4.002 | 4.156 | 4.309 | 4.461 | 4.613 | 4.917 | 5.521 | 6.123 |
| (381.79) | h | 1342.8 | 1353.1 | 1363.4 | 1373.6 | 1399.2 | 1424.8 | 1450.4 | 1476.2 | 1502.0 | 1528.0 | 1554.2 | 1580.5 | 1633.7 | 1742.6 | 1855.0 |
| | s | 1.6959 | 1.7052 | 1.7143 | 1.7232 | 1.7448 | 1.7655 | 1.7855 | 1.8048 | 1.8234 | 1.8415 | 1.8592 | 1.8763 | 1.9094 | 1.9713 | 2.0287 |
| **202** | v | 3.156 | 3.220 | 3.283 | 3.346 | 3.501 | 3.656 | 3.809 | 3.962 | 4.114 | 4.266 | 4.417 | 4.567 | 4.868 | 5.466 | 6.062 |
| (382.62) | h | 1342.7 | 1353.0 | 1363.3 | 1373.5 | 1399.1 | 1424.7 | 1450.4 | 1476.1 | 1502.0 | 1528.0 | 1554.1 | 1580.5 | 1633.7 | 1742.6 | 1855.0 |
| | s | 1.6947 | 1.7040 | 1.7131 | 1.7220 | 1.7436 | 1.7644 | 1.7843 | 1.8036 | 1.8223 | 1.8404 | 1.8580 | 1.8752 | 1.9083 | 1.9702 | 2.0276 |
| **204** | v | 3.124 | 3.187 | 3.250 | 3.312 | 3.466 | 3.619 | 3.771 | 3.923 | 4.073 | 4.223 | 4.373 | 4.522 | 4.820 | 5.412 | 6.002 |
| (383.45) | h | 1342.6 | 1352.9 | 1363.2 | 1373.4 | 1399.0 | 1424.6 | 1450.3 | 1476.0 | 1501.9 | 1527.9 | 1554.1 | 1580.4 | 1633.6 | 1742.5 | 1854.9 |
| | s | 1.6935 | 1.7028 | 1.7120 | 1.7209 | 1.7425 | 1.7632 | 1.7832 | 1.8025 | 1.8212 | 1.8393 | 1.8569 | 1.8741 | 1.9072 | 1.9691 | 2.0265 |
| **206** | v | 3.093 | 3.155 | 3.218 | 3.279 | 3.432 | 3.584 | 3.734 | 3.884 | 4.033 | 4.182 | 4.330 | 4.478 | 4.773 | 5.359 | 5.944 |
| (384.27) | h | 1342.4 | 1352.8 | 1363.1 | 1373.3 | 1399.0 | 1424.6 | 1450.2 | 1476.0 | 1501.8 | 1527.8 | 1554.0 | 1580.4 | 1633.6 | 1742.5 | 1854.9 |
| | s | 1.6924 | 1.7017 | 1.7108 | 1.7197 | 1.7414 | 1.7621 | 1.7821 | 1.8014 | 1.8201 | 1.8382 | 1.8558 | 1.8730 | 1.9061 | 1.9680 | 2.0254 |
| **208** | v | 3.063 | 3.124 | 3.186 | 3.247 | 3.398 | 3.549 | 3.698 | 3.846 | 3.994 | 4.141 | 4.288 | 4.435 | 4.726 | 5.308 | 5.887 |
| (385.09) | h | 1342.3 | 1352.6 | 1362.9 | 1373.2 | 1398.9 | 1424.5 | 1450.2 | 1475.9 | 1501.8 | 1527.8 | 1554.0 | 1580.3 | 1633.6 | 1742.5 | 1854.9 |
| | s | 1.6912 | 1.7005 | 1.7097 | 1.7186 | 1.7402 | 1.7610 | 1.7810 | 1.8003 | 1.8190 | 1.8371 | 1.8547 | 1.8719 | 1.9050 | 1.9669 | 2.0243 |
| **210** | v | 3.033 | 3.094 | 3.155 | 3.215 | 3.365 | 3.514 | 3.662 | 3.809 | 3.956 | 4.102 | 4.247 | 4.392 | 4.681 | 5.257 | 5.830 |
| (385.90) | h | 1342.2 | 1352.5 | 1362.8 | 1373.1 | 1398.8 | 1424.4 | 1450.1 | 1475.8 | 1501.7 | 1527.7 | 1553.9 | 1580.3 | 1633.5 | 1742.5 | 1854.9 |
| | s | 1.6901 | 1.6994 | 1.7085 | 1.7175 | 1.7391 | 1.7599 | 1.7799 | 1.7992 | 1.8179 | 1.8360 | 1.8537 | 1.8708 | 1.9039 | 1.9659 | 2.0233 |
| **212** | v | 3.003 | 3.064 | 3.124 | 3.184 | 3.333 | 3.481 | 3.627 | 3.773 | 3.918 | 4.062 | 4.207 | 4.350 | 4.637 | 5.207 | 5.775 |
| (386.70) | h | 1342.1 | 1352.4 | 1362.7 | 1373.0 | 1398.7 | 1424.3 | 1450.0 | 1475.8 | 1501.7 | 1527.7 | 1553.9 | 1580.2 | 1633.5 | 1742.4 | 1854.8 |
| | s | 1.6890 | 1.6983 | 1.7074 | 1.7164 | 1.7380 | 1.7588 | 1.7788 | 1.7981 | 1.8168 | 1.8350 | 1.8526 | 1.8698 | 1.9029 | 1.9648 | 2.0222 |

# Table 3. Superheated Vapor

| Abs. Press. Lb./Sq. In. (Sat. Temp.) | | Sat. Liquid | Sat. Vapor | Temperature—Degrees Fahrenheit 390° | 400° | 420° | 440° | 460° | 480° | 500° | 520° | 540° | 560° | 580° | 600° | 620° | 640° |
|---|---|---|---|---|---|---|---|---|---|---|---|---|---|---|---|---|---|
| | v | 0.018 | 2.144 | 2.153 | 2.191 | 2.265 | 2.336 | 2.405 | 2.472 | 2.538 | 2.603 | 2.666 | 2.729 | 2.791 | 2.853 | 2.914 | 2.974 |
| **214** | h | 361.5 | 1199.2 | 1200.9 | 1207.6 | 1220.5 | 1232.8 | 1244.6 | 1256.1 | 1267.4 | 1278.4 | 1289.3 | 1300.0 | 1310.6 | 1321.1 | 1331.5 | 1341.9 |
| (387.50) | s | 0.5507 | 1.5396 | 1.5416 | 1.5494 | 1.5642 | 1.5780 | 1.5911 | 1.6034 | 1.6153 | 1.6267 | 1.6376 | 1.6482 | 1.6585 | 1.6685 | 1.6783 | 1.6878 |
| | v | 0.018 | 2.125 | 2.131 | 2.169 | 2.242 | 2.313 | 2.381 | 2.448 | 2.513 | 2.577 | 2.640 | 2.703 | 2.764 | 2.825 | 2.886 | 2.946 |
| **216** | h | 362.3 | 1199.4 | 1200.5 | 1207.2 | 1220.2 | 1232.5 | 1244.4 | 1255.9 | 1267.2 | 1278.2 | 1289.1 | 1299.8 | 1310.4 | 1321.0 | 1331.4 | 1341.8 |
| (388.29) | s | 0.5517 | 1.5388 | 1.5402 | 1.5480 | 1.5629 | 1.5768 | 1.5898 | 1.6022 | 1.6141 | 1.6255 | 1.6365 | 1.6471 | 1.6574 | 1.6674 | 1.6772 | 1.6867 |
| | v | 0.018 | 2.106 | 2.109 | 2.147 | 2.220 | 2.290 | 2.358 | 2.424 | 2.489 | 2.552 | 2.615 | 2.677 | 2.738 | 2.798 | 2.858 | 2.918 |
| **218** | h | 363.2 | 1199.5 | 1200.1 | 1206.9 | 1219.8 | 1232.2 | 1244.1 | 1255.7 | 1266.9 | 1278.0 | 1288.9 | 1299.6 | 1310.3 | 1320.8 | 1331.3 | 1341.7 |
| (389.08) | s | 0.5527 | 1.5380 | 1.5388 | 1.5467 | 1.5616 | 1.5755 | 1.5886 | 1.6010 | 1.6129 | 1.6243 | 1.6353 | 1.6459 | 1.6562 | 1.6663 | 1.6761 | 1.6856 |
| | v | 0.019 | 2.087 | 2.087 | 2.125 | 2.198 | 2.267 | 2.335 | 2.400 | 2.465 | 2.528 | 2.590 | 2.652 | 2.712 | 2.772 | 2.831 | 2.891 |
| **220** | h | 364.0 | 1199.6 | 1199.7 | 1206.5 | 1219.5 | 1231.9 | 1243.8 | 1255.4 | 1266.7 | 1277.8 | 1288.7 | 1299.5 | 1310.1 | 1320.7 | 1331.1 | 1341.6 |
| (389.86) | s | 0.5537 | 1.5372 | 1.5373 | 1.5453 | 1.5603 | 1.5742 | 1.5874 | 1.5998 | 1.6117 | 1.6231 | 1.6342 | 1.6448 | 1.6551 | 1.6652 | 1.6750 | 1.6845 |
| | v | 0.019 | 2.069 | | 2.104 | 2.176 | 2.245 | 2.312 | 2.377 | 2.441 | 2.504 | 2.566 | 2.627 | 2.687 | 2.746 | 2.805 | 2.864 |
| **222** | h | 364.8 | 1199.7 | | 1206.1 | 1219.1 | 1231.6 | 1243.6 | 1255.2 | 1266.5 | 1277.6 | 1288.5 | 1299.3 | 1310.0 | 1320.5 | 1331.0 | 1341.4 |
| (390.63) | s | 0.5547 | 1.5365 | | 1.5439 | 1.5590 | 1.5729 | 1.5861 | 1.5986 | 1.6105 | 1.6220 | 1.6330 | 1.6437 | 1.6540 | 1.6641 | 1.6739 | 1.6835 |
| | v | 0.019 | 2.051 | | 2.083 | 2.154 | 2.223 | 2.290 | 2.355 | 2.418 | 2.480 | 2.542 | 2.602 | 2.662 | 2.721 | 2.779 | 2.838 |
| **224** | h | 365.7 | 1199.8 | | 1205.7 | 1218.8 | 1231.3 | 1243.3 | 1254.9 | 1266.3 | 1277.4 | 1288.3 | 1299.1 | 1309.8 | 1320.4 | 1330.9 | 1341.3 |
| (391.40) | s | 0.5556 | 1.5357 | | 1.5426 | 1.5577 | 1.5717 | 1.5849 | 1.5974 | 1.6094 | 1.6208 | 1.6319 | 1.6426 | 1.6529 | 1.6630 | 1.6728 | 1.6824 |
| | v | 0.019 | 2.033 | | 2.062 | 2.133 | 2.202 | 2.268 | 2.333 | 2.396 | 2.457 | 2.518 | 2.578 | 2.637 | 2.696 | 2.754 | 2.812 |
| **226** | h | 366.5 | 1199.9 | | 1205.3 | 1218.5 | 1231.0 | 1243.0 | 1254.7 | 1266.1 | 1277.2 | 1288.2 | 1299.0 | 1309.6 | 1320.2 | 1330.7 | 1341.2 |
| (392.17) | s | 0.5566 | 1.5349 | | 1.5412 | 1.5564 | 1.5704 | 1.5837 | 1.5962 | 1.6082 | 1.6197 | 1.6308 | 1.6415 | 1.6518 | 1.6619 | 1.6717 | 1.6813 |
| | v | 0.019 | 2.016 | | 2.042 | 2.113 | 2.181 | 2.247 | 2.311 | 2.373 | 2.434 | 2.495 | 2.554 | 2.613 | 2.671 | 2.729 | 2.786 |
| **228** | h | 367.3 | 1200.0 | | 1204.9 | 1218.1 | 1230.7 | 1242.8 | 1254.5 | 1265.9 | 1277.0 | 1288.0 | 1298.8 | 1309.5 | 1320.1 | 1330.6 | 1341.0 |
| (392.93) | s | 0.5575 | 1.5342 | | 1.5399 | 1.5551 | 1.5692 | 1.5825 | 1.5951 | 1.6071 | 1.6186 | 1.6297 | 1.6404 | 1.6507 | 1.6608 | 1.6706 | 1.6802 |
| | v | 0.0185 | 1.9992 | | 2.022 | 2.093 | 2.160 | 2.226 | 2.289 | 2.351 | 2.412 | 2.472 | 2.531 | 2.589 | 2.647 | 2.704 | 2.761 |
| **230** | h | 368.1 | 1200.1 | | 1204.5 | 1217.8 | 1230.4 | 1242.5 | 1254.2 | 1265.6 | 1276.8 | 1287.8 | 1298.6 | 1309.3 | 1319.9 | 1330.5 | 1340.9 |
| (393.68) | s | 0.5585 | 1.5334 | | 1.5385 | 1.5538 | 1.5680 | 1.5813 | 1.5939 | 1.6059 | 1.6174 | 1.6286 | 1.6393 | 1.6497 | 1.6598 | 1.6696 | 1.6792 |
| | v | 0.0186 | 1.9825 | | 2.003 | 2.073 | 2.140 | 2.205 | 2.268 | 2.330 | 2.390 | 2.450 | 2.508 | 2.566 | 2.623 | 2.680 | 2.736 |
| **232** | h | 368.9 | 1200.2 | | 1204.1 | 1217.4 | 1230.1 | 1242.2 | 1254.0 | 1265.4 | 1276.6 | 1287.6 | 1298.5 | 1309.2 | 1319.8 | 1330.3 | 1340.8 |
| (394.43) | s | 0.5594 | 1.5327 | | 1.5372 | 1.5525 | 1.5667 | 1.5801 | 1.5927 | 1.6048 | 1.6163 | 1.6275 | 1.6382 | 1.6486 | 1.6587 | 1.6685 | 1.6782 |
| | v | 0.0186 | 1.9661 | | 1.9834 | 2.053 | 2.120 | 2.185 | 2.247 | 2.309 | 2.369 | 2.428 | 2.486 | 2.543 | 2.600 | 2.656 | 2.712 |
| **234** | h | 369.7 | 1200.3 | | 1203.7 | 1217.1 | 1229.8 | 1241.9 | 1253.7 | 1265.2 | 1276.4 | 1287.4 | 1298.3 | 1309.0 | 1319.6 | 1330.2 | 1340.7 |
| (395.17) | s | 0.5603 | 1.5319 | | 1.5359 | 1.5512 | 1.5655 | 1.5789 | 1.5916 | 1.6037 | 1.6152 | 1.6264 | 1.6371 | 1.6475 | 1.6577 | 1.6675 | 1.6771 |
| | v | 0.0186 | 1.9499 | | 1.9645 | 2.034 | 2.100 | 2.165 | 2.227 | 2.288 | 2.348 | 2.406 | 2.464 | 2.521 | 2.577 | 2.633 | 2.688 |
| **236** | h | 370.5 | 1200.4 | | 1203.3 | 1216.7 | 1229.5 | 1241.7 | 1253.5 | 1265.0 | 1276.2 | 1287.2 | 1298.1 | 1308.9 | 1319.5 | 1330.0 | 1340.5 |
| (395.91) | s | 0.5613 | 1.5312 | | 1.5345 | 1.5500 | 1.5643 | 1.5778 | 1.5904 | 1.6025 | 1.6141 | 1.6253 | 1.6360 | 1.6465 | 1.6566 | 1.6665 | 1.6761 |
| | v | 0.0186 | 1.9340 | | 1.9459 | 2.015 | 2.081 | 2.145 | 2.207 | 2.267 | 2.327 | 2.385 | 2.442 | 2.499 | 2.555 | 2.610 | 2.665 |
| **238** | h | 371.3 | 1200.5 | | 1202.9 | 1216.4 | 1229.1 | 1241.4 | 1253.2 | 1264.7 | 1276.0 | 1287.1 | 1297.9 | 1308.7 | 1319.3 | 1329.9 | 1340.4 |
| (396.64) | s | 0.5622 | 1.5305 | | 1.5332 | 1.5487 | 1.5631 | 1.5766 | 1.5893 | 1.6014 | 1.6130 | 1.6242 | 1.6350 | 1.6454 | 1.6556 | 1.6654 | 1.6751 |
| | v | 0.0186 | 1.9183 | | 1.9276 | 1.9964 | 2.062 | 2.126 | 2.187 | 2.247 | 2.306 | 2.364 | 2.421 | 2.477 | 2.533 | 2.587 | 2.642 |
| **240** | h | 372.1 | 1200.6 | | 1202.5 | 1216.0 | 1228.8 | 1241.1 | 1253.0 | 1264.5 | 1275.8 | 1286.9 | 1297.8 | 1308.5 | 1319.2 | 1329.8 | 1340.3 |
| (397.37) | s | 0.5631 | 1.5298 | | 1.5319 | 1.5475 | 1.5619 | 1.5754 | 1.5882 | 1.6003 | 1.6120 | 1.6232 | 1.6339 | 1.6444 | 1.6546 | 1.6644 | 1.6741 |
| | v | 0.0186 | 1.9030 | | 1.9096 | 1.9780 | 2.043 | 2.107 | 2.168 | 2.227 | 2.286 | 2.343 | 2.400 | 2.456 | 2.511 | 2.565 | 2.620 |
| **242** | h | 372.9 | 1200.7 | | 1202.1 | 1215.7 | 1228.5 | 1240.8 | 1252.7 | 1264.3 | 1275.6 | 1286.7 | 1297.6 | 1308.4 | 1319.0 | 1329.6 | 1340.1 |
| (398.10) | s | 0.5640 | 1.5291 | | 1.5306 | 1.5463 | 1.5607 | 1.5743 | 1.5871 | 1.5992 | 1.6109 | 1.6221 | 1.6329 | 1.6434 | 1.6535 | 1.6634 | 1.6731 |
| | v | 0.0186 | 1.8879 | | 1.8919 | 1.9600 | 2.025 | 2.088 | 2.149 | 2.208 | 2.266 | 2.323 | 2.379 | 2.435 | 2.489 | 2.543 | 2.598 |
| **244** | h | 373.7 | 1200.8 | | 1201.6 | 1215.3 | 1228.2 | 1240.6 | 1252.5 | 1264.1 | 1275.4 | 1286.5 | 1297.4 | 1308.2 | 1318.9 | 1329.5 | 1340.0 |
| (398.82) | s | 0.5649 | 1.5284 | | 1.5293 | 1.5450 | 1.5595 | 1.5731 | 1.5859 | 1.5982 | 1.6098 | 1.6210 | 1.6319 | 1.6423 | 1.6525 | 1.6624 | 1.6721 |
| | v | 0.0186 | 1.8730 | | 1.8745 | 1.9423 | 2.007 | 2.069 | 2.130 | 2.189 | 2.247 | 2.303 | 2.359 | 2.414 | 2.468 | 2.522 | 2.576 |
| **246** | h | 374.5 | 1200.9 | | 1201.2 | 1214.9 | 1227.9 | 1240.3 | 1252.2 | 1263.8 | 1275.2 | 1286.3 | 1297.3 | 1308.1 | 1318.8 | 1329.4 | 1339.9 |
| (399.53) | s | 0.5658 | 1.5277 | | 1.5280 | 1.5438 | 1.5584 | 1.5720 | 1.5848 | 1.5971 | 1.6088 | 1.6200 | 1.6308 | 1.6413 | 1.6515 | 1.6614 | 1.6711 |
| | v | 0.0186 | 1.8583 | | | 1.9249 | 1.9892 | 2.051 | 2.111 | 2.170 | 2.227 | 2.283 | 2.339 | 2.394 | 2.448 | 2.501 | 2.554 |
| **248** | h | 375.2 | 1201.0 | | | 1214.6 | 1227.6 | 1240.0 | 1252.0 | 1263.6 | 1275.0 | 1286.1 | 1297.1 | 1307.9 | 1318.6 | 1329.2 | 1339.8 |
| (400.24) | s | 0.5667 | 1.5270 | | | 1.5426 | 1.5572 | 1.5709 | 1.5837 | 1.5960 | 1.6077 | 1.6190 | 1.6298 | 1.6403 | 1.6505 | 1.6604 | 1.6701 |

Temperature—Degrees Fahrenheit

| 660° | 680° | 700° | 720° | 740° | 760° | 780° | 800° | 850° | 900° | 950° | 1000° | 1100° | 1200° | 1400° | 1600° | | Abs. Press. Lb./Sq. In. (Sat. Temp.) |
|---|---|---|---|---|---|---|---|---|---|---|---|---|---|---|---|---|---|
| 3.034 | 3.094 | 3.154 | 3.213 | 3.272 | 3.331 | 3.389 | 3.448 | 3.593 | 3.737 | 3.881 | 4.024 | 4.309 | 4.593 | 5.158 | 5.721 | v | |
| 1352.3 | 1362.6 | 1372.9 | 1383.2 | 1393.4 | 1403.7 | 1414.0 | 1424.2 | 1449.9 | 1475.7 | 1501.6 | 1527.6 | 1580.2 | 1633.4 | 1742.4 | 1854.8 | h | **214** |
| 1.6972 | 1.7063 | 1.7153 | 1.7241 | 1.7327 | 1.7412 | 1.7495 | 1.7577 | 1.7777 | 1.7970 | 1.8157 | 1.8339 | 1.8687 | 1.9018 | 1.9638 | 2.0212 | s | (387.50) |
| 3.005 | 3.065 | 3.124 | 3.183 | 3.241 | 3.299 | 3.357 | 3.415 | 3.559 | 3.702 | 3.845 | 3.986 | 4.269 | 4.550 | 5.110 | 5.668 | v | |
| 1352.2 | 1362.5 | 1372.8 | 1383.1 | 1393.3 | 1403.6 | 1413.9 | 1424.1 | 1449.9 | 1475.6 | 1501.5 | 1527.6 | 1580.1 | 1633.4 | 1742.4 | 1854.8 | h | **216** |
| 1.6961 | 1.7052 | 1.7142 | 1.7230 | 1.7316 | 1.7401 | 1.7484 | 1.7567 | 1.7767 | 1.7960 | 1.8147 | 1.8328 | 1.8677 | 1.9008 | 1.9627 | 2.0201 | s | (388.29) |
| 2.977 | 3.036 | 3.094 | 3.153 | 3.211 | 3.268 | 3.326 | 3.383 | 3.526 | 3.668 | 3.809 | 3.950 | 4.230 | 4.508 | 5.063 | 5.616 | v | |
| 1352.0 | 1362.4 | 1372.7 | 1383.0 | 1393.2 | 1403.5 | 1413.8 | 1424.1 | 1449.8 | 1475.6 | 1501.5 | 1527.5 | 1580.1 | 1633.4 | 1742.3 | 1854.8 | h | **218** |
| 1.6950 | 1.7041 | 1.7131 | 1.7219 | 1.7305 | 1.7390 | 1.7474 | 1.7556 | 1.7756 | 1.7949 | 1.8136 | 1.8318 | 1.8666 | 1.8997 | 1.9617 | 2.0191 | s | (389.08) |
| 2.949 | 3.008 | 3.066 | 3.123 | 3.181 | 3.238 | 3.295 | 3.352 | 3.493 | 3.634 | 3.774 | 3.913 | 4.191 | 4.467 | 5.017 | 5.565 | v | |
| 1351.9 | 1362.3 | 1372.6 | 1382.9 | 1393.1 | 1403.4 | 1413.7 | 1424.0 | 1449.7 | 1475.5 | 1501.4 | 1527.5 | 1580.0 | 1633.3 | 1742.3 | 1854.7 | h | **220** |
| 1.6939 | 1.7030 | 1.7120 | 1.7208 | 1.7295 | 1.7380 | 1.7463 | 1.7545 | 1.7746 | 1.7939 | 1.8126 | 1.8308 | 1.8656 | 1.8987 | 1.9607 | 2.0181 | s | (389.86) |
| 2.922 | 2.980 | 3.037 | 3.094 | 3.151 | 3.208 | 3.265 | 3.321 | 3.461 | 3.601 | 3.739 | 3.878 | 4.153 | 4.427 | 4.972 | 5.515 | v | |
| 1351.8 | 1362.2 | 1372.5 | 1382.8 | 1393.0 | 1403.3 | 1413.6 | 1423.9 | 1449.6 | 1475.4 | 1501.4 | 1527.4 | 1580.0 | 1633.3 | 1742.3 | 1854.7 | h | **222** |
| 1.6928 | 1.7020 | 1.7110 | 1.7198 | 1.7284 | 1.7369 | 1.7453 | 1.7535 | 1.7735 | 1.7928 | 1.8116 | 1.8297 | 1.8646 | 1.8977 | 1.9597 | 2.0171 | s | (390.63) |
| 2.895 | 2.952 | 3.009 | 3.066 | 3.122 | 3.179 | 3.235 | 3.291 | 3.430 | 3.568 | 3.706 | 3.843 | 4.115 | 4.387 | 4.927 | 5.465 | v | |
| 1351.7 | 1362.0 | 1372.4 | 1382.7 | 1393.0 | 1403.2 | 1413.5 | 1423.8 | 1449.6 | 1475.4 | 1501.3 | 1527.3 | 1579.9 | 1633.2 | 1742.2 | 1854.7 | h | **224** |
| 1.6917 | 1.7009 | 1.7099 | 1.7187 | 1.7274 | 1.7359 | 1.7442 | 1.7524 | 1.7725 | 1.7918 | 1.8105 | 1.8287 | 1.8635 | 1.8967 | 1.9587 | 2.0161 | s | (391.40) |
| 2.869 | 2.925 | 2.982 | 3.038 | 3.094 | 3.150 | 3.206 | 3.261 | 3.399 | 3.536 | 3.673 | 3.808 | 4.079 | 4.348 | 4.883 | 5.417 | v | |
| 1351.6 | 1361.9 | 1372.3 | 1382.6 | 1392.9 | 1403.1 | 1413.4 | 1423.7 | 1449.5 | 1475.3 | 1501.2 | 1527.3 | 1579.9 | 1633.2 | 1742.2 | 1854.7 | h | **226** |
| 1.6907 | 1.6999 | 1.7088 | 1.7177 | 1.7263 | 1.7348 | 1.7432 | 1.7514 | 1.7715 | 1.7908 | 1.8095 | 1.8277 | 1.8625 | 1.8957 | 1.9577 | 2.0151 | s | (392.17) |
| 2.843 | 2.899 | 2.955 | 3.011 | 3.066 | 3.122 | 3.177 | 3.232 | 3.369 | 3.505 | 3.640 | 3.775 | 4.043 | 4.309 | 4.840 | 5.369 | v | |
| 1351.5 | 1361.8 | 1372.2 | 1382.5 | 1392.8 | 1403.1 | 1413.4 | 1423.7 | 1449.4 | 1475.2 | 1501.2 | 1527.2 | 1579.8 | 1633.1 | 1742.2 | 1854.6 | h | **228** |
| 1.6896 | 1.6988 | 1.7078 | 1.7166 | 1.7253 | 1.7338 | 1.7422 | 1.7504 | 1.7705 | 1.7898 | 1.8085 | 1.8267 | 1.8616 | 1.8947 | 1.9567 | 2.0141 | s | (392.93) |
| 2.817 | 2.873 | 2.929 | 2.984 | 3.039 | 3.094 | 3.149 | 3.203 | 3.339 | 3.474 | 3.608 | 3.741 | 4.007 | 4.272 | 4.798 | 5.322 | v | |
| 1351.3 | 1361.7 | 1372.0 | 1382.4 | 1392.7 | 1403.0 | 1413.3 | 1423.6 | 1449.3 | 1475.2 | 1501.1 | 1527.2 | 1579.8 | 1633.1 | 1742.1 | 1854.6 | h | **230** |
| 1.6886 | 1.6978 | 1.7068 | 1.7156 | 1.7243 | 1.7328 | 1.7412 | 1.7494 | 1.7695 | 1.7888 | 1.8075 | 1.8257 | 1.8606 | 1.8937 | 1.9557 | 2.0131 | s | (393.68) |
| 2.792 | 2.848 | 2.903 | 2.958 | 3.012 | 3.067 | 3.121 | 3.175 | 3.310 | 3.443 | 3.576 | 3.709 | 3.972 | 4.235 | 4.757 | 5.276 | v | |
| 1351.2 | 1361.6 | 1371.9 | 1382.3 | 1392.6 | 1402.9 | 1413.2 | 1423.5 | 1449.3 | 1475.1 | 1501.0 | 1527.1 | 1579.7 | 1633.1 | 1742.1 | 1854.6 | h | **232** |
| 1.6876 | 1.6967 | 1.7057 | 1.7146 | 1.7233 | 1.7318 | 1.7401 | 1.7484 | 1.7685 | 1.7878 | 1.8065 | 1.8247 | 1.8596 | 1.8927 | 1.9548 | 2.0122 | s | (394.43) |
| 2.768 | 2.823 | 2.878 | 2.932 | 2.986 | 3.040 | 3.094 | 3.148 | 3.281 | 3.414 | 3.545 | 3.677 | 3.938 | 4.198 | 4.716 | 5.231 | v | |
| 1351.1 | 1361.5 | 1371.8 | 1382.2 | 1392.5 | 1402.8 | 1413.1 | 1423.4 | 1449.2 | 1475.0 | 1501.0 | 1527.1 | 1579.7 | 1633.0 | 1742.1 | 1854.6 | h | **234** |
| 1.6865 | 1.6957 | 1.7047 | 1.7136 | 1.7223 | 1.7308 | 1.7391 | 1.7474 | 1.7675 | 1.7868 | 1.8056 | 1.8238 | 1.8586 | 1.8918 | 1.9538 | 2.0112 | s | (395.17) |
| 2.744 | 2.798 | 2.853 | 2.906 | 2.960 | 3.014 | 3.067 | 3.121 | 3.253 | 3.384 | 3.515 | 3.645 | 3.905 | 4.162 | 4.676 | 5.187 | v | |
| 1351.0 | 1361.4 | 1371.7 | 1382.1 | 1392.4 | 1402.7 | 1413.0 | 1423.3 | 1449.1 | 1475.0 | 1500.9 | 1527.0 | 1579.7 | 1633.0 | 1742.0 | 1854.5 | h | **236** |
| 1.6855 | 1.6947 | 1.7037 | 1.7126 | 1.7213 | 1.7298 | 1.7382 | 1.7464 | 1.7665 | 1.7858 | 1.8046 | 1.8228 | 1.8577 | 1.8908 | 1.9528 | 2.0103 | s | (395.91) |
| 2.720 | 2.774 | 2.828 | 2.881 | 2.935 | 2.988 | 3.041 | 3.094 | 3.225 | 3.355 | 3.485 | 3.614 | 3.871 | 4.127 | 4.636 | 5.143 | v | |
| 1350.9 | 1361.2 | 1371.6 | 1382.0 | 1392.3 | 1402.6 | 1412.9 | 1423.2 | 1449.0 | 1474.9 | 1500.9 | 1527.0 | 1579.6 | 1632.9 | 1742.0 | 1854.5 | h | **238** |
| 1.6845 | 1.6937 | 1.7027 | 1.7116 | 1.7203 | 1.7288 | 1.7372 | 1.7454 | 1.7655 | 1.7849 | 1.8036 | 1.8218 | 1.8567 | 1.8899 | 1.9519 | 2.0093 | s | (396.64) |
| 2.696 | 2.750 | 2.804 | 2.857 | 2.910 | 2.963 | 3.015 | 3.068 | 3.198 | 3.327 | 3.456 | 3.584 | 3.839 | 4.093 | 4.597 | 5.100 | v | |
| 1350.7 | 1361.1 | 1371.5 | 1381.9 | 1392.2 | 1402.5 | 1412.8 | 1423.2 | 1449.0 | 1474.8 | 1500.8 | 1526.9 | 1579.6 | 1632.9 | 1742.0 | 1854.5 | h | **240** |
| 1.6835 | 1.6927 | 1.7017 | 1.7106 | 1.7193 | 1.7278 | 1.7362 | 1.7444 | 1.7645 | 1.7839 | 1.8027 | 1.8209 | 1.8558 | 1.8889 | 1.9510 | 2.0084 | s | (397.37) |
| 2.673 | 2.727 | 2.780 | 2.833 | 2.885 | 2.938 | 2.990 | 3.042 | 3.171 | 3.299 | 3.427 | 3.554 | 3.807 | 4.059 | 4.559 | 5.058 | v | |
| 1350.6 | 1361.0 | 1371.4 | 1381.8 | 1392.1 | 1402.4 | 1412.7 | 1423.1 | 1448.9 | 1474.8 | 1500.7 | 1526.8 | 1579.5 | 1632.9 | 1742.0 | 1854.5 | h | **242** |
| 1.6825 | 1.6917 | 1.7008 | 1.7096 | 1.7183 | 1.7268 | 1.7352 | 1.7435 | 1.7636 | 1.7830 | 1.8017 | 1.8199 | 1.8548 | 1.8880 | 1.9500 | 2.0075 | s | (398.10) |
| 2.651 | 2.704 | 2.756 | 2.809 | 2.861 | 2.913 | 2.965 | 3.016 | 3.144 | 3.272 | 3.398 | 3.524 | 3.775 | 4.025 | 4.522 | 5.016 | v | |
| 1350.5 | 1360.9 | 1371.3 | 1381.7 | 1392.0 | 1402.3 | 1412.7 | 1423.0 | 1448.8 | 1474.7 | 1500.7 | 1526.8 | 1579.5 | 1632.8 | 1741.9 | 1854.4 | h | **244** |
| 1.6815 | 1.6907 | 1.6998 | 1.7086 | 1.7173 | 1.7259 | 1.7343 | 1.7425 | 1.7626 | 1.7820 | 1.8008 | 1.8190 | 1.8539 | 1.8871 | 1.9491 | 2.0066 | s | (398.82) |
| 2.629 | 2.681 | 2.733 | 2.785 | 2.837 | 2.889 | 2.940 | 2.991 | 3.118 | 3.245 | 3.370 | 3.495 | 3.744 | 3.992 | 4.485 | 4.975 | v | |
| 1350.4 | 1360.8 | 1371.2 | 1381.6 | 1391.9 | 1402.2 | 1412.6 | 1422.9 | 1448.7 | 1474.6 | 1500.6 | 1526.7 | 1579.4 | 1632.8 | 1741.9 | 1854.4 | h | **246** |
| 1.6805 | 1.6898 | 1.6988 | 1.7077 | 1.7164 | 1.7249 | 1.7333 | 1.7416 | 1.7617 | 1.7811 | 1.7999 | 1.8181 | 1.8530 | 1.8861 | 1.9482 | 2.0057 | s | (399.53) |
| 2.607 | 2.659 | 2.711 | 2.762 | 2.814 | 2.865 | 2.916 | 2.967 | 3.093 | 3.218 | 3.343 | 3.467 | 3.714 | 3.960 | 4.448 | 4.935 | v | |
| 1350.3 | 1360.7 | 1371.1 | 1381.5 | 1391.8 | 1402.2 | 1412.5 | 1422.8 | 1448.7 | 1474.6 | 1500.6 | 1526.7 | 1579.4 | 1632.7 | 1741.9 | 1854.4 | h | **248** |
| 1.6796 | 1.6888 | 1.6979 | 1.7067 | 1.7154 | 1.7240 | 1.7324 | 1.7407 | 1.7608 | 1.7802 | 1.7989 | 1.8172 | 1.8521 | 1.8852 | 1.9473 | 2.0048 | s | (400.24) |

# Table 3. Superheated Vapor

| Abs. Press. Lb./Sq. In. (Sat. Temp.) | | Sat. Liquid | Sat. Vapor | Temperature—Degrees Fahrenheit 420° | 440° | 460° | 480° | 500° | 520° | 540° | 560° | 580° | 600° | 620° | 640° | 660° | 680° |
|---|---|---|---|---|---|---|---|---|---|---|---|---|---|---|---|---|---|
| **250** | v | 0.0187 | 1.8438 | 1.9077 | 1.9717 | 2.033 | 2.093 | 2.151 | 2.208 | 2.264 | 2.319 | 2.374 | 2.427 | 2.480 | 2.533 | 2.585 | 2.637 |
| (400.95) | h | 376.0 | 1201.1 | 1214.2 | 1227.3 | 1239.7 | 1251.7 | 1263.4 | 1274.8 | 1285.9 | 1296.9 | 1307.7 | 1318.5 | 1329.1 | 1339.6 | 1350.1 | 1360.6 |
| | s | 0.5675 | 1.5263 | 1.5414 | 1.5560 | 1.5697 | 1.5826 | 1.5949 | 1.6067 | 1.6179 | 1.6288 | 1.6393 | 1.6495 | 1.6595 | 1.6691 | 1.6786 | 1.6878 |
| **255** | v | 0.0187 | 1.8086 | 1.8659 | 1.9292 | 1.9900 | 2.049 | 2.106 | 2.162 | 2.217 | 2.272 | 2.325 | 2.378 | 2.430 | 2.482 | 2.533 | 2.584 |
| (402.70) | h | 377.9 | 1201.3 | 1213.3 | 1226.5 | 1239.0 | 1251.1 | 1262.8 | 1274.3 | 1285.5 | 1296.5 | 1307.3 | 1318.1 | 1328.7 | 1339.3 | 1349.8 | 1360.3 |
| | s | 0.5697 | 1.5246 | 1.5384 | 1.5531 | 1.5670 | 1.5800 | 1.5923 | 1.6041 | 1.6154 | 1.6263 | 1.6368 | 1.6471 | 1.6570 | 1.6667 | 1.6762 | 1.6855 |
| **260** | v | 0.0187 | 1.7748 | 1.8257 | 1.8882 | 1.9483 | 2.006 | 2.063 | 2.118 | 2.172 | 2.226 | 2.278 | 2.330 | 2.381 | 2.432 | 2.482 | 2.532 |
| (404.42) | h | 379.8 | 1201.5 | 1212.4 | 1225.7 | 1238.3 | 1250.5 | 1262.3 | 1273.8 | 1285.0 | 1296.0 | 1306.9 | 1317.7 | 1328.4 | 1339.0 | 1349.5 | 1360.0 |
| | s | 0.5719 | 1.5229 | 1.5354 | 1.5503 | 1.5642 | 1.5773 | 1.5897 | 1.6015 | 1.6129 | 1.6238 | 1.6344 | 1.6447 | 1.6547 | 1.6644 | 1.6739 | 1.6832 |
| **265** | v | 0.0187 | 1.7422 | 1.7870 | 1.8488 | 1.9081 | 1.9654 | 2.021 | 2.076 | 2.129 | 2.181 | 2.233 | 2.284 | 2.334 | 2.384 | 2.434 | 2.483 |
| (406.11) | h | 381.6 | 1201.7 | 1211.5 | 1224.9 | 1237.6 | 1249.9 | 1261.7 | 1273.2 | 1284.5 | 1295.6 | 1306.5 | 1317.3 | 1328.0 | 1338.7 | 1349.2 | 1359.7 |
| | s | 0.5740 | 1.5212 | 1.5324 | 1.5475 | 1.5615 | 1.5747 | 1.5871 | 1.5990 | 1.6104 | 1.6214 | 1.6320 | 1.6423 | 1.6523 | 1.6621 | 1.6716 | 1.6809 |
| **270** | v | 0.0188 | 1.7107 | 1.7497 | 1.8108 | 1.8694 | 1.9260 | 1.9809 | 2.035 | 2.087 | 2.139 | 2.190 | 2.240 | 2.289 | 2.338 | 2.387 | 2.436 |
| (407.78) | h | 383.4 | 1201.9 | 1210.6 | 1224.1 | 1236.9 | 1249.2 | 1261.1 | 1272.7 | 1284.0 | 1295.2 | 1306.1 | 1317.0 | 1327.7 | 1338.3 | 1348.9 | 1359.4 |
| | s | 0.5760 | 1.5196 | 1.5295 | 1.5447 | 1.5588 | 1.5721 | 1.5846 | 1.5965 | 1.6080 | 1.6190 | 1.6297 | 1.6400 | 1.6500 | 1.6598 | 1.6693 | 1.6786 |
| **275** | v | 0.0188 | 1.6804 | 1.7137 | 1.7742 | 1.8320 | 1.8879 | 1.9421 | 1.9950 | 2.047 | 2.098 | 2.148 | 2.197 | 2.246 | 2.294 | 2.342 | 2.390 |
| (409.43) | h | 385.2 | 1202.1 | 1209.6 | 1223.3 | 1236.2 | 1248.6 | 1260.5 | 1272.2 | 1283.6 | 1294.7 | 1305.7 | 1316.6 | 1327.3 | 1338.0 | 1348.6 | 1359.1 |
| | s | 0.5781 | 1.5180 | 1.5266 | 1.5420 | 1.5562 | 1.5695 | 1.5821 | 1.5941 | 1.6056 | 1.6167 | 1.6274 | 1.6377 | 1.6477 | 1.6575 | 1.6671 | 1.6764 |
| **280** | v | 0.0188 | 1.6511 | 1.6789 | 1.7388 | 1.7960 | 1.8512 | 1.9047 | 1.9569 | 2.008 | 2.058 | 2.107 | 2.156 | 2.204 | 2.252 | 2.299 | 2.346 |
| (411.05) | h | 387.0 | 1202.3 | 1208.7 | 1222.4 | 1235.4 | 1247.9 | 1260.0 | 1271.7 | 1283.1 | 1294.3 | 1305.3 | 1316.2 | 1327.0 | 1337.7 | 1348.3 | 1358.9 |
| | s | 0.5801 | 1.5164 | 1.5238 | 1.5392 | 1.5536 | 1.5670 | 1.5796 | 1.5917 | 1.6032 | 1.6143 | 1.6251 | 1.6354 | 1.6455 | 1.6553 | 1.6649 | 1.6742 |
| **285** | v | 0.0188 | 1.6228 | 1.6454 | 1.7047 | 1.7613 | 1.8158 | 1.8686 | 1.9201 | 1.9704 | 2.020 | 2.068 | 2.116 | 2.164 | 2.211 | 2.257 | 2.303 |
| (412.65) | h | 388.7 | 1202.4 | 1207.7 | 1221.6 | 1234.7 | 1247.3 | 1259.4 | 1271.1 | 1282.6 | 1293.9 | 1304.9 | 1315.8 | 1326.6 | 1337.4 | 1348.0 | 1358.6 |
| | s | 0.5821 | 1.5149 | 1.5209 | 1.5365 | 1.5510 | 1.5645 | 1.5772 | 1.5893 | 1.6009 | 1.6121 | 1.6228 | 1.6332 | 1.6433 | 1.6531 | 1.6627 | 1.6721 |
| **290** | v | 0.0188 | 1.5954 | 1.6130 | 1.6717 | 1.7277 | 1.7816 | 1.8338 | 1.8846 | 1.9342 | 1.9829 | 2.031 | 2.078 | 2.125 | 2.171 | 2.217 | 2.262 |
| (414.23) | h | 390.5 | 1202.6 | 1206.8 | 1220.8 | 1234.0 | 1246.6 | 1258.8 | 1270.6 | 1282.1 | 1293.4 | 1304.5 | 1315.5 | 1326.3 | 1337.0 | 1347.7 | 1358.3 |
| | s | 0.5841 | 1.5133 | 1.5181 | 1.5339 | 1.5484 | 1.5620 | 1.5748 | 1.5870 | 1.5986 | 1.6098 | 1.6206 | 1.6310 | 1.6411 | 1.6510 | 1.6606 | 1.6700 |
| **295** | v | 0.0189 | 1.5689 | 1.5816 | 1.6398 | 1.6952 | 1.7485 | 1.8001 | 1.8502 | 1.8992 | 1.9472 | 1.9945 | 2.041 | 2.087 | 2.133 | 2.178 | 2.222 |
| (415.79) | h | 392.2 | 1202.7 | 1205.8 | 1219.9 | 1233.3 | 1246.0 | 1258.2 | 1270.1 | 1281.6 | 1293.0 | 1304.1 | 1315.1 | 1325.9 | 1336.7 | 1347.4 | 1358.0 |
| | s | 0.5860 | 1.5118 | 1.5154 | 1.5312 | 1.5459 | 1.5596 | 1.5725 | 1.5847 | 1.5964 | 1.6076 | 1.6184 | 1.6289 | 1.6390 | 1.6489 | 1.6585 | 1.6679 |
| **300** | v | 0.0189 | 1.5433 | 1.5513 | 1.6090 | 1.6638 | 1.7165 | 1.7675 | 1.8170 | 1.8654 | 1.9128 | 1.9594 | 2.005 | 2.051 | 2.095 | 2.140 | 2.184 |
| (417.33) | h | 393.8 | 1202.8 | 1204.8 | 1219.1 | 1232.5 | 1245.3 | 1257.6 | 1269.5 | 1281.1 | 1292.5 | 1303.7 | 1314.7 | 1325.6 | 1336.4 | 1347.1 | 1357.7 |
| | s | 0.5879 | 1.5104 | 1.5126 | 1.5286 | 1.5434 | 1.5572 | 1.5701 | 1.5824 | 1.5941 | 1.6054 | 1.6163 | 1.6268 | 1.6369 | 1.6468 | 1.6565 | 1.6659 |
| **305** | v | 0.0189 | 1.5184 | 1.5219 | 1.5791 | 1.6334 | 1.6856 | 1.7360 | 1.7849 | 1.8327 | 1.8794 | 1.9254 | 1.9707 | 2.015 | 2.060 | 2.103 | 2.147 |
| (418.85) | h | 395.4 | 1203.0 | 1203.8 | 1218.2 | 1231.8 | 1244.7 | 1257.0 | 1269.0 | 1280.7 | 1292.1 | 1303.3 | 1314.3 | 1325.2 | 1336.0 | 1346.8 | 1357.4 |
| | s | 0.5898 | 1.5089 | 1.5099 | 1.5260 | 1.5409 | 1.5548 | 1.5678 | 1.5802 | 1.5919 | 1.6032 | 1.6141 | 1.6247 | 1.6349 | 1.6448 | 1.6544 | 1.6639 |
| **310** | v | 0.0189 | 1.4944 | | 1.5502 | 1.6040 | 1.6556 | 1.7054 | 1.7538 | 1.8010 | 1.8472 | 1.8925 | 1.9372 | 1.9813 | 2.025 | 2.068 | 2.111 |
| (420.35) | h | 397.1 | 1203.1 | | 1217.4 | 1231.0 | 1244.0 | 1256.4 | 1268.5 | 1280.2 | 1291.6 | 1302.9 | 1313.9 | 1324.9 | 1335.7 | 1346.4 | 1357.1 |
| | s | 0.5916 | 1.5075 | | 1.5235 | 1.5385 | 1.5524 | 1.5655 | 1.5780 | 1.5898 | 1.6011 | 1.6120 | 1.6226 | 1.6328 | 1.6428 | 1.6524 | 1.6619 |
| **315** | v | 0.0190 | 1.4711 | | 1.5222 | 1.5755 | 1.6266 | 1.6759 | 1.7237 | 1.7703 | 1.8159 | 1.8607 | 1.9048 | 1.9483 | 1.9912 | 2.034 | 2.076 |
| (421.83) | h | 398.8 | 1203.2 | | 1216.5 | 1230.3 | 1243.3 | 1255.8 | 1267.9 | 1279.7 | 1291.2 | 1302.5 | 1313.6 | 1324.5 | 1335.4 | 1346.1 | 1356.8 |
| | s | 0.5934 | 1.5060 | | 1.5209 | 1.5361 | 1.5501 | 1.5633 | 1.5758 | 1.5876 | 1.5990 | 1.6100 | 1.6206 | 1.6308 | 1.6408 | 1.6505 | 1.6599 |
| **320** | v | 0.0190 | 1.4485 | | 1.4950 | 1.5479 | 1.5985 | 1.6472 | 1.6945 | 1.7406 | 1.7856 | 1.8298 | 1.8734 | 1.9163 | 1.9586 | 2.001 | 2.042 |
| (423.29) | h | 400.4 | 1203.4 | | 1215.6 | 1229.5 | 1242.6 | 1255.2 | 1267.4 | 1279.2 | 1290.7 | 1302.0 | 1313.2 | 1324.2 | 1335.0 | 1345.8 | 1356.5 |
| | s | 0.5952 | 1.5046 | | 1.5184 | 1.5337 | 1.5478 | 1.5611 | 1.5736 | 1.5855 | 1.5969 | 1.6079 | 1.6186 | 1.6288 | 1.6388 | 1.6485 | 1.6580 |
| **325** | v | 0.0190 | 1.4266 | | 1.4687 | 1.5211 | 1.5712 | 1.6194 | 1.6662 | 1.7118 | 1.7562 | 1.7999 | 1.8429 | 1.8852 | 1.9271 | 1.9685 | 2.009 |
| (424.74) | h | 402.0 | 1203.5 | | 1214.8 | 1228.7 | 1241.9 | 1254.6 | 1266.8 | 1278.7 | 1290.3 | 1301.6 | 1312.8 | 1323.8 | 1334.7 | 1345.5 | 1356.2 |
| | s | 0.5970 | 1.5033 | | 1.5159 | 1.5313 | 1.5455 | 1.5588 | 1.5714 | 1.5834 | 1.5949 | 1.6059 | 1.6166 | 1.6269 | 1.6369 | 1.6466 | 1.6561 |
| **330** | v | 0.0190 | 1.4053 | | 1.4431 | 1.4951 | 1.5448 | 1.5925 | 1.6388 | 1.6838 | 1.7278 | 1.7709 | 1.8134 | 1.8552 | 1.8965 | 1.9373 | 1.9777 |
| (426.16) | h | 403.6 | 1203.6 | | 1213.9 | 1228.0 | 1241.3 | 1254.0 | 1266.3 | 1278.2 | 1289.8 | 1301.2 | 1312.4 | 1323.5 | 1334.4 | 1345.2 | 1356.0 |
| | s | 0.5988 | 1.5019 | | 1.5134 | 1.5289 | 1.5432 | 1.5567 | 1.5693 | 1.5814 | 1.5929 | 1.6039 | 1.6146 | 1.6249 | 1.6350 | 1.6447 | 1.6542 |
| **335** | v | 0.0191 | 1.3846 | | 1.4182 | 1.4699 | 1.5191 | 1.5664 | 1.6121 | 1.6567 | 1.7001 | 1.7428 | 1.7847 | 1.8260 | 1.8667 | 1.9071 | 1.9470 |
| (427.58) | h | 405.1 | 1203.6 | | 1213.0 | 1227.2 | 1240.6 | 1253.4 | 1265.7 | 1277.7 | 1289.4 | 1300.8 | 1312.0 | 1323.1 | 1334.0 | 1344.9 | 1355.7 |
| | s | 0.6005 | 1.5005 | | 1.5110 | 1.5266 | 1.5410 | 1.5545 | 1.5672 | 1.5793 | 1.5909 | 1.6020 | 1.6127 | 1.6230 | 1.6331 | 1.6428 | 1.6524 |

Temperature—Degrees Fahrenheit

| 700° | 720° | 740° | 760° | 780° | 800° | 850° | 900° | 950° | 1000° | 1050° | 1100° | 1200° | 1400° | 1600° | | Abs. Press. Lb./Sq. In. (Sat. Temp.) |
|---|---|---|---|---|---|---|---|---|---|---|---|---|---|---|---|---|
| 2.688 | 2.740 | 2.791 | 2.841 | 2.892 | 2.942 | 3.068 | 3.192 | 3.316 | 3.439 | 3.562 | 3.684 | 3.928 | 4.413 | 4.896 | v | |
| 1371.0 | 1381.4 | 1391.7 | 1402.1 | 1412.4 | 1422.7 | 1448.6 | 1474.5 | 1500.5 | 1526.6 | 1552.9 | 1579.3 | 1632.7 | 1741.8 | 1854.4 | h | **250** |
| 1.6969 | 1.7058 | 1.7145 | 1.7230 | 1.7315 | 1.7397 | 1.7598 | 1.7793 | 1.7980 | 1.8162 | 1.8339 | 1.8512 | 1.8843 | 1.9464 | 2.0039 | s | (400.95) |
| 2.634 | 2.684 | 2.734 | 2.784 | 2.834 | 2.884 | 3.006 | 3.128 | 3.250 | 3.371 | 3.491 | 3.611 | 3.850 | 4.326 | 4.799 | v | |
| 1370.7 | 1381.1 | 1391.5 | 1401.8 | 1412.2 | 1422.5 | 1448.4 | 1474.3 | 1500.3 | 1526.5 | 1552.8 | 1579.2 | 1632.6 | 1741.8 | 1854.3 | h | **255** |
| 1.6945 | 1.7034 | 1.7122 | 1.7207 | 1.7291 | 1.7374 | 1.7575 | 1.7770 | 1.7958 | 1.8140 | 1.8317 | 1.8489 | 1.8821 | 1.9442 | 2.0017 | s | (402.70) |
| 2.582 | 2.631 | 2.681 | 2.729 | 2.778 | 2.827 | 2.947 | 3.067 | 3.187 | 3.305 | 3.423 | 3.541 | 3.776 | 4.242 | 4.707 | v | |
| 1370.4 | 1380.8 | 1391.2 | 1401.6 | 1412.0 | 1422.3 | 1448.2 | 1474.2 | 1500.2 | 1526.3 | 1552.6 | 1579.1 | 1632.5 | 1741.7 | 1854.2 | h | **260** |
| 1.6922 | 1.7011 | 1.7099 | 1.7185 | 1.7269 | 1.7352 | 1.7553 | 1.7748 | 1.7936 | 1.8118 | 1.8295 | 1.8467 | 1.8799 | 1.9420 | 1.9995 | s | (404.42) |
| 2.532 | 2.580 | 2.629 | 2.677 | 2.725 | 2.772 | 2.891 | 3.009 | 3.126 | 3.242 | 3.358 | 3.474 | 3.704 | 4.162 | 4.618 | v | |
| 1370.2 | 1380.6 | 1391.0 | 1401.4 | 1411.7 | 1422.1 | 1448.0 | 1474.0 | 1500.0 | 1526.2 | 1552.5 | 1579.0 | 1632.4 | 1741.6 | 1854.2 | h | **265** |
| 1.6900 | 1.6989 | 1.7076 | 1.7162 | 1.7246 | 1.7329 | 1.7531 | 1.7726 | 1.7914 | 1.8096 | 1.8273 | 1.8446 | 1.8778 | 1.9399 | 1.9974 | s | (406.11) |
| 2.484 | 2.531 | 2.579 | 2.626 | 2.673 | 2.720 | 2.836 | 2.952 | 3.067 | 3.181 | 3.295 | 3.409 | 3.635 | 4.085 | 4.532 | v | |
| 1369.9 | 1380.3 | 1390.7 | 1401.1 | 1411.5 | 1421.9 | 1447.8 | 1473.8 | 1499.9 | 1526.1 | 1552.4 | 1578.8 | 1632.3 | 1741.5 | 1854.1 | h | **270** |
| 1.6877 | 1.6967 | 1.7054 | 1.7140 | 1.7224 | 1.7308 | 1.7509 | 1.7704 | 1.7892 | 1.8075 | 1.8252 | 1.8425 | 1.8757 | 1.9378 | 1.9953 | s | (407.78) |
| 2.437 | 2.484 | 2.530 | 2.577 | 2.623 | 2.669 | 2.784 | 2.897 | 3.010 | 3.123 | 3.235 | 3.346 | 3.568 | 4.010 | 4.449 | v | |
| 1369.6 | 1380.1 | 1390.5 | 1400.9 | 1411.3 | 1421.7 | 1447.7 | 1473.7 | 1499.7 | 1525.9 | 1552.3 | 1578.7 | 1632.2 | 1741.4 | 1854.0 | h | **275** |
| 1.6855 | 1.6945 | 1.7032 | 1.7118 | 1.7203 | 1.7286 | 1.7488 | 1.7683 | 1.7871 | 1.8054 | 1.8231 | 1.8404 | 1.8736 | 1.9357 | 1.9932 | s | (409.43) |
| 2.392 | 2.438 | 2.484 | 2.530 | 2.575 | 2.621 | 2.733 | 2.845 | 2.956 | 3.066 | 3.176 | 3.286 | 3.504 | 3.938 | 4.370 | v | |
| 1369.4 | 1379.8 | 1390.3 | 1400.7 | 1411.1 | 1421.5 | 1447.5 | 1473.5 | 1499.6 | 1525.8 | 1552.1 | 1578.6 | 1632.1 | 1741.4 | 1854.0 | h | **280** |
| 1.6834 | 1.6923 | 1.7011 | 1.7097 | 1.7182 | 1.7265 | 1.7467 | 1.7662 | 1.7851 | 1.8033 | 1.8211 | 1.8383 | 1.8716 | 1.9337 | 1.9912 | s | (411.05) |
| 2.349 | 2.394 | 2.439 | 2.484 | 2.529 | 2.574 | 2.684 | 2.794 | 2.903 | 3.012 | 3.120 | 3.228 | 3.442 | 3.869 | 4.293 | v | |
| 1369.1 | 1379.6 | 1390.0 | 1400.4 | 1410.9 | 1421.3 | 1447.3 | 1473.3 | 1499.4 | 1525.6 | 1552.0 | 1578.5 | 1632.0 | 1741.3 | 1853.9 | h | **285** |
| 1.6812 | 1.6902 | 1.6990 | 1.7076 | 1.7161 | 1.7244 | 1.7447 | 1.7642 | 1.7830 | 1.8013 | 1.8191 | 1.8363 | 1.8696 | 1.9317 | 1.9892 | s | (412.65) |
| 2.307 | 2.352 | 2.396 | 2.440 | 2.484 | 2.528 | 2.637 | 2.745 | 2.852 | 2.959 | 3.065 | 3.171 | 3.382 | 3.802 | 4.219 | v | |
| 1368.8 | 1379.3 | 1389.8 | 1400.2 | 1410.6 | 1421.1 | 1447.1 | 1473.2 | 1499.3 | 1525.5 | 1551.9 | 1578.4 | 1631.9 | 1741.2 | 1853.9 | h | **290** |
| 1.6791 | 1.6881 | 1.6969 | 1.7056 | 1.7140 | 1.7224 | 1.7426 | 1.7622 | 1.7810 | 1.7993 | 1.8171 | 1.8343 | 1.8676 | 1.9298 | 1.9873 | s | (414.23) |
| 2.267 | 2.310 | 2.354 | 2.398 | 2.441 | 2.484 | 2.591 | 2.698 | 2.803 | 2.908 | 3.013 | 3.117 | 3.325 | 3.737 | 4.147 | v | |
| 1368.5 | 1379.1 | 1389.5 | 1400.0 | 1410.4 | 1420.9 | 1446.9 | 1473.0 | 1499.1 | 1525.4 | 1551.7 | 1578.3 | 1631.8 | 1741.1 | 1853.8 | h | **295** |
| 1.6771 | 1.6861 | 1.6949 | 1.7035 | 1.7120 | 1.7204 | 1.7407 | 1.7602 | 1.7791 | 1.7974 | 1.8151 | 1.8324 | 1.8657 | 1.9279 | 1.9854 | s | (415.79) |
| 2.227 | 2.271 | 2.314 | 2.357 | 2.399 | 2.442 | 2.547 | 2.652 | 2.756 | 2.859 | 2.962 | 3.065 | 3.269 | 3.674 | 4.078 | v | |
| 1368.3 | 1378.8 | 1389.3 | 1399.8 | 1410.2 | 1420.6 | 1446.7 | 1472.8 | 1499.0 | 1525.2 | 1551.6 | 1578.1 | 1631.7 | 1741.0 | 1853.7 | h | **300** |
| 1.6751 | 1.6841 | 1.6929 | 1.7015 | 1.7100 | 1.7184 | 1.7387 | 1.7582 | 1.7771 | 1.7954 | 1.8132 | 1.8305 | 1.8638 | 1.9260 | 1.9835 | s | (417.33) |
| 2.190 | 2.232 | 2.275 | 2.317 | 2.359 | 2.401 | 2.505 | 2.608 | 2.710 | 2.812 | 2.913 | 3.014 | 3.215 | 3.614 | 4.011 | v | |
| 1368.0 | 1378.5 | 1389.0 | 1399.5 | 1410.0 | 1420.4 | 1446.5 | 1472.7 | 1498.8 | 1525.1 | 1551.5 | 1578.0 | 1631.6 | 1741.0 | 1853.7 | h | **305** |
| 1.6731 | 1.6821 | 1.6909 | 1.6996 | 1.7081 | 1.7165 | 1.7368 | 1.7563 | 1.7752 | 1.7935 | 1.8113 | 1.8286 | 1.8619 | 1.9241 | 1.9817 | s | (418.85) |
| 2.153 | 2.195 | 2.237 | 2.279 | 2.320 | 2.361 | 2.463 | 2.565 | 2.666 | 2.766 | 2.865 | 2.965 | 3.162 | 3.555 | 3.946 | v | |
| 1367.7 | 1378.3 | 1388.8 | 1399.3 | 1409.8 | 1420.2 | 1446.4 | 1472.5 | 1498.7 | 1524.9 | 1551.4 | 1577.9 | 1631.5 | 1740.9 | 1853.6 | h | **310** |
| 1.6711 | 1.6801 | 1.6890 | 1.6977 | 1.7062 | 1.7145 | 1.7349 | 1.7545 | 1.7734 | 1.7917 | 1.8095 | 1.8268 | 1.8601 | 1.9223 | 1.9799 | s | (420.35) |
| 2.118 | 2.159 | 2.200 | 2.241 | 2.282 | 2.323 | 2.423 | 2.523 | 2.622 | 2.721 | 2.819 | 2.917 | 3.112 | 3.498 | 3.883 | v | |
| 1367.4 | 1378.0 | 1388.6 | 1399.1 | 1409.5 | 1420.0 | 1446.2 | 1472.3 | 1498.5 | 1524.8 | 1551.2 | 1577.8 | 1631.4 | 1740.8 | 1853.5 | h | **315** |
| 1.6692 | 1.6782 | 1.6871 | 1.6958 | 1.7043 | 1.7127 | 1.7330 | 1.7526 | 1.7715 | 1.7899 | 1.8077 | 1.8250 | 1.8583 | 1.9205 | 1.9781 | s | (421.83) |
| 2.083 | 2.124 | 2.165 | 2.205 | 2.245 | 2.285 | 2.385 | 2.483 | 2.581 | 2.678 | 2.775 | 2.871 | 3.063 | 3.443 | 3.822 | v | |
| 1367.2 | 1377.8 | 1388.3 | 1398.8 | 1409.3 | 1419.8 | 1446.0 | 1472.1 | 1498.4 | 1524.7 | 1551.1 | 1577.7 | 1631.3 | 1740.7 | 1853.5 | h | **320** |
| 1.6673 | 1.6763 | 1.6852 | 1.6939 | 1.7024 | 1.7108 | 1.7312 | 1.7508 | 1.7697 | 1.7880 | 1.8059 | 1.8232 | 1.8565 | 1.9187 | 1.9763 | s | (423.29) |
| 2.050 | 2.090 | 2.130 | 2.170 | 2.210 | 2.249 | 2.347 | 2.444 | 2.540 | 2.636 | 2.731 | 2.826 | 3.015 | 3.390 | 3.763 | v | |
| 1366.9 | 1377.5 | 1388.1 | 1398.6 | 1409.1 | 1419.6 | 1445.8 | 1472.0 | 1498.2 | 1524.5 | 1551.0 | 1577.5 | 1631.2 | 1740.7 | 1853.4 | h | **325** |
| 1.6654 | 1.6744 | 1.6833 | 1.6920 | 1.7006 | 1.7090 | 1.7294 | 1.7490 | 1.7679 | 1.7863 | 1.8041 | 1.8214 | 1.8547 | 1.9170 | 1.9746 | s | (424.74) |
| 2.018 | 2.058 | 2.097 | 2.136 | 2.175 | 2.214 | 2.311 | 2.406 | 2.501 | 2.596 | 2.690 | 2.783 | 2.969 | 3.339 | 3.706 | v | |
| 1366.6 | 1377.2 | 1387.8 | 1398.4 | 1408.9 | 1419.4 | 1445.6 | 1471.8 | 1498.0 | 1524.4 | 1550.8 | 1577.4 | 1631.1 | 1740.6 | 1853.4 | h | **330** |
| 1.6635 | 1.6726 | 1.6815 | 1.6902 | 1.6988 | 1.7072 | 1.7276 | 1.7472 | 1.7662 | 1.7845 | 1.8023 | 1.8197 | 1.8530 | 1.9153 | 1.9729 | s | (426.16) |
| 1.9866 | 2.026 | 2.065 | 2.103 | 2.142 | 2.180 | 2.275 | 2.370 | 2.463 | 2.556 | 2.649 | 2.741 | 2.924 | 3.288 | 3.650 | v | |
| 1366.3 | 1377.0 | 1387.6 | 1398.1 | 1408.7 | 1419.2 | 1445.4 | 1471.6 | 1497.9 | 1524.2 | 1550.7 | 1577.3 | 1631.0 | 1740.5 | 1853.3 | h | **335** |
| 1.6617 | 1.6708 | 1.6797 | 1.6884 | 1.6970 | 1.7054 | 1.7258 | 1.7454 | 1.7644 | 1.7828 | 1.8006 | 1.8179 | 1.8513 | 1.9136 | 1.9712 | s | (427.58) |

## Table 3. Superheated Vapor

| Abs. Press. Lb./Sq. In. (Sat. Temp.) | | Sat. Liquid | Sat. Vapor | Temperature—Degrees Fahrenheit 430° | 440° | 450° | 460° | 470° | 480° | 490° | 500° | 520° | 540° | 560° | 580° | 600° | 620° |
|---|---|---|---|---|---|---|---|---|---|---|---|---|---|---|---|---|---|
| | v | 0.0191 | 1.3645 | 1.3673 | 1.3941 | 1.4201 | 1.4454 | 1.4700 | 1.4941 | 1.5178 | 1.5410 | 1.5863 | 1.6303 | 1.6733 | 1.7155 | 1.7569 | 1.7976 |
| **340** | h | 406.7 | 1203.7 | 1204.5 | 1212.1 | 1219.4 | 1226.4 | 1233.2 | 1239.9 | 1246.4 | 1252.8 | 1265.2 | 1277.2 | 1288.9 | 1300.4 | 1311.6 | 1322.7 |
| (428.97) | s | 0.6022 | 1.4992 | 1.5000 | 1.5085 | 1.5166 | 1.5243 | 1.5317 | 1.5388 | 1.5457 | 1.5524 | 1.5652 | 1.5773 | 1.5889 | 1.6000 | 1.6108 | 1.6211 |
| | v | 0.0191 | 1.3450 | | 1.3707 | 1.3964 | 1.4215 | 1.4460 | 1.4699 | 1.4933 | 1.5163 | 1.5612 | 1.6047 | 1.6472 | 1.6889 | 1.7298 | 1.7701 |
| **345** | h | 408.2 | 1203.8 | | 1211.2 | 1218.5 | 1225.6 | 1232.5 | 1239.2 | 1245.8 | 1252.1 | 1264.6 | 1276.7 | 1288.4 | 1299.9 | 1311.2 | 1322.4 |
| (430.35) | s | 0.6039 | 1.4979 | | 1.5061 | 1.5142 | 1.5220 | 1.5294 | 1.5366 | 1.5435 | 1.5502 | 1.5631 | 1.5753 | 1.5869 | 1.5981 | 1.6089 | 1.6193 |
| | v | 0.0191 | 1.3260 | | 1.3478 | 1.3734 | 1.3984 | 1.4226 | 1.4463 | 1.4696 | 1.4923 | 1.5368 | 1.5799 | 1.6220 | 1.6631 | 1.7036 | 1.7434 |
| **350** | h | 409.7 | 1203.9 | | 1210.3 | 1217.7 | 1224.8 | 1231.7 | 1238.5 | 1245.1 | 1251.5 | 1264.1 | 1276.2 | 1288.0 | 1299.5 | 1310.9 | 1322.0 |
| (431.72) | s | 0.6056 | 1.4966 | | 1.5037 | 1.5119 | 1.5197 | 1.5272 | 1.5344 | 1.5414 | 1.5481 | 1.5611 | 1.5733 | 1.5850 | 1.5962 | 1.6070 | 1.6174 |
| | v | 0.0191 | 1.3075 | | 1.3257 | 1.3511 | 1.3759 | 1.3999 | 1.4234 | 1.4465 | 1.4690 | 1.5130 | 1.5557 | 1.5973 | 1.6381 | 1.6781 | 1.7174 |
| **355** | h | 411.2 | 1204.0 | | 1209.3 | 1216.8 | 1224.0 | 1231.0 | 1237.8 | 1244.4 | 1250.9 | 1263.5 | 1275.7 | 1287.5 | 1299.1 | 1310.5 | 1321.7 |
| (433.06) | s | 0.6073 | 1.4953 | | 1.5013 | 1.5095 | 1.5174 | 1.5250 | 1.5323 | 1.5393 | 1.5461 | 1.5591 | 1.5714 | 1.5831 | 1.5943 | 1.6052 | 1.6156 |
| | v | 0.0192 | 1.2895 | | 1.3041 | 1.3294 | 1.3539 | 1.3778 | 1.4012 | 1.4240 | 1.4464 | 1.4900 | 1.5322 | 1.5734 | 1.6137 | 1.6533 | 1.6922 |
| **360** | h | 412.7 | 1204.1 | | 1208.4 | 1215.9 | 1223.2 | 1230.2 | 1237.1 | 1243.8 | 1250.3 | 1263.0 | 1275.2 | 1287.1 | 1298.7 | 1310.1 | 1321.3 |
| (434.40) | s | 0.6090 | 1.4941 | | 1.4989 | 1.5072 | 1.5152 | 1.5228 | 1.5301 | 1.5372 | 1.5440 | 1.5571 | 1.5694 | 1.5812 | 1.5925 | 1.6033 | 1.6138 |
| | v | 0.0192 | 1.2720 | | 1.2831 | 1.3082 | 1.3326 | 1.3563 | 1.3795 | 1.4021 | 1.4243 | 1.4675 | 1.5094 | 1.5501 | 1.5900 | 1.6291 | 1.6676 |
| **365** | h | 414.1 | 1204.1 | | 1207.5 | 1215.1 | 1222.4 | 1229.5 | 1236.4 | 1243.1 | 1249.6 | 1262.4 | 1274.7 | 1286.6 | 1298.2 | 1309.7 | 1320.9 |
| (435.72) | s | 0.6106 | 1.4928 | | 1.4965 | 1.5049 | 1.5129 | 1.5206 | 1.5280 | 1.5351 | 1.5420 | 1.5551 | 1.5675 | 1.5793 | 1.5906 | 1.6015 | 1.6120 |
| | v | 0.0192 | 1.2550 | | 1.2626 | 1.2876 | 1.3118 | 1.3354 | 1.3584 | 1.3808 | 1.4028 | 1.4457 | 1.4871 | 1.5275 | 1.5670 | 1.6057 | 1.6437 |
| **370** | h | 415.6 | 1204.2 | | 1206.5 | 1214.2 | 1221.6 | 1228.7 | 1235.7 | 1242.4 | 1249.0 | 1261.8 | 1274.1 | 1286.1 | 1297.8 | 1309.3 | 1320.6 |
| (437.03) | s | 0.6122 | 1.4916 | | 1.4942 | 1.5027 | 1.5107 | 1.5185 | 1.5259 | 1.5330 | 1.5399 | 1.5531 | 1.5656 | 1.5775 | 1.5888 | 1.5997 | 1.6103 |
| | v | 0.0192 | 1.2384 | | 1.2427 | 1.2675 | 1.2916 | 1.3150 | 1.3378 | 1.3601 | 1.3819 | 1.4244 | 1.4655 | 1.5054 | 1.5445 | 1.5828 | 1.6205 |
| **375** | h | 417.0 | 1204.3 | | 1205.6 | 1213.3 | 1220.7 | 1227.9 | 1234.9 | 1241.8 | 1248.4 | 1261.2 | 1273.6 | 1285.6 | 1297.4 | 1308.9 | 1320.2 |
| (438.32) | s | 0.6137 | 1.4904 | | 1.4918 | 1.5004 | 1.5085 | 1.5163 | 1.5238 | 1.5310 | 1.5379 | 1.5512 | 1.5637 | 1.5756 | 1.5870 | 1.5980 | 1.6086 |
| | v | 0.0192 | 1.2222 | | 1.2232 | 1.2479 | 1.2719 | 1.2951 | 1.3178 | 1.3399 | 1.3616 | 1.4037 | 1.4444 | 1.4840 | 1.5226 | 1.5605 | 1.5978 |
| **380** | h | 418.5 | 1204.3 | | 1204.6 | 1212.4 | 1219.9 | 1227.2 | 1234.2 | 1241.1 | 1247.7 | 1260.7 | 1273.1 | 1285.2 | 1297.0 | 1308.5 | 1319.8 |
| (439.60) | s | 0.6153 | 1.4891 | | 1.4895 | 1.4981 | 1.5063 | 1.5142 | 1.5217 | 1.5289 | 1.5359 | 1.5493 | 1.5618 | 1.5738 | 1.5852 | 1.5962 | 1.6068 |
| | v | 0.0193 | 1.2064 | | | 1.2289 | 1.2527 | 1.2758 | 1.2983 | 1.3202 | 1.3417 | 1.3835 | 1.4238 | 1.4830 | 1.5013 | 1.5389 | 1.5757 |
| **385** | h | 419.9 | 1204.3 | | | 1211.5 | 1219.1 | 1226.4 | 1233.5 | 1240.4 | 1247.1 | 1260.1 | 1272.6 | 1284.7 | 1296.5 | 1308.1 | 1319.5 |
| (440.86) | s | 0.6169 | 1.4879 | | | 1.4959 | 1.5042 | 1.5121 | 1.5196 | 1.5269 | 1.5340 | 1.5474 | 1.5600 | 1.5720 | 1.5835 | 1.5945 | 1.6051 |
| | v | 0.0193 | 1.1910 | | | 1.2102 | 1.2339 | 1.2569 | 1.2793 | 1.3011 | 1.3224 | 1.3638 | 1.4038 | 1.4426 | 1.4806 | 1.5177 | 1.5542 |
| **390** | h | 421.3 | 1204.4 | | | 1210.6 | 1218.2 | 1225.6 | 1232.8 | 1239.7 | 1246.4 | 1259.5 | 1272.1 | 1284.2 | 1296.1 | 1307.7 | 1319.1 |
| (442.12) | s | 0.6184 | 1.4867 | | | 1.4937 | 1.5020 | 1.5100 | 1.5176 | 1.5249 | 1.5320 | 1.5455 | 1.5582 | 1.5702 | 1.5817 | 1.5928 | 1.6034 |
| | v | 0.0193 | 1.1760 | | | 1.1921 | 1.2156 | 1.2385 | 1.2607 | 1.2824 | 1.3035 | 1.3446 | 1.3843 | 1.4228 | 1.4603 | 1.4971 | 1.5332 |
| **395** | h | 422.7 | 1204.4 | | | 1209.7 | 1217.4 | 1224.8 | 1232.0 | 1239.0 | 1245.8 | 1258.9 | 1271.5 | 1283.8 | 1295.7 | 1307.3 | 1318.7 |
| (443.36) | s | 0.6199 | 1.4856 | | | 1.4914 | 1.4998 | 1.5079 | 1.5156 | 1.5230 | 1.5301 | 1.5436 | 1.5564 | 1.5685 | 1.5800 | 1.5911 | 1.6018 |
| | v | 0.0193 | 1.1613 | | | 1.1744 | 1.1978 | 1.2205 | 1.2426 | 1.2641 | 1.2851 | 1.3259 | 1.3652 | 1.4034 | 1.4406 | 1.4770 | 1.5128 |
| **400** | h | 424.0 | 1204.5 | | | 1208.8 | 1216.5 | 1224.0 | 1231.3 | 1238.3 | 1245.1 | 1258.3 | 1271.0 | 1283.3 | 1295.2 | 1306.9 | 1318.3 |
| (444.59) | s | 0.6214 | 1.4844 | | | 1.4892 | 1.4977 | 1.5058 | 1.5135 | 1.5210 | 1.5281 | 1.5417 | 1.5546 | 1.5667 | 1.5783 | 1.5894 | 1.6001 |
| | v | 0.0193 | 1.1470 | | | 1.1571 | 1.1804 | 1.2030 | 1.2249 | 1.2463 | 1.2672 | 1.3076 | 1.3466 | 1.3845 | 1.4214 | 1.4574 | 1.4928 |
| **405** | h | 425.4 | 1204.5 | | | 1207.9 | 1215.7 | 1223.2 | 1230.5 | 1237.6 | 1244.5 | 1257.8 | 1270.5 | 1282.8 | 1294.8 | 1306.5 | 1318.0 |
| (445.81) | s | 0.6229 | 1.4832 | | | 1.4870 | 1.4956 | 1.5037 | 1.5115 | 1.5190 | 1.5262 | 1.5399 | 1.5528 | 1.5650 | 1.5766 | 1.5878 | 1.5985 |
| | v | 0.0194 | 1.1330 | | | 1.1401 | 1.1634 | 1.1859 | 1.2077 | 1.2289 | 1.2496 | 1.2898 | 1.3285 | 1.3660 | 1.4026 | 1.4383 | 1.4734 |
| **410** | h | 426.8 | 1204.5 | | | 1206.9 | 1214.8 | 1222.4 | 1229.8 | 1236.9 | 1243.8 | 1257.2 | 1270.0 | 1282.3 | 1294.3 | 1306.1 | 1317.6 |
| (447.01) | s | 0.6243 | 1.4821 | | | 1.4848 | 1.4935 | 1.5017 | 1.5095 | 1.5171 | 1.5243 | 1.5381 | 1.5510 | 1.5632 | 1.5749 | 1.5861 | 1.5969 |
| | v | 0.0194 | 1.1194 | | | 1.1236 | 1.1468 | 1.1692 | 1.1908 | 1.2119 | 1.2325 | 1.2724 | 1.3108 | 1.3480 | 1.3842 | 1.4196 | 1.4544 |
| **415** | h | 428.1 | 1204.6 | | | 1206.0 | 1214.0 | 1221.6 | 1229.0 | 1236.2 | 1243.2 | 1256.6 | 1269.4 | 1281.8 | 1293.9 | 1305.7 | 1317.2 |
| (448.21) | s | 0.6258 | 1.4810 | | | 1.4826 | 1.4913 | 1.4996 | 1.5075 | 1.5151 | 1.5224 | 1.5363 | 1.5492 | 1.5615 | 1.5733 | 1.5845 | 1.5953 |
| | v | 0.0194 | 1.1061 | | | 1.1075 | 1.1305 | 1.1528 | 1.1744 | 1.1953 | 1.2158 | 1.2554 | 1.2935 | 1.3304 | 1.3663 | 1.4014 | 1.4359 |
| **420** | h | 429.4 | 1204.6 | | | 1205.0 | 1213.1 | 1220.8 | 1228.3 | 1235.5 | 1242.5 | 1256.0 | 1268.9 | 1281.4 | 1293.5 | 1305.3 | 1316.9 |
| (449.39) | s | 0.6272 | 1.4799 | | | 1.4804 | 1.4892 | 1.4976 | 1.5056 | 1.5132 | 1.5205 | 1.5345 | 1.5475 | 1.5598 | 1.5716 | 1.5829 | 1.5937 |
| | v | 0.0194 | 1.0931 | | | | 1.1146 | 1.1368 | 1.1583 | 1.1791 | 1.1995 | 1.2388 | 1.2766 | 1.3132 | 1.3488 | 1.3836 | 1.4177 |
| **425** | h | 430.7 | 1204.6 | | | | 1212.2 | 1220.0 | 1227.5 | 1234.8 | 1241.8 | 1255.4 | 1268.3 | 1280.9 | 1293.0 | 1304.9 | 1316.5 |
| (450.57) | s | 0.6287 | 1.4788 | | | | 1.4872 | 1.4956 | 1.5036 | 1.5113 | 1.5187 | 1.5327 | 1.5458 | 1 5582 | 1.5700 | 1.5813 | 1.5921 |

**Temperature—Degrees Fahrenheit**

| 640° | 660° | 680° | 700° | 720° | 740° | 760° | 780° | 800° | 850° | 900° | 950° | 1000° | 1200° | 1400° | 1600° | | Abs. Press. Lb./Sq. In. (Sat. Temp.) |
|---|---|---|---|---|---|---|---|---|---|---|---|---|---|---|---|---|---|
| 1.8379 | 1.8777 | 1.9171 | 1.9562 | 1.9949 | 2.033 | 2.072 | 2.110 | 2.147 | 2.241 | 2.334 | 2.426 | 2.518 | 2.881 | 3.240 | 3.596 | v | |
| 1333.7 | 1344.6 | 1355.4 | 1366.1 | 1376.7 | 1387.3 | 1397.9 | 1408.4 | 1419.0 | 1445.2 | 1471.5 | 1497.7 | 1524.1 | 1630.9 | 1740.4 | 1853.2 | h | **340** |
| 1.6312 | 1.6410 | 1.6505 | 1.6599 | 1.6690 | 1.6779 | 1.6866 | 1.6952 | 1.7036 | 1.7241 | 1.7437 | 1.7627 | 1.7811 | 1.8496 | 1.9119 | 1.9695 | s | (428.97) |
| 1.8099 | 1.8492 | 1.8881 | 1.9267 | 1.9650 | 2.003 | 2.041 | 2.079 | 2.115 | 2.208 | 2.300 | 2.391 | 2.481 | 2.839 | 3.193 | 3.544 | v | |
| 1333.4 | 1344.3 | 1355.1 | 1365.8 | 1376.5 | 1387.1 | 1397.7 | 1408.2 | 1418.8 | 1445.0 | 1471.3 | 1497.6 | 1524.0 | 1630.8 | 1740.3 | 1853.2 | h | **345** |
| 1.6294 | 1.6392 | 1.6487 | 1.6581 | 1.6672 | 1.6761 | 1.6849 | 1.6935 | 1.7019 | 1.7223 | 1.7420 | 1.7610 | 1.7794 | 1.8480 | 1.9103 | 1.9679 | s | (430.35) |
| 1.7827 | 1.8215 | 1.8600 | 1.8980 | 1.9358 | 1.9732 | 2.010 | 2.048 | 2.084 | 2.176 | 2.266 | 2.356 | 2.445 | 2.798 | 3.147 | 3.493 | v | |
| 1333.0 | 1343.9 | 1354.8 | 1365.5 | 1376.2 | 1386.8 | 1397.4 | 1408.0 | 1418.5 | 1444.8 | 1471.1 | 1497.4 | 1523.8 | 1630.7 | 1740.3 | 1853.1 | h | **350** |
| 1.6275 | 1.6374 | 1.6470 | 1.6563 | 1.6654 | 1.6744 | 1.6831 | 1.6917 | 1.7002 | 1.7206 | 1.7403 | 1.7594 | 1.7777 | 1.8463 | 1.9086 | 1.9663 | s | (431.72) |
| 1.7563 | 1.7946 | 1.8326 | 1.8702 | 1.9074 | 1.9444 | 1.9812 | 2.018 | 2.054 | 2.145 | 2.234 | 2.322 | 2.410 | 2.758 | 3.102 | 3.444 | v | |
| 1332.7 | 1343.6 | 1354.5 | 1365.2 | 1375.9 | 1386.6 | 1397.2 | 1407.8 | 1418.3 | 1444.7 | 1471.0 | 1497.3 | 1523.7 | 1630.6 | 1740.2 | 1853.1 | h | **355** |
| 1.6258 | 1.6356 | 1.6452 | 1.6546 | 1.6637 | 1.6727 | 1.6814 | 1.6900 | 1.6985 | 1.7190 | 1.7387 | 1.7577 | 1.7761 | 1.8447 | 1.9071 | 1.9647 | s | (433.06) |
| 1.7306 | 1.7685 | 1.8060 | 1.8431 | 1.8799 | 1.9164 | 1.9527 | 1.9888 | 2.025 | 2.114 | 2.202 | 2.289 | 2.376 | 2.719 | 3.059 | 3.396 | v | |
| 1332.4 | 1343.3 | 1354.2 | 1365.0 | 1375.7 | 1386.3 | 1397.0 | 1407.5 | 1418.1 | 1444.5 | 1470.8 | 1497.1 | 1523.5 | 1630.5 | 1740.1 | 1853.0 | h | **360** |
| 1.6240 | 1.6339 | 1.6435 | 1.6529 | 1.6620 | 1.6710 | 1.6798 | 1.6884 | 1.6968 | 1.7173 | 1.7371 | 1.7561 | 1.7745 | 1.8431 | 1.9055 | 1.9631 | s | (434.40) |
| 1.7056 | 1.7430 | 1.7801 | 1.8168 | 1.8531 | 1.8892 | 1.9250 | 1.9606 | 1.9961 | 2.084 | 2.171 | 2.257 | 2.343 | 2.682 | 3.017 | 3.349 | v | |
| 1332.0 | 1343.0 | 1353.9 | 1364.7 | 1375.4 | 1386.1 | 1396.7 | 1407.3 | 1417.9 | 1444.3 | 1470.6 | 1497.0 | 1523.4 | 1630.4 | 1740.0 | 1852.9 | h | **365** |
| 1.6222 | 1.6321 | 1.6417 | 1.6511 | 1.6603 | 1.6693 | 1.6781 | 1.6867 | 1.6952 | 1.7157 | 1.7354 | 1.7545 | 1.7729 | 1.8416 | 1.9039 | 1.9616 | s | (435.72) |
| 1.6812 | 1.7182 | 1.7549 | 1.7911 | 1.8271 | 1.8627 | 1.8981 | 1.9333 | 1.9683 | 2.055 | 2.141 | 2.226 | 2.311 | 2.645 | 2.976 | 3.304 | v | |
| 1331.7 | 1342.7 | 1353.6 | 1364.4 | 1375.1 | 1385.8 | 1396.5 | 1407.1 | 1417.7 | 1444.1 | 1470.4 | 1496.8 | 1523.3 | 1630.3 | 1740.0 | 1852.9 | h | **370** |
| 1.6205 | 1.6304 | 1.6401 | 1.6495 | 1.6587 | 1.6677 | 1.6765 | 1.6851 | 1.6936 | 1.7141 | 1.7339 | 1.7529 | 1.7713 | 1.8400 | 1.9024 | 1.9601 | s | (437.03) |
| 1.6575 | 1.6941 | 1.7304 | 1.7662 | 1.8017 | 1.8369 | 1.8719 | 1.9067 | 1.9413 | 2.027 | 2.112 | 2.196 | 2.279 | 2.610 | 2.936 | 3.260 | v | |
| 1331.4 | 1342.4 | 1353.3 | 1364.1 | 1374.9 | 1385.6 | 1396.2 | 1406.9 | 1417.5 | 1443.9 | 1470.3 | 1496.7 | 1523.1 | 1630.2 | 1739.9 | 1852.8 | h | **375** |
| 1.6188 | 1.6287 | 1.6384 | 1.6478 | 1.6570 | 1.6660 | 1.6748 | 1.6835 | 1.6919 | 1.7125 | 1.7323 | 1.7514 | 1.7698 | 1.8385 | 1.9009 | 1.9586 | s | (438.32) |
| 1.6345 | 1.6707 | 1.7065 | 1.7419 | 1.7770 | 1.8118 | 1.8464 | 1.8808 | 1.9149 | 1.9996 | 2.083 | 2.166 | 2.249 | 2.575 | 2.897 | 3.217 | v | |
| 1331.0 | 1342.0 | 1353.0 | 1363.8 | 1374.6 | 1385.3 | 1396.0 | 1406.6 | 1417.3 | 1443.7 | 1470.1 | 1496.5 | 1523.0 | 1630.0 | 1739.8 | 1852.7 | h | **380** |
| 1.6171 | 1.6270 | 1.6367 | 1.6462 | 1.6554 | 1.6644 | 1.6732 | 1.6819 | 1.6904 | 1.7110 | 1.7307 | 1.7498 | 1.7683 | 1.8370 | 1.8994 | 1.9571 | s | (439.60) |
| 1.6120 | 1.6478 | 1.6832 | 1.7183 | 1.7529 | 1.7873 | 1.8215 | 1.8555 | 1.8893 | 1.9729 | 2.056 | 2.138 | 2.219 | 2.541 | 2.859 | 3.175 | v | |
| 1330.7 | 1341.7 | 1352.7 | 1363.6 | 1374.3 | 1385.1 | 1395.8 | 1406.4 | 1417.0 | 1443.5 | 1469.9 | 1496.4 | 1522.8 | 1629.9 | 1739.7 | 1852.7 | h | **385** |
| 1.6154 | 1.6254 | 1.6351 | 1.6445 | 1.6538 | 1.6628 | 1.6716 | 1.6803 | 1.6888 | 1.7094 | 1.7292 | 1.7483 | 1.7667 | 1.8355 | 1.8979 | 1.9556 | s | (440.86) |
| 1.5901 | 1.6255 | 1.6606 | 1.6952 | 1.7295 | 1.7635 | 1.7973 | 1.8309 | 1.8643 | 1.9469 | 2.029 | 2.110 | 2.190 | 2.508 | 2.822 | 3.134 | v | |
| 1330.3 | 1341.4 | 1352.4 | 1363.3 | 1374.1 | 1384.8 | 1395.5 | 1406.2 | 1416.8 | 1443.3 | 1469.8 | 1496.2 | 1522.7 | 1629.8 | 1739.6 | 1852.6 | h | **390** |
| 1.6138 | 1.6237 | 1.6335 | 1.6429 | 1.6522 | 1.6612 | 1.6701 | 1.6788 | 1.6873 | 1.7079 | 1.7277 | 1.7468 | 1.7653 | 1.8340 | 1.8965 | 1.9542 | s | (442.12) |
| 1.5688 | 1.6038 | 1.6385 | 1.6727 | 1.7067 | 1.7403 | 1.7737 | 1.8069 | 1.8399 | 1.9215 | 2.002 | 2.083 | 2.162 | 2.476 | 2.786 | 3.094 | v | |
| 1330.0 | 1341.1 | 1352.1 | 1363.0 | 1373.8 | 1384.6 | 1395.3 | 1406.0 | 1416.6 | 1443.1 | 1469.6 | 1496.0 | 1522.6 | 1629.7 | 1739.6 | 1852.6 | h | **395** |
| 1.6121 | 1.6221 | 1.6319 | 1.6413 | 1.6506 | 1.6597 | 1.6685 | 1.6772 | 1.6857 | 1.7064 | 1.7262 | 1.7453 | 1.7638 | 1.8326 | 1.8950 | 1.9527 | s | (443.36) |
| 1.5480 | 1.5827 | 1.6169 | 1.6508 | 1.6844 | 1.7177 | 1.7507 | 1.7835 | 1.8161 | 1.8968 | 1.9767 | 2.056 | 2.134 | 2.445 | 2.751 | 3.055 | v | |
| 1329.6 | 1340.8 | 1351.8 | 1362.7 | 1373.6 | 1384.3 | 1395.1 | 1405.8 | 1416.4 | 1442.9 | 1469.4 | 1495.9 | 1522.4 | 1629.6 | 1739.5 | 1852.5 | h | **400** |
| 1.6105 | 1.6205 | 1.6303 | 1.6398 | 1.6491 | 1.6581 | 1.6670 | 1.6757 | 1.6842 | 1.7049 | 1.7247 | 1.7438 | 1.7623 | 1.8311 | 1.8936 | 1.9513 | s | (444.59) |
| 1.5277 | 1.5620 | 1.5959 | 1.6294 | 1.6626 | 1.6956 | 1.7282 | 1.7607 | 1.7929 | 1.8727 | 1.9517 | 2.030 | 2.108 | 2.414 | 2.717 | 3.017 | v | |
| 1329.3 | 1340.4 | 1351.5 | 1362.4 | 1373.3 | 1384.1 | 1394.8 | 1405.5 | 1416.2 | 1442.8 | 1469.2 | 1495.7 | 1522.3 | 1629.5 | 1739.4 | 1852.4 | h | **405** |
| 1.6089 | 1.6189 | 1.6287 | 1.6382 | 1.6475 | 1.6566 | 1.6655 | 1.6742 | 1.6827 | 1.7034 | 1.7232 | 1.7424 | 1.7609 | 1.8297 | 1.8922 | 1.9499 | s | (445.81) |
| 1.5079 | 1.5419 | 1.5754 | 1.6086 | 1.6414 | 1.6740 | 1.7063 | 1.7384 | 1.7703 | 1.8492 | 1.9273 | 2.005 | 2.081 | 2.384 | 2.684 | 2.980 | v | |
| 1328.9 | 1340.1 | 1351.2 | 1362.2 | 1373.0 | 1383.8 | 1394.6 | 1405.3 | 1416.0 | 1442.6 | 1469.1 | 1495.6 | 1522.1 | 1629.4 | 1739.3 | 1852.4 | h | **410** |
| 1.6073 | 1.6174 | 1.6272 | 1.6367 | 1.6460 | 1.6551 | 1.6640 | 1.6727 | 1.6812 | 1.7019 | 1.7218 | 1.7409 | 1.7594 | 1.8283 | 1.8908 | 1.9485 | s | (447.01) |
| 1.4885 | 1.5222 | 1.5554 | 1.5883 | 1.6207 | 1.6529 | 1.6849 | 1.7167 | 1.7482 | 1.8263 | 1.9035 | 1.9800 | 2.056 | 2.355 | 2.651 | 2.944 | v | |
| 1328.6 | 1339.8 | 1350.9 | 1361.9 | 1372.8 | 1383.6 | 1394.4 | 1405.1 | 1415.8 | 1442.4 | 1468.9 | 1495.4 | 1522.0 | 1629.3 | 1739.2 | 1852.3 | h | **415** |
| 1.6057 | 1.6158 | 1.6256 | 1.6352 | 1.6445 | 1.6536 | 1.6625 | 1.6712 | 1.6798 | 1.7005 | 1.7204 | 1.7395 | 1.7580 | 1.8269 | 1.8894 | 1.9472 | s | (448.21) |
| 1.4697 | 1.5030 | 1.5359 | 1.5684 | 1.6005 | 1.6324 | 1.6641 | 1.6955 | 1.7267 | 1.8039 | 1.8802 | 1.9558 | 2.031 | 2.327 | 2.619 | 2.909 | v | |
| 1328.3 | 1339.5 | 1350.6 | 1361.6 | 1372.5 | 1383.3 | 1394.1 | 1404.8 | 1415.5 | 1442.2 | 1468.7 | 1495.3 | 1521.9 | 1629.2 | 1739.2 | 1852.2 | h | **420** |
| 1.6042 | 1.6143 | 1.6241 | 1.6337 | 1.6430 | 1.6521 | 1.6610 | 1.6697 | 1.6783 | 1.6990 | 1.7189 | 1.7381 | 1.7566 | 1.8256 | 1.8881 | 1.9458 | s | (449.39) |
| 1.4512 | 1.4842 | 1.5168 | 1.5490 | 1.5808 | 1.6124 | 1.6437 | 1.6748 | 1.7056 | 1.7820 | 1.8576 | 1.9322 | 2.007 | 2.299 | 2.588 | 2.875 | v | |
| 1327.9 | 1339.2 | 1350.3 | 1361.3 | 1372.2 | 1383.1 | 1393.9 | 1404.6 | 1415.3 | 1442.0 | 1468.6 | 1495.1 | 1521.7 | 1629.1 | 1739.1 | 1852.2 | h | **425** |
| 1.6026 | 1.6127 | 1.6226 | 1.6322 | 1.6415 | 1.6507 | 1.6596 | 1.6683 | 1.6769 | 1.6976 | 1.7175 | 1.7367 | 1.7553 | 1.8242 | 1.8868 | 1.9445 | s | (450.57) |

## Table 3. Superheated Vapor

| Abs. Press. Lb./Sq. In. (Sat. Temp.) | | Sat. Liquid | Sat. Vapor | Temperature—Degrees Fahrenheit 460° | 470° | 480° | 490° | 500° | 520° | 540° | 560° | 580° | 600° | 620° | 640° | 660° | 680° |
|---|---|---|---|---|---|---|---|---|---|---|---|---|---|---|---|---|---|
| 430 | v | 0.0194 | 1.0803 | 1.0991 | 1.1212 | 1.1425 | 1.1633 | 1.1835 | 1.2226 | 1.2601 | 1.2965 | 1.3318 | 1.3662 | 1.4000 | 1.4332 | 1.4659 | 1.4982 |
| (451.73) | h | 432.1 | 1204.6 | 1211.3 | 1219.2 | 1226.7 | 1234.0 | 1241.1 | 1254.8 | 1267.8 | 1280.4 | 1292.6 | 1304.4 | 1316.1 | 1327.6 | 1338.8 | 1350.0 |
| | s | 0.6301 | 1.4777 | 1.4851 | 1.4936 | 1.5017 | 1.5094 | 1.5168 | 1.5309 | 1.5441 | 1.5565 | 1.5684 | 1.5797 | 1.5906 | 1.6011 | 1.6112 | 1.6211 |
| 435 | v | 0.0195 | 1.0678 | 1.0839 | 1.1059 | 1.1272 | 1.1479 | 1.1679 | 1.2067 | 1.2440 | 1.2800 | 1.3151 | 1.3493 | 1.3827 | 1.4157 | 1.4480 | 1.4800 |
| (452.88) | h | 433.4 | 1204.6 | 1210.4 | 1218.3 | 1225.9 | 1233.3 | 1240.5 | 1254.2 | 1267.3 | 1279.9 | 1292.1 | 1304.0 | 1315.7 | 1327.2 | 1338.5 | 1349.7 |
| | s | 0.6315 | 1.4766 | 1.4830 | 1.4916 | 1.4997 | 1.5075 | 1.5150 | 1.5291 | 1.5424 | 1.5549 | 1.5667 | 1.5781 | 1.5890 | 1.5996 | 1.6097 | 1.6196 |
| 440 | v | 0.0195 | 1.0556 | 1.0691 | 1.0910 | 1.1121 | 1.1327 | 1.1526 | 1.1912 | 1.2282 | 1.2640 | 1.2988 | 1.3327 | 1.3658 | 1.3984 | 1.4306 | 1.4622 |
| (454.02) | h | 434.6 | 1204.6 | 1209.5 | 1217.5 | 1225.2 | 1232.6 | 1239.8 | 1253.6 | 1266.7 | 1279.4 | 1291.7 | 1303.6 | 1315.4 | 1326.9 | 1338.2 | 1349.4 |
| | s | 0.6329 | 1.4755 | 1.4809 | 1.4896 | 1.4978 | 1.5056 | 1.5132 | 1.5274 | 1.5407 | 1.5532 | 1.5652 | 1.5766 | 1.5875 | 1.5981 | 1.6083 | 1.6182 |
| 445 | v | 0.0195 | 1.0437 | 1.0546 | 1.0764 | 1.0974 | 1.1179 | 1.1377 | 1.1760 | 1.2128 | 1.2483 | 1.2828 | 1.3164 | 1.3493 | 1.3816 | 1.4134 | 1.4448 |
| (455.15) | h | 435.9 | 1204.6 | 1208.6 | 1216.7 | 1224.4 | 1231.8 | 1239.1 | 1253.0 | 1266.2 | 1278.9 | 1291.2 | 1303.2 | 1315.0 | 1326.5 | 1337.9 | 1349.1 |
| | s | 0.6343 | 1.4745 | 1.4789 | 1.4876 | 1.4959 | 1.5038 | 1.5113 | 1.5256 | 1.5390 | 1.5516 | 1.5636 | 1.5750 | 1.5860 | 1.5966 | 1.6068 | 1.6167 |
| 450 | v | 0.0195 | 1.0320 | 1.0403 | 1.0621 | 1.0830 | 1.1033 | 1.1231 | 1.1612 | 1.1977 | 1.2330 | 1.2672 | 1.3005 | 1.3332 | 1.3652 | 1.3967 | 1.4278 |
| (456.28) | h | 437.2 | 1204.6 | 1207.7 | 1215.8 | 1223.6 | 1231.1 | 1238.4 | 1252.3 | 1265.6 | 1278.4 | 1290.8 | 1302.8 | 1314.6 | 1326.2 | 1337.5 | 1348.8 |
| | s | 0.6356 | 1.4734 | 1.4768 | 1.4856 | 1.4939 | 1.5019 | 1.5095 | 1.5239 | 1.5373 | 1.5500 | 1.5620 | 1.5735 | 1.5845 | 1.5951 | 1.6054 | 1.6153 |
| 455 | v | 0.0195 | 1.0206 | 1.0264 | 1.0481 | 1.0689 | 1.0892 | 1.1088 | 1.1467 | 1.1830 | 1.2180 | 1.2519 | 1.2850 | 1.3174 | 1.3491 | 1.3804 | 1.4111 |
| (457.39) | h | 438.5 | 1204.6 | 1206.8 | 1215.0 | 1222.8 | 1230.4 | 1237.7 | 1251.7 | 1265.1 | 1277.9 | 1290.3 | 1302.4 | 1314.2 | 1325.8 | 1337.2 | 1348.5 |
| | s | 0.6370 | 1.4724 | 1.4748 | 1.4836 | 1.4920 | 1.5000 | 1.5077 | 1.5222 | 1.5357 | 1.5484 | 1.5605 | 1.5720 | 1.5830 | 1.5936 | 1.6039 | 1.6139 |
| 460 | v | 0.0196 | 1.0094 | 1.0128 | 1.0344 | 1.0551 | 1.0753 | 1.0948 | 1.1325 | 1.1685 | 1.2033 | 1.2370 | 1.2698 | 1.3019 | 1.3334 | 1.3644 | 1.3949 |
| (458.50) | h | 439.7 | 1204.6 | 1205.8 | 1214.1 | 1222.0 | 1229.6 | 1237.0 | 1251.1 | 1264.5 | 1277.4 | 1289.9 | 1302.0 | 1313.8 | 1325.4 | 1336.9 | 1348.1 |
| | s | 0.6383 | 1.4713 | 1.4728 | 1.4817 | 1.4901 | 1.4982 | 1.5059 | 1.5205 | 1.5340 | 1.5468 | 1.5589 | 1.5705 | 1.5815 | 1.5922 | 1.6025 | 1.6125 |
| 465 | v | 0.0196 | 0.9985 | 0.9994 | 1.0210 | 1.0416 | 1.0617 | 1.0811 | 1.1186 | 1.1544 | 1.1889 | 1.2224 | 1.2550 | 1.2868 | 1.3180 | 1.3488 | 1.3789 |
| (459.59) | h | 440.9 | 1204.6 | 1204.9 | 1213.2 | 1221.2 | 1228.9 | 1236.3 | 1250.5 | 1264.0 | 1276.9 | 1289.4 | 1301.6 | 1313.5 | 1325.1 | 1336.5 | 1347.8 |
| | s | 0.6397 | 1.4703 | 1.4707 | 1.4797 | 1.4882 | 1.4964 | 1.5042 | 1.5188 | 1.5324 | 1.5452 | 1.5574 | 1.5690 | 1.5800 | 1.5907 | 1.6011 | 1.6111 |
| 470 | v | 0.0196 | 0.9878 | | 1.0078 | 1.0284 | 1.0483 | 1.0677 | 1.1050 | 1.1406 | 1.1749 | 1.2081 | 1.2404 | 1.2720 | 1.3029 | 1.3335 | 1.3633 |
| (460.68) | h | 442.2 | 1204.6 | | 1212.4 | 1220.4 | 1228.1 | 1235.6 | 1249.9 | 1263.4 | 1276.4 | 1289.0 | 1301.2 | 1313.1 | 1324.7 | 1336.2 | 1347.5 |
| | s | 0.6410 | 1.4693 | | 1.4778 | 1.4864 | 1.4946 | 1.5024 | 1.5171 | 1.5308 | 1.5437 | 1.5559 | 1.5675 | 1.5786 | 1.5893 | 1.5997 | 1.6097 |
| 475 | v | 0.0196 | 0.9773 | | 0.9949 | 1.0154 | 1.0353 | 1.0546 | 1.0916 | 1.1270 | 1.1611 | 1.1941 | 1.2261 | 1.2575 | 1.2882 | 1.3185 | 1.3481 |
| (461.75) | h | 443.4 | 1204.5 | | 1211.5 | 1219.6 | 1227.3 | 1234.9 | 1249.2 | 1262.8 | 1275.9 | 1288.5 | 1300.7 | 1312.7 | 1324.4 | 1335.9 | 1347.2 |
| | s | 0.6423 | 1.4683 | | 1.4758 | 1.4845 | 1.4927 | 1.5006 | 1.5154 | 1.5292 | 1.5421 | 1.5543 | 1.5660 | 1.5772 | 1.5879 | 1.5983 | 1.6083 |
| 480 | v | 0.0197 | 0.9670 | | 0.9822 | 1.0027 | 1.0225 | 1.0417 | 1.0786 | 1.1138 | 1.1476 | 1.1803 | 1.2122 | 1.2433 | 1.2737 | 1.3038 | 1.3331 |
| (462.82) | h | 444.6 | 1204.5 | | 1210.6 | 1218.8 | 1226.6 | 1234.2 | 1248.6 | 1262.3 | 1275.4 | 1288.0 | 1300.3 | 1312.3 | 1324.0 | 1335.6 | 1346.9 |
| | s | 0.6436 | 1.4673 | | 1.4739 | 1.4826 | 1.4909 | 1.4989 | 1.5138 | 1.5276 | 1.5406 | 1.5528 | 1.5645 | 1.5757 | 1.5865 | 1.5969 | 1.6069 |
| 485 | v | 0.0197 | 0.9569 | | 0.9698 | 0.9902 | 1.0100 | 1.0291 | 1.0658 | 1.1007 | 1.1344 | 1.1669 | 1.1985 | 1.2294 | 1.2596 | 1.2895 | 1.3185 |
| (463.88) | h | 445.8 | 1204.5 | | 1209.7 | 1217.9 | 1225.8 | 1233.4 | 1248.0 | 1261.7 | 1274.9 | 1287.6 | 1299.9 | 1311.9 | 1323.7 | 1335.2 | 1346.6 |
| | s | 0.6449 | 1.4663 | | 1.4720 | 1.4808 | 1.4891 | 1.4971 | 1.5121 | 1.5260 | 1.5390 | 1.5514 | 1.5631 | 1.5743 | 1.5851 | 1.5955 | 1.6056 |
| 490 | v | 0.0197 | 0.9470 | | 0.9576 | 0.9780 | 0.9977 | 1.0167 | 1.0532 | 1.0880 | 1.1214 | 1.1537 | 1.1851 | 1.2157 | 1.2457 | 1.2754 | 1.3042 |
| (464.93) | h | 447.0 | 1204.5 | | 1208.8 | 1217.1 | 1225.0 | 1232.7 | 1247.3 | 1261.2 | 1274.4 | 1287.1 | 1299.5 | 1311.5 | 1323.3 | 1334.9 | 1346.3 |
| | s | 0.6462 | 1.4653 | | 1.4700 | 1.4789 | 1.4874 | 1.4954 | 1.5104 | 1.5244 | 1.5375 | 1.5499 | 1.5617 | 1.5729 | 1.5837 | 1.5942 | 1.6043 |
| 495 | v | 0.0197 | 0.9373 | | 0.9457 | 0.9660 | 0.9856 | 1.0046 | 1.0409 | 1.0755 | 1.1087 | 1.1408 | 1.1720 | 1.2024 | 1.2321 | 1.2616 | 1.2901 |
| (465.97) | h | 448.2 | 1204.4 | | 1207.9 | 1216.3 | 1224.3 | 1232.0 | 1246.7 | 1260.6 | 1273.9 | 1286.7 | 1299.0 | 1311.1 | 1322.9 | 1334.6 | 1346.0 |
| | s | 0.6475 | 1.4644 | | 1.4681 | 1.4771 | 1.4856 | 1.4936 | 1.5088 | 1.5229 | 1.5360 | 1.5484 | 1.5602 | 1.5715 | 1.5824 | 1.5928 | 1.6029 |
| 500 | v | 0.0197 | 0.9278 | | 0.9340 | 0.9543 | 0.9738 | 0.9927 | 1.0289 | 1.0633 | 1.0963 | 1.1282 | 1.1591 | 1.1893 | 1.2188 | 1.2478 | 1.2763 |
| (467.01) | h | 449.4 | 1204.4 | | 1207.0 | 1215.4 | 1223.5 | 1231.3 | 1246.0 | 1260.0 | 1273.4 | 1286.2 | 1298.6 | 1310.7 | 1322.6 | 1334.2 | 1345.7 |
| | s | 0.6487 | 1.4634 | | 1.4662 | 1.4752 | 1.4838 | 1.4919 | 1.5072 | 1.5213 | 1.5345 | 1.5470 | 1.5588 | 1.5701 | 1.5810 | 1.5915 | 1.6016 |
| 510 | v | 0.0198 | 0.9093 | | 0.9113 | 0.9314 | 0.9508 | 0.9696 | 1.0054 | 1.0395 | 1.0721 | 1.1036 | 1.1341 | 1.1639 | 1.1930 | 1.2215 | 1.2496 |
| (469.05) | h | 451.8 | 1204.3 | | 1205.2 | 1213.7 | 1221.9 | 1229.8 | 1244.7 | 1258.9 | 1272.4 | 1285.3 | 1297.8 | 1310.0 | 1321.9 | 1333.6 | 1345.0 |
| | s | 0.6512 | 1.4615 | | 1.4624 | 1.4716 | 1.4802 | 1.4885 | 1.5039 | 1.5182 | 1.5315 | 1.5441 | 1.5560 | 1.5674 | 1.5783 | 1.5889 | 1.5990 |
| 520 | v | 0.0198 | 0.8915 | | | 0.9094 | 0.9287 | 0.9473 | 0.9829 | 1.0166 | 1.0488 | 1.0799 | 1.1101 | 1.1394 | 1.1681 | 1.1962 | 1.2239 |
| (471.07) | h | 454.1 | 1204.2 | | | 1212.0 | 1220.3 | 1228.3 | 1243.4 | 1257.7 | 1271.3 | 1284.3 | 1296.9 | 1309.2 | 1321.1 | 1332.9 | 1344.4 |
| | s | 0.6536 | 1.4596 | | | 1.4679 | 1.4767 | 1.4851 | 1.5007 | 1.5151 | 1.5286 | 1.5412 | 1.5532 | 1.5647 | 1.5757 | 1.5863 | 1.5965 |
| 530 | v | 0.0199 | 0.8743 | | | 0.8882 | 0.9074 | 0.9259 | 0.9611 | 0.9945 | 1.0264 | 1.0572 | 1.0869 | 1.1159 | 1.1442 | 1.1719 | 1.1991 |
| (473.06) | h | 456.4 | 1204.1 | | | 1210.3 | 1218.7 | 1226.8 | 1242.1 | 1256.5 | 1270.2 | 1283.4 | 1296.1 | 1308.4 | 1320.4 | 1332.2 | 1343.8 |
| | s | 0.6560 | 1.4578 | | | 1.4643 | 1.4732 | 1.4817 | 1.4975 | 1.5121 | 1.5257 | 1.5384 | 1.5505 | 1.5620 | 1.5731 | 1.5837 | 1.5940 |

Temperature—Degrees Fahrenheit

| 700° | 720° | 740° | 760° | 780° | 800° | 850° | 900° | 950° | 1000° | 1050° | 1100° | 1200° | 1400° | 1600° | | Abs. Press. Lb./Sq. In. (Sat. Temp.) |
|---|---|---|---|---|---|---|---|---|---|---|---|---|---|---|---|---|
| 1.5300 | 1.5616 | 1.5928 | 1.6238 | 1.6545 | 1.6851 | 1.7607 | 1.8354 | 1.9093 | 1.9827 | 2.056 | 2.128 | 2.272 | 2.558 | 2.841 | v | |
| 1361.0 | 1372.0 | 1382.8 | 1393.6 | 1404.4 | 1415.1 | 1441.8 | 1468.4 | 1495.0 | 1521.6 | 1548.3 | 1575.1 | 1629.0 | 1739.0 | 1852.1 | h | 430 |
| 1.6307 | 1.6401 | 1.6492 | 1.6581 | 1.6669 | 1.6755 | 1.6963 | 1.7162 | 1.7354 | 1.7539 | 1.7719 | 1.7893 | 1.8229 | 1.8854 | 1.9432 | s | (451.73) |
| 1.5115 | 1.5427 | 1.5736 | 1.6043 | 1.6348 | 1.6650 | 1.7398 | 1.8137 | 1.8868 | 1.9595 | 2.032 | 2.103 | 2.246 | 2.528 | 2.808 | v | |
| 1360.7 | 1371.7 | 1382.6 | 1393.4 | 1404.2 | 1414.9 | 1441.6 | 1468.2 | 1494.8 | 1521.4 | 1548.1 | 1574.9 | 1628.9 | 1738.9 | 1852.1 | h | 435 |
| 1.6292 | 1.6386 | 1.6478 | 1.6567 | 1.6655 | 1.6741 | 1.6949 | 1.7148 | 1.7340 | 1.7526 | 1.7706 | 1.7880 | 1.8216 | 1.8841 | 1.9419 | s | (452.88) |
| 1.4934 | 1.5243 | 1.5549 | 1.5853 | 1.6155 | 1.6454 | 1.7194 | 1.7925 | 1.8649 | 1.9368 | 2.008 | 2.079 | 2.220 | 2.499 | 2.776 | v | |
| 1360.4 | 1371.4 | 1382.3 | 1393.2 | 1403.9 | 1414.7 | 1441.4 | 1468.1 | 1494.7 | 1521.3 | 1548.0 | 1574.8 | 1628.8 | 1738.9 | 1852.0 | h | 440 |
| 1.6278 | 1.6372 | 1.6464 | 1.6553 | 1.6641 | 1.6727 | 1.6935 | 1.7135 | 1.7327 | 1.7512 | 1.7692 | 1.7867 | 1.8203 | 1.8828 | 1.9406 | s | (454.02) |
| 1.4757 | 1.5063 | 1.5366 | 1.5667 | 1.5966 | 1.6262 | 1.6995 | 1.7718 | 1.8435 | 1.9146 | 1.9851 | 2.055 | 2.195 | 2.471 | 2.745 | v | |
| 1360.2 | 1371.2 | 1382.1 | 1392.9 | 1403.7 | 1414.5 | 1441.2 | 1467.9 | 1494.5 | 1521.2 | 1547.9 | 1574.7 | 1628.7 | 1738.8 | 1851.9 | h | 445 |
| 1.6264 | 1.6358 | 1.6450 | 1.6539 | 1.6627 | 1.6713 | 1.6921 | 1.7121 | 1.7313 | 1.7499 | 1.7679 | 1.7854 | 1.8190 | 1.8816 | 1.9393 | s | (455.15) |
| 1.4584 | 1.4888 | 1.5188 | 1.5486 | 1.5781 | 1.6074 | 1.6800 | 1.7516 | 1.8225 | 1.8928 | 1.9627 | 2.032 | 2.170 | 2.443 | 2.714 | v | |
| 1359.9 | 1370.9 | 1381.8 | 1392.7 | 1403.5 | 1414.3 | 1441.0 | 1467.7 | 1494.3 | 1521.0 | 1547.7 | 1574.6 | 1628.6 | 1738.7 | 1851.9 | h | 450 |
| 1.6250 | 1.6344 | 1.6436 | 1.6525 | 1.6613 | 1.6699 | 1.6908 | 1.7108 | 1.7300 | 1.7486 | 1.7666 | 1.7841 | 1.8177 | 1.8803 | 1.9381 | s | (456.28) |
| 1.4415 | 1.4716 | 1.5013 | 1.5308 | 1.5601 | 1.5891 | 1.6610 | 1.7318 | 1.8020 | 1.8716 | 1.9407 | 2.009 | 2.146 | 2.416 | 2.784 | v | |
| 1359.6 | 1370.6 | 1381.6 | 1392.4 | 1403.3 | 1414.0 | 1440.8 | 1467.5 | 1494.2 | 1520.9 | 1547.6 | 1574.5 | 1628.5 | 1738.6 | 1851.8 | h | 455 |
| 1.6236 | 1.6330 | 1.6422 | 1.6512 | 1.6600 | 1.6686 | 1.6895 | 1.7095 | 1.7287 | 1.7473 | 1.7653 | 1.7828 | 1.8164 | 1.8790 | 1.9368 | s | (457.39) |
| 1.4250 | 1.4547 | 1.4842 | 1.5134 | 1.5424 | 1.5711 | 1.6423 | 1.7124 | 1.7819 | 1.8508 | 1.9192 | 1.9873 | 2.122 | 2.390 | 2.655 | v | |
| 1359.3 | 1370.3 | 1381.3 | 1392.2 | 1403.0 | 1413.8 | 1440.7 | 1467.4 | 1494.0 | 1520.7 | 1547.5 | 1574.3 | 1628.4 | 1738.5 | 1851.7 | h | 460 |
| 1.6222 | 1.6316 | 1.6408 | 1.6498 | 1.6586 | 1.6673 | 1.6882 | 1.7082 | 1.7274 | 1.7460 | 1.7641 | 1.7816 | 1.8152 | 1.8778 | 1.9356 | s | (458.50) |
| 1.4088 | 1.4382 | 1.4675 | 1.4964 | 1.5251 | 1.5536 | 1.6240 | 1.6935 | 1.7623 | 1.8305 | 1.8982 | 1.9655 | 2.099 | 2.364 | 2.626 | v | |
| 1359.0 | 1370.1 | 1381.1 | 1392.0 | 1402.8 | 1413.6 | 1440.5 | 1467.2 | 1493.9 | 1520.6 | 1547.4 | 1574.2 | 1628.3 | 1738.5 | 1851.7 | h | 465 |
| 1.6208 | 1.6302 | 1.6395 | 1.6485 | 1.6573 | 1.6660 | 1.6869 | 1.7069 | 1.7262 | 1.7448 | 1.7628 | 1.7803 | 1.8139 | 1.8766 | 1.9344 | s | (459.59) |
| 1.3929 | 1.4221 | 1.4511 | 1.4797 | 1.5082 | 1.5364 | 1.6061 | 1.6750 | 1.7431 | 1.8106 | 1.8776 | 1.9443 | 2.077 | 2.339 | 2.598 | v | |
| 1358.7 | 1369.8 | 1380.8 | 1391.7 | 1402.6 | 1413.4 | 1440.3 | 1467.0 | 1493.7 | 1520.4 | 1547.2 | 1574.1 | 1628.2 | 1738.4 | 1851.6 | h | 470 |
| 1.6194 | 1.6289 | 1.6382 | 1.6472 | 1.6560 | 1.6646 | 1.6856 | 1.7056 | 1.7249 | 1.7435 | 1.7616 | 1.7791 | 1.8127 | 1.8754 | 1.9332 | s | (460.68) |
| 1.3774 | 1.4063 | 1.4350 | 1.4634 | 1.4916 | 1.5196 | 1.5886 | 1.6568 | 1.7243 | 1.7911 | 1.8575 | 1.9235 | 2.054 | 2.314 | 2.571 | v | |
| 1358.4 | 1369.5 | 1380.5 | 1391.5 | 1402.3 | 1413.2 | 1440.1 | 1466.8 | 1493.6 | 1520.3 | 1547.1 | 1574.0 | 1628.1 | 1738.3 | 1851.6 | h | 475 |
| 1.6181 | 1.6276 | 1.6368 | 1.6458 | 1.6547 | 1.6633 | 1.6843 | 1.7044 | 1.7237 | 1.7423 | 1.7603 | 1.7779 | 1.8115 | 1.8742 | 1.9320 | s | (461.75) |
| 1.3622 | 1.3909 | 1.4193 | 1.4475 | 1.4754 | 1.5031 | 1.5715 | 1.6390 | 1.7058 | 1.7720 | 1.8378 | 1.9031 | 2.033 | 2.290 | 2.544 | v | |
| 1358.2 | 1369.3 | 1380.3 | 1391.2 | 1402.1 | 1412.9 | 1439.9 | 1466.7 | 1493.4 | 1520.2 | 1547.0 | 1573.9 | 1628.0 | 1738.2 | 1851.5 | h | 480 |
| 1.6167 | 1.6262 | 1.6355 | 1.6445 | 1.6534 | 1.6621 | 1.6830 | 1.7031 | 1.7224 | 1.7411 | 1.7591 | 1.7767 | 1.8103 | 1.8730 | 1.9308 | s | (462.82) |
| 1.3473 | 1.3758 | 1.4040 | 1.4318 | 1.4595 | 1.4869 | 1.5547 | 1.6216 | 1.6878 | 1.7533 | 1.8185 | 1.8831 | 2.012 | 2.266 | 2.518 | v | |
| 1357.9 | 1369.0 | 1380.0 | 1391.0 | 1401.9 | 1412.7 | 1439.7 | 1466.5 | 1493.3 | 1520.0 | 1546.8 | 1573.7 | 1627.9 | 1738.1 | 1851.4 | h | 485 |
| 1.6154 | 1.6249 | 1.6342 | 1.6433 | 1.6521 | 1.6608 | 1.6818 | 1.7019 | 1.7212 | 1.7399 | 1.7579 | 1.7755 | 1.8091 | 1.8718 | 1.9297 | s | (463.88) |
| 1.3327 | 1.3610 | 1.3890 | 1.4165 | 1.4439 | 1.4711 | 1.5383 | 1.6046 | 1.6701 | 1.7350 | 1.7995 | 1.8636 | 1.9907 | 2.242 | 2.492 | v | |
| 1357.6 | 1368.7 | 1379.8 | 1390.7 | 1401.7 | 1412.5 | 1439.5 | 1466.3 | 1493.1 | 1519.9 | 1546.7 | 1573.6 | 1627.8 | 1738.1 | 1851.4 | h | 490 |
| 1.6141 | 1.6236 | 1.6329 | 1.6420 | 1.6509 | 1.6595 | 1.6805 | 1.7007 | 1.7200 | 1.7387 | 1.7567 | 1.7743 | 1.8079 | 1.8706 | 1.9285 | s | (464.93) |
| 1.3184 | 1.3465 | 1.3742 | 1.4015 | 1.4286 | 1.4557 | 1.5222 | 1.5879 | 1.6528 | 1.7171 | 1.7810 | 1.8444 | 1.9704 | 2.219 | 2.466 | v | |
| 1357.3 | 1368.5 | 1379.5 | 1390.5 | 1401.4 | 1412.3 | 1439.3 | 1466.2 | 1493.0 | 1519.7 | 1546.6 | 1573.5 | 1627.7 | 1738.0 | 1851.3 | h | 495 |
| 1.6128 | 1.6223 | 1.6316 | 1.6407 | 1.6496 | 1.6583 | 1.6793 | 1.6994 | 1.7188 | 1.7375 | 1.7555 | 1.7731 | 1.8068 | 1.8695 | 1.9274 | s | (465.97) |
| 1.3044 | 1.3322 | 1.3596 | 1.3868 | 1.4137 | 1.4405 | 1.5065 | 1.5715 | 1.6358 | 1.6996 | 1.7628 | 1.8256 | 1.9504 | 2.197 | 2.442 | v | |
| 1357.0 | 1368.2 | 1379.3 | 1390.3 | 1401.2 | 1412.1 | 1439.1 | 1466.0 | 1492.8 | 1519.6 | 1546.4 | 1573.4 | 1627.6 | 1737.9 | 1851.3 | h | 500 |
| 1.6115 | 1.6210 | 1.6304 | 1.6395 | 1.6483 | 1.6571 | 1.6781 | 1.6982 | 1.7176 | 1.7363 | 1.7544 | 1.7719 | 1.8056 | 1.8683 | 1.9262 | s | (467.01) |
| 1.2772 | 1.3046 | 1.3315 | 1.3583 | 1.3847 | 1.4110 | 1.4758 | 1.5397 | 1.6029 | 1.6654 | 1.7275 | 1.7892 | 1.9116 | 2.154 | 2.394 | v | |
| 1356.4 | 1367.6 | 1378.7 | 1389.8 | 1400.7 | 1411.6 | 1438.7 | 1465.6 | 1492.5 | 1519.3 | 1546.2 | 1573.2 | 1627.4 | 1737.7 | 1851.1 | h | 510 |
| 1.6089 | 1.6185 | 1.6279 | 1.6370 | 1.6459 | 1.6546 | 1.6757 | 1.6959 | 1.7153 | 1.7340 | 1.7521 | 1.7696 | 1.8033 | 1.8661 | 1.9240 | s | (469.05) |
| 1.2511 | 1.2780 | 1.3045 | 1.3308 | 1.3568 | 1.3826 | 1.4464 | 1.5091 | 1.5712 | 1.6326 | 1.6936 | 1.7542 | 1.8743 | 2.112 | 2.347 | v | |
| 1355.8 | 1367.1 | 1378.2 | 1389.3 | 1400.3 | 1411.2 | 1438.3 | 1465.3 | 1492.2 | 1519.0 | 1545.9 | 1572.9 | 1627.2 | 1737.6 | 1851.0 | h | 520 |
| 1.6064 | 1.6160 | 1.6254 | 1.6345 | 1.6435 | 1.6522 | 1.6733 | 1.6935 | 1.7130 | 1.7317 | 1.7498 | 1.7674 | 1.8011 | 1.8639 | 1.9218 | s | (471.07) |
| 1.2259 | 1.2524 | 1.2785 | 1.3043 | 1.3299 | 1.3554 | 1.4180 | 1.4797 | 1.5407 | 1.6011 | 1.6609 | 1.7205 | 1.8384 | 2.072 | 2.303 | v | |
| 1355.2 | 1366.5 | 1377.7 | 1388.8 | 1399.8 | 1410.8 | 1438.0 | 1465.0 | 1491.9 | 1518.7 | 1545.7 | 1572.7 | 1627.0 | 1737.4 | 1850.9 | h | 530 |
| 1.6039 | 1.6136 | 1.6230 | 1.6321 | 1.6411 | 1.6498 | 1.6710 | 1.6912 | 1.7107 | 1.7294 | 1.7476 | 1.7652 | 1.7989 | 1.8617 | 1.9196 | s | (473.06) |

# Table 3. Superheated Vapor

| Abs. Press. Lb./Sq. In. (Sat. Temp.) | | Sat. Liquid | Sat. Vapor | Temperature—Degrees Fahrenheit 480° | 490° | 500° | 510° | 520° | 530° | 540° | 550° | 560° | 570° | 580° | 590° | 600° | 620° |
|---|---|---|---|---|---|---|---|---|---|---|---|---|---|---|---|---|---|
| | v | 0.0199 | 0.8578 | 0.8676 | 0.8868 | 0.9052 | 0.9229 | 0.9402 | 0.9569 | 0.9733 | 0.9892 | 1.0049 | 1.0202 | 1.0352 | 1.0500 | 1.0646 | 1.0932 |
| 540 | h | 458.6 | 1204.0 | 1208.5 | 1217.1 | 1225.3 | 1233.1 | 1240.8 | 1248.2 | 1255.4 | 1262.3 | 1269.2 | 1275.9 | 1282.4 | 1288.8 | 1295.2 | 1307.6 |
| (475.01) | s | 0.6584 | 1.4560 | 1.4607 | 1.4698 | 1.4784 | 1.4866 | 1.4944 | 1.5019 | 1.5091 | 1.5161 | 1.5228 | 1.5293 | 1.5356 | 1.5418 | 1.5478 | 1.5594 |
| | v | 0.0199 | 0.8419 | 0.8478 | 0.8669 | 0.8852 | 0.9028 | 0.9199 | 0.9365 | 0.9528 | 0.9686 | 0.9840 | 0.9992 | 1.0141 | 1.0287 | 1.0431 | 1.0714 |
| 550 | h | 460.8 | 1203.9 | 1206.7 | 1215.4 | 1223.7 | 1231.7 | 1239.4 | 1246.9 | 1254.2 | 1261.2 | 1268.1 | 1274.9 | 1281.5 | 1287.9 | 1294.3 | 1306.8 |
| (476.94) | s | 0.6608 | 1.4542 | 1.4571 | 1.4663 | 1.4751 | 1.4834 | 1.4913 | 1.4988 | 1.5061 | 1.5131 | 1.5199 | 1.5265 | 1.5329 | 1.5391 | 1.5451 | 1.5568 |
| | v | 0.0200 | 0.8265 | 0.8287 | 0.8477 | 0.8659 | 0.8834 | 0.9004 | 0.9169 | 0.9330 | 0.9486 | 0.9639 | 0.9789 | 0.9937 | 1.0081 | 1.0224 | 1.0503 |
| 560 | h | 463.0 | 1203.8 | 1204.9 | 1213.7 | 1222.2 | 1230.3 | 1238.1 | 1245.6 | 1253.0 | 1260.1 | 1267.1 | 1273.8 | 1280.5 | 1287.0 | 1293.4 | 1306.0 |
| (478.85) | s | 0.6631 | 1.4524 | 1.4535 | 1.4629 | 1.4717 | 1.4802 | 1.4882 | 1.4958 | 1.5032 | 1.5103 | 1.5171 | 1.5237 | 1.5302 | 1.5364 | 1.5425 | 1.5542 |
| | v | 0.0200 | 0.8117 | | 0.8291 | 0.8472 | 0.8646 | 0.8815 | 0.8979 | 0.9138 | 0.9293 | 0.9445 | 0.9594 | 0.9740 | 0.9883 | 1.0024 | 1.0299 |
| 570 | h | 465.2 | 1203.6 | | 1212.0 | 1220.6 | 1228.8 | 1236.7 | 1244.3 | 1251.8 | 1259.0 | 1266.0 | 1272.8 | 1279.5 | 1286.1 | 1292.6 | 1305.2 |
| (480.73) | s | 0.6654 | 1.4506 | | 1.4594 | 1.4684 | 1.4770 | 1.4851 | 1.4928 | 1.5003 | 1.5074 | 1.5143 | 1.5210 | 1.5275 | 1.5338 | 1.5399 | 1.5517 |
| | v | 0.0201 | 0.7973 | | 0.8111 | 0.8291 | 0.8465 | 0.8633 | 0.8795 | 0.8954 | 0.9107 | 0.9258 | 0.9405 | 0.9549 | 0.9691 | 0.9830 | 1.0103 |
| 580 | h | 467.4 | 1203.5 | | 1210.3 | 1219.0 | 1227.3 | 1235.3 | 1243.0 | 1250.5 | 1257.8 | 1264.9 | 1271.8 | 1278.5 | 1285.2 | 1291.7 | 1304.4 |
| (482.58) | s | 0.6676 | 1.4489 | | 1.4560 | 1.4652 | 1.4738 | 1.4820 | 1.4898 | 1.4974 | 1.5046 | 1.5116 | 1.5183 | 1.5248 | 1.5312 | 1.5373 | 1.5492 |
| | v | 0.0201 | 0.7833 | | 0.7937 | 0.8117 | 0.8290 | 0.8457 | 0.8618 | 0.8775 | 0.8927 | 0.9077 | 0.9222 | 0.9365 | 0.9506 | 0.9643 | 0.9913 |
| 590 | h | 469.5 | 1203.4 | | 1208.5 | 1217.4 | 1225.8 | 1233.9 | 1241.7 | 1249.3 | 1256.6 | 1263.8 | 1270.8 | 1277.6 | 1284.3 | 1290.8 | 1303.6 |
| (484.41) | s | 0.6698 | 1.4472 | | 1.4526 | 1.4619 | 1.4706 | 1.4790 | 1.4869 | 1.4945 | 1.5018 | 1.5088 | 1.5156 | 1.5222 | 1.5286 | 1.5348 | 1.5467 |
| | v | 0.0201 | 0.7698 | | 0.7768 | 0.7947 | 0.8120 | 0.8286 | 0.8446 | 0.8602 | 0.8753 | 0.8901 | 0.9046 | 0.9187 | 0.9326 | 0.9463 | 0.9729 |
| 600 | h | 471.6 | 1203.2 | | 1206.7 | 1215.7 | 1224.3 | 1232.5 | 1240.4 | 1248.1 | 1255.5 | 1262.7 | 1269.7 | 1276.6 | 1283.3 | 1289.9 | 1302.7 |
| (486.21) | s | 0.6720 | 1.4454 | | 1.4492 | 1.4586 | 1.4675 | 1.4759 | 1.4839 | 1.4916 | 1.4990 | 1.5061 | 1.5130 | 1.5196 | 1.5260 | 1.5323 | 1.5443 |
| | v | 0.0202 | 0.7567 | | 0.7604 | 0.7783 | 0.7955 | 0.8120 | 0.8279 | 0.8434 | 0.8585 | 0.8732 | 0.8875 | 0.9015 | 0.9153 | 0.9288 | 0.9551 |
| 610 | h | 473.7 | 1203.1 | | 1205.0 | 1214.1 | 1222.8 | 1231.1 | 1239.1 | 1246.8 | 1254.3 | 1261.6 | 1268.6 | 1275.6 | 1282.3 | 1289.0 | 1301.9 |
| (487.99) | s | 0.6742 | 1.4437 | | 1.4458 | 1.4553 | 1.4644 | 1.4729 | 1.4810 | 1.4888 | 1.4963 | 1.5034 | 1.5103 | 1.5170 | 1.5235 | 1.5298 | 1.5419 |
| | v | 0.0202 | 0.7440 | | 0.7445 | 0.7624 | 0.7795 | 0.7960 | 0.8118 | 0.8272 | 0.8421 | 0.8567 | 0.8709 | 0.8848 | 0.8984 | 0.9118 | 0.9379 |
| 620 | h | 475.7 | 1202.9 | | 1203.1 | 1212.4 | 1221.2 | 1229.6 | 1237.7 | 1245.5 | 1253.1 | 1260.4 | 1267.6 | 1274.6 | 1281.4 | 1288.1 | 1301.1 |
| (489.75) | s | 0.6763 | 1.4421 | | 1.4424 | 1.4521 | 1.4613 | 1.4699 | 1.4781 | 1.4860 | 1.4935 | 1.5007 | 1.5077 | 1.5145 | 1.5210 | 1.5273 | 1.5395 |
| | v | 0.0202 | 0.7317 | | | 0.7469 | 0.7640 | 0.7804 | 0.7962 | 0.8115 | 0.8263 | 0.8408 | 0.8549 | 0.8687 | 0.8821 | 0.8954 | 0.9213 |
| 630 | h | 477.8 | 1202.7 | | | 1210.7 | 1219.6 | 1228.2 | 1236.4 | 1244.3 | 1251.9 | 1259.3 | 1266.5 | 1273.5 | 1280.4 | 1287.1 | 1300.2 |
| (491.49) | s | 0.6784 | 1.4405 | | | 1.4489 | 1.4582 | 1.4669 | 1.4753 | 1.4832 | 1.4908 | 1.4981 | 1.5051 | 1.5119 | 1.5185 | 1.5249 | 1.5371 |
| | v | 0.0203 | 0.7198 | | | 0.7319 | 0.7490 | 0.7653 | 0.7810 | 0.7962 | 0.8110 | 0.8253 | 0.8393 | 0.8530 | 0.8664 | 0.8795 | 0.9051 |
| 640 | h | 479.8 | 1202.5 | | | 1209.0 | 1218.0 | 1226.7 | 1235.0 | 1243.0 | 1250.7 | 1258.2 | 1265.4 | 1272.5 | 1279.4 | 1286.2 | 1299.4 |
| (493.21) | s | 0.6805 | 1.4389 | | | 1.4456 | 1.4551 | 1.4640 | 1.4724 | 1.4804 | 1.4881 | 1.4955 | 1.5026 | 1.5094 | 1.5160 | 1.5225 | 1.5348 |
| | v | 0.0203 | 0.7083 | | | 0.7173 | 0.7343 | 0.7506 | 0.7663 | 0.7814 | 0.7961 | 0.8103 | 0.8242 | 0.8378 | 0.8511 | 0.8641 | 0.8894 |
| 650 | h | 481.8 | 1202.3 | | | 1207.2 | 1216.4 | 1225.2 | 1233.6 | 1241.7 | 1249.5 | 1257.0 | 1264.4 | 1271.5 | 1278.5 | 1285.3 | 1298.6 |
| (494.90) | s | 0.6825 | 1.4373 | | | 1.4424 | 1.4520 | 1.4610 | 1.4695 | 1.4776 | 1.4854 | 1.4928 | 1.5000 | 1.5069 | 1.5136 | 1.5201 | 1.5325 |
| | v | 0.0204 | 0.6971 | | | 0.7032 | 0.7201 | 0.7363 | 0.7520 | 0.7670 | 0.7816 | 0.7958 | 0.8096 | 0.8230 | 0.8362 | 0.8491 | 0.8742 |
| 660 | h | 483.8 | 1202.1 | | | 1205.4 | 1214.8 | 1223.7 | 1232.2 | 1240.4 | 1248.2 | 1255.9 | 1263.3 | 1270.5 | 1277.5 | 1284.4 | 1297.7 |
| (496.58) | s | 0.6846 | 1.4358 | | | 1.4392 | 1.4489 | 1.4581 | 1.4667 | 1.4749 | 1.4828 | 1.4903 | 1.4975 | 1.5045 | 1.5112 | 1.5177 | 1.5302 |
| | v | 0.0204 | 0.6862 | | | 0.6894 | 0.7063 | 0.7225 | 0.7381 | 0.7530 | 0.7675 | 0.7816 | 0.7953 | 0.8087 | 0.8218 | 0.8346 | 0.8595 |
| 670 | h | 485.8 | 1201.9 | | | 1203.6 | 1213.1 | 1222.2 | 1230.8 | 1239.0 | 1247.0 | 1254.7 | 1262.2 | 1269.4 | 1276.5 | 1283.4 | 1296.9 |
| (498.24) | s | 0.6866 | 1.4342 | | | 1.4360 | 1.4459 | 1.4551 | 1.4639 | 1.4722 | 1.4801 | 1.4877 | 1.4950 | 1.5020 | 1.5088 | 1.5154 | 1.5279 |
| | v | 0.0204 | 0.6757 | | | 0.6759 | 0.6929 | 0.7090 | 0.7245 | 0.7395 | 0.7539 | 0.7679 | 0.7815 | 0.7948 | 0.8078 | 0.8205 | 0.8451 |
| 680 | h | 487.7 | 1201.7 | | | 1201.8 | 1211.4 | 1220.6 | 1229.3 | 1237.7 | 1245.7 | 1253.5 | 1261.1 | 1268.4 | 1275.5 | 1282.5 | 1296.0 |
| (499.88) | s | 0.6886 | 1.4327 | | | 1.4328 | 1.4428 | 1.4522 | 1.4611 | 1.4695 | 1.4775 | 1.4851 | 1.4925 | 1.4996 | 1.5064 | 1.5130 | 1.5257 |
| | v | 0.0205 | 0.6654 | | | | 0.6798 | 0.6959 | 0.7114 | 0.7262 | 0.7406 | 0.7545 | 0.7681 | 0.7813 | 0.7941 | 0.8067 | 0.8312 |
| 690 | h | 489.6 | 1201.4 | | | | 1209.7 | 1219.0 | 1227.9 | 1236.3 | 1244.5 | 1252.3 | 1259.9 | 1267.3 | 1274.5 | 1281.5 | 1295.2 |
| (501.50) | s | 0.6905 | 1.4311 | | | | 1.4398 | 1.4493 | 1.4583 | 1.4667 | 1.4748 | 1.4826 | 1.4900 | 1.4971 | 1.5040 | 1.5107 | 1.5234 |
| | v | 0.0205 | 0.6554 | | | | 0.6670 | 0.6832 | 0.6986 | 0.7134 | 0.7277 | 0.7416 | 0.7550 | 0.7681 | 0.7809 | 0.7934 | 0.8177 |
| 700 | h | 491.5 | 1201.2 | | | | 1208.0 | 1217.5 | 1226.4 | 1235.0 | 1243.2 | 1251.1 | 1258.8 | 1266.3 | 1273.5 | 1280.6 | 1294.3 |
| (503.10) | s | 0.6925 | 1.4296 | | | | 1.4367 | 1.4463 | 1.4554 | 1.4640 | 1.4722 | 1.4800 | 1.4875 | 1.4947 | 1.5017 | 1.5084 | 1.5212 |
| | v | 0.0205 | 0.6457 | | | | 0.6546 | 0.6707 | 0.6861 | 0.7009 | 0.7151 | 0.7289 | 0.7423 | 0.7553 | 0.7680 | 0.7804 | 0.8045 |
| 710 | h | 493.4 | 1201.0 | | | | 1206.3 | 1215.9 | 1224.9 | 1233.6 | 1241.9 | 1249.9 | 1257.7 | 1265.2 | 1272.5 | 1279.6 | 1293.4 |
| (504.68) | s | 0.6944 | 1.4281 | | | | 1.4336 | 1.4434 | 1.4527 | 1.4614 | 1.4696 | 1.4775 | 1.4851 | 1.4924 | 1.4994 | 1.5061 | 1.5190 |

| 640° | 660° | 680° | 700° | 720° | 740° | 760° | 780° | 800° | 850° | 900° | 950° | 1000° | 1200° | 1400° | 1600° | | Abs. Press. Lb./Sq. In. (Sat. Temp.) |
|---|---|---|---|---|---|---|---|---|---|---|---|---|---|---|---|---|---|
| | | | | | | | | **Temperature—Degrees Fahrenheit** | | | | | | | | | |
| 1.1211 | 1.1485 | 1.1753 | 1.2017 | 1.2277 | 1.2535 | 1.2789 | 1.3041 | 1.3291 | 1.3907 | 1.4514 | 1.5113 | 1.5707 | 1.8039 | 2.033 | 2.260 | v | |
| 1319.7 | 1331.5 | 1343.2 | 1354.6 | 1366.0 | 1377.2 | 1388.3 | 1399.4 | 1410.3 | 1437.6 | 1464.6 | 1491.5 | 1518.5 | 1626.8 | 1737.3 | 1850.7 | h | **540** |
| 1.5705 | 1.5812 | 1.5915 | 1.6015 | 1.6111 | 1.6206 | 1.6298 | 1.6388 | 1.6475 | 1.6687 | 1.6890 | 1.7085 | 1.7272 | 1.7968 | 1.8596 | 1.9175 | s | (475.01) |
| 1.0989 | 1.1259 | 1.1523 | 1.1783 | 1.2040 | 1.2293 | 1.2544 | 1.2792 | 1.3038 | 1.3644 | 1.4241 | 1.4830 | 1.5414 | 1.7706 | 1.9957 | 2.219 | v | |
| 1318.9 | 1330.8 | 1342.5 | 1354.0 | 1365.4 | 1376.7 | 1387.8 | 1398.9 | 1409.9 | 1437.2 | 1464.3 | 1491.2 | 1518.2 | 1626.6 | 1737.1 | 1850.6 | h | **550** |
| 1.5680 | 1.5787 | 1.5890 | 1.5991 | 1.6088 | 1.6182 | 1.6274 | 1.6365 | 1.6452 | 1.6665 | 1.6868 | 1.7063 | 1.7250 | 1.7946 | 1.8575 | 1.9155 | s | (476.94) |
| 1.0775 | 1.1041 | 1.1302 | 1.1558 | 1.1811 | 1.2060 | 1.2307 | 1.2552 | 1.2794 | 1.3390 | 1.3978 | 1.4557 | 1.5132 | 1.7385 | 1.9597 | 2.179 | v | |
| 1318.2 | 1330.2 | 1341.9 | 1353.5 | 1364.9 | 1376.1 | 1387.3 | 1398.4 | 1409.4 | 1436.8 | 1463.9 | 1490.9 | 1517.9 | 1626.4 | 1737.0 | 1850.5 | h | **560** |
| 1.5655 | 1.5762 | 1.5866 | 1.5967 | 1.6064 | 1.6159 | 1.6251 | 1.6342 | 1.6430 | 1.6643 | 1.6846 | 1.7041 | 1.7229 | 1.7926 | 1.8555 | 1.9134 | s | (478.85) |
| 1.0568 | 1.0830 | 1.1088 | 1.1341 | 1.1590 | 1.1835 | 1.2079 | 1.2320 | 1.2558 | 1.3145 | 1.3724 | 1.4294 | 1.4859 | 1.7075 | 1.9250 | 2.140 | v | |
| 1317.5 | 1329.5 | 1341.3 | 1352.9 | 1364.3 | 1375.6 | 1386.8 | 1398.0 | 1409.0 | 1436.4 | 1463.6 | 1490.6 | 1517.6 | 1626.2 | 1736.8 | 1850.4 | h | **570** |
| 1.5630 | 1.5738 | 1.5842 | 1.5943 | 1.6041 | 1.6136 | 1.6229 | 1.6320 | 1.6408 | 1.6621 | 1.6825 | 1.7020 | 1.7208 | 1.7905 | 1.8534 | 1.9114 | s | (480.73) |
| 1.0368 | 1.0627 | 1.0881 | 1.1131 | 1.1376 | 1.1619 | 1.1859 | 1.2096 | 1.2331 | 1.2909 | 1.3479 | 1.4040 | 1.4596 | 1.6776 | 1.8915 | 2.103 | v | |
| 1316.7 | 1328.8 | 1340.6 | 1352.3 | 1363.7 | 1375.1 | 1386.3 | 1397.5 | 1408.6 | 1436.0 | 1463.2 | 1490.3 | 1517.3 | 1626.0 | 1736.6 | 1850.2 | h | **580** |
| 1.5605 | 1.5714 | 1.5819 | 1.5920 | 1.6018 | 1.6114 | 1.6207 | 1.6298 | 1.6386 | 1.6600 | 1.6804 | 1.6999 | 1.7188 | 1.7885 | 1.8515 | 1.9095 | s | (482.58) |
| 1.0175 | 1.0431 | 1.0682 | 1.0928 | 1.1170 | 1.1410 | 1.1646 | 1.1879 | 1.2111 | 1.2681 | 1.3242 | 1.3794 | 1.4342 | 1.6487 | 1.8592 | 2.067 | v | |
| 1316.0 | 1328.1 | 1340.0 | 1351.7 | 1363.2 | 1374.6 | 1385.8 | 1397.0 | 1408.1 | 1435.6 | 1462.9 | 1490.0 | 1517.0 | 1625.7 | 1736.5 | 1850.1 | h | **590** |
| 1.5581 | 1.5691 | 1.5796 | 1.5898 | 1.5996 | 1.6092 | 1.6185 | 1.6276 | 1.6365 | 1.6579 | 1.6783 | 1.6979 | 1.7167 | 1.7865 | 1.8495 | 1.9075 | s | (484.41) |
| 0.9988 | 1.0241 | 1.0489 | 1.0732 | 1.0971 | 1.1207 | 1.1440 | 1.1670 | 1.1899 | 1.2460 | 1.3013 | 1.3557 | 1.4096 | 1.6208 | 1.8279 | 2.033 | v | |
| 1315.2 | 1327.4 | 1339.3 | 1351.1 | 1362.6 | 1374.0 | 1385.3 | 1396.5 | 1407.7 | 1435.2 | 1462.5 | 1489.7 | 1516.7 | 1625.5 | 1736.3 | 1850.0 | h | **600** |
| 1.5558 | 1.5667 | 1.5773 | 1.5875 | 1.5974 | 1.6070 | 1.6163 | 1.6254 | 1.6343 | 1.6558 | 1.6762 | 1.6958 | 1.7147 | 1.7846 | 1.8476 | 1.9056 | s | (486.21) |
| 0.9807 | 1.0057 | 1.0302 | 1.0542 | 1.0778 | 1.1011 | 1.1241 | 1.1468 | 1.1693 | 1.2247 | 1.2791 | 1.3327 | 1.3858 | 1.5938 | 1.7976 | 1.9991 | v | |
| 1314.5 | 1326.7 | 1338.7 | 1350.5 | 1362.1 | 1373.5 | 1384.8 | 1396.1 | 1407.2 | 1434.8 | 1462.2 | 1489.3 | 1515.5 | 1625.3 | 1736.2 | 1849.9 | h | **610** |
| 1.5534 | 1.5644 | 1.5750 | 1.5853 | 1.5952 | 1.6048 | 1.6142 | 1.6233 | 1.6323 | 1.6538 | 1.6742 | 1.6939 | 1.7128 | 1.7827 | 1.8457 | 1.9038 | s | (487.99) |
| 0.9633 | 0.9880 | 1.0121 | 1.0358 | 1.0591 | 1.0821 | 1.1048 | 1.1272 | 1.1494 | 1.2040 | 1.2577 | 1.3105 | 1.3628 | 1.5676 | 1.7683 | 1.9667 | v | |
| 1313.7 | 1326.0 | 1338.0 | 1349.9 | 1361.5 | 1373.0 | 1384.3 | 1395.6 | 1406.8 | 1434.4 | 1461.8 | 1489.0 | 1516.2 | 1625.1 | 1736.0 | 1849.8 | h | **620** |
| 1.5511 | 1.5622 | 1.5728 | 1.5831 | 1.5931 | 1.6027 | 1.6121 | 1.6213 | 1.6302 | 1.6517 | 1.6722 | 1.6919 | 1.7108 | 1.7808 | 1.8438 | 1.9019 | s | (489.75) |
| 0.9463 | 0.9707 | 0.9946 | 1.0180 | 1.0410 | 1.0637 | 1.0861 | 1.1083 | 1.1302 | 1.1840 | 1.2369 | 1.2890 | 1.3405 | 1.5423 | 1.7399 | 1.9353 | v | |
| 1312.9 | 1325.3 | 1337.4 | 1349.2 | 1360.9 | 1372.4 | 1383.8 | 1395.1 | 1406.3 | 1434.0 | 1461.5 | 1488.7 | 1515.9 | 1624.9 | 1735.9 | 1849.6 | h | **630** |
| 1.5488 | 1.5599 | 1.5706 | 1.5809 | 1.5909 | 1.6006 | 1.6100 | 1.6192 | 1.6282 | 1.6498 | 1.6703 | 1.6900 | 1.7089 | 1.7789 | 1.8420 | 1.9001 | s | (491.49) |
| 0.9299 | 0.9541 | 0.9777 | 1.0008 | 1.0235 | 1.0459 | 1.0680 | 1.0899 | 1.1115 | 1.1647 | 1.2168 | 1.2681 | 1.3190 | 1.5178 | 1.7125 | 1.9048 | v | |
| 1312.2 | 1324.6 | 1336.7 | 1348.6 | 1360.4 | 1371.9 | 1383.3 | 1394.7 | 1405.9 | 1433.7 | 1461.1 | 1488.4 | 1515.6 | 1624.7 | 1735.7 | 1849.5 | h | **640** |
| 1.5465 | 1.5577 | 1.5684 | 1.5788 | 1.5888 | 1.5985 | 1.6080 | 1.6172 | 1.6262 | 1.6478 | 1.6684 | 1.6881 | 1.7070 | 1.7771 | 1.8402 | 1.8983 | s | (493.21) |
| 0.9140 | 0.9379 | 0.9612 | 0.9841 | 1.0065 | 1.0286 | 1.0505 | 1.0721 | 1.0934 | 1.1459 | 1.1973 | 1.2479 | 1.2981 | 1.4940 | 1.6858 | 1.8753 | v | |
| 1311.4 | 1323.9 | 1336.1 | 1348.0 | 1359.8 | 1371.4 | 1382.8 | 1394.2 | 1405.4 | 1433.3 | 1460.8 | 1488.1 | 1515.3 | 1624.5 | 1735.5 | 1849.4 | h | **650** |
| 1.5443 | 1.5555 | 1.5663 | 1.5767 | 1.5867 | 1.5965 | 1.6060 | 1.6152 | 1.6242 | 1.6458 | 1.6665 | 1.6862 | 1.7052 | 1.7753 | 1.8384 | 1.8966 | s | (494.90) |
| 0.8985 | 0.9222 | 0.9453 | 0.9679 | 0.9901 | 1.0119 | 1.0335 | 1.0548 | 1.0759 | 1.1277 | 1.1784 | 1.2283 | 1.2778 | 1.4709 | 1.6600 | 1.8467 | v | |
| 1310.6 | 1323.2 | 1335.4 | 1347.4 | 1359.2 | 1370.8 | 1382.3 | 1393.7 | 1405.0 | 1432.9 | 1460.4 | 1487.8 | 1515.0 | 1624.3 | 1735.4 | 1849.3 | h | **660** |
| 1.5420 | 1.5533 | 1.5642 | 1.5746 | 1.5847 | 1.5945 | 1.6040 | 1.6132 | 1.6222 | 1.6439 | 1.6646 | 1.6843 | 1.7034 | 1.7735 | 1.8367 | 1.8948 | s | (496.58) |
| 0.8835 | 0.9070 | 0.9298 | 0.9522 | 0.9741 | 0.9957 | 1.0170 | 1.0380 | 1.0589 | 1.1100 | 1.1601 | 1.2093 | 1.2581 | 1.4486 | 1.6349 | 1.8190 | v | |
| 1309.9 | 1322.4 | 1334.8 | 1346.8 | 1358.6 | 1370.3 | 1381.8 | 1393.2 | 1404.5 | 1432.5 | 1460.0 | 1487.4 | 1514.7 | 1624.1 | 1735.2 | 1849.1 | h | **670** |
| 1.5398 | 1.5512 | 1.5621 | 1.5726 | 1.5827 | 1.5925 | 1.6020 | 1.6113 | 1.6203 | 1.6421 | 1.6627 | 1.6825 | 1.7016 | 1.7718 | 1.8350 | 1.8931 | s | (498.24) |
| 0.8690 | 0.8922 | 0.9148 | 0.9369 | 0.9586 | 0.9800 | 1.0010 | 1.0218 | 1.0424 | 1.0928 | 1.1423 | 1.1909 | 1.2390 | 1.4269 | 1.6106 | 1.7920 | v | |
| 1309.1 | 1321.7 | 1334.1 | 1346.2 | 1358.1 | 1369.8 | 1381.3 | 1392.8 | 1404.1 | 1432.1 | 1459.7 | 1487.1 | 1514.5 | 1623.9 | 1735.1 | 1849.0 | h | **680** |
| 1.5376 | 1.5491 | 1.5600 | 1.5705 | 1.5807 | 1.5905 | 1.6001 | 1.6094 | 1.6184 | 1.6402 | 1.6609 | 1.6807 | 1.6998 | 1.7700 | 1.8333 | 1.8914 | s | (499.88) |
| 0.8549 | 0.8778 | 0.9002 | 0.9221 | 0.9435 | 0.9647 | 0.9856 | 1.0060 | 1.0263 | 1.0761 | 1.1250 | 1.1730 | 1.2205 | 1.4058 | 1.5870 | 1.7659 | v | |
| 1308.3 | 1321.0 | 1333.4 | 1345.6 | 1357.5 | 1369.2 | 1380.8 | 1392.3 | 1403.6 | 1431.7 | 1459.3 | 1486.8 | 1514.2 | 1623.7 | 1734.9 | 1848.9 | h | **690** |
| 1.5355 | 1.5469 | 1.5579 | 1.5685 | 1.5787 | 1.5886 | 1.5981 | 1.6075 | 1.6165 | 1.6384 | 1.6591 | 1.6789 | 1.6980 | 1.7683 | 1.8316 | 1.8898 | s | (501.50) |
| 0.8411 | 0.8639 | 0.8860 | 0.9077 | 0.9289 | 0.9498 | 0.9704 | 0.9907 | 1.0108 | 1.0600 | 1.1082 | 1.1556 | 1.2024 | 1.3853 | 1.5641 | 1.7405 | v | |
| 1307.5 | 1320.3 | 1332.8 | 1345.0 | 1356.9 | 1368.7 | 1380.3 | 1391.8 | 1403.2 | 1431.3 | 1459.0 | 1486.5 | 1513.9 | 1623.5 | 1734.8 | 1848.8 | h | **700** |
| 1.5333 | 1.5449 | 1.5559 | 1.5665 | 1.5767 | 1.5866 | 1.5962 | 1.6056 | 1.6147 | 1.6366 | 1.6573 | 1.6771 | 1.6963 | 1.7666 | 1.8299 | 1.8881 | s | (503.10) |
| 0.8278 | 0.8503 | 0.8722 | 0.8937 | 0.9147 | 0.9353 | 0.9557 | 0.9758 | 0.9956 | 1.0442 | 1.0919 | 1.1386 | 1.1849 | 1.3654 | 1.5418 | 1.7158 | v | |
| 1306.7 | 1319.6 | 1332.1 | 1344.3 | 1356.3 | 1368.1 | 1379.8 | 1391.3 | 1402.7 | 1430.9 | 1458.6 | 1486.2 | 1513.6 | 1623.3 | 1734.6 | 1848.6 | h | **710** |
| 1.5312 | 1.5428 | 1.5539 | 1.5645 | 1.5748 | 1.5847 | 1.5944 | 1.6037 | 1.6129 | 1.6348 | 1.6556 | 1.6754 | 1.6946 | 1.7650 | 1.8283 | 1.8865 | s | (504.68) |

# Table 3. Superheated Vapor

| Abs. Press. Lb./Sq. In. (Sat. Temp.) | | Sat. Liquid | Sat. Vapor | 510° | 520° | 530° | 540° | 550° | 560° | 570° | 580° | 590° | 600° | 620° | 640° | 660° | 680° |
|---|---|---|---|---|---|---|---|---|---|---|---|---|---|---|---|---|---|
| | v | 0.0206 | 0.6362 | 0.6425 | 0.6586 | 0.6740 | 0.6887 | 0.7029 | 0.7166 | 0.7299 | 0.7429 | 0.7555 | 0.7678 | 0.7917 | 0.8148 | 0.8371 | 0.8588 |
| 720 | h | 495.3 | 1200.7 | 1204.5 | 1214.2 | 1223.4 | 1232.2 | 1240.6 | 1248.7 | 1256.5 | 1264.1 | 1271.5 | 1278.7 | 1292.5 | 1305.9 | 1318.8 | 1331.4 |
| (506.25) | s | 0.6963 | 1.4266 | 1.4306 | 1.4405 | 1.4499 | 1.4587 | 1.4670 | 1.4750 | 1.4827 | 1.4900 | 1.4971 | 1.5039 | 1.5168 | 1.5291 | 1.5408 | 1.5519 |
| | v | 0.0206 | 0.6270 | 0.6307 | 0.6468 | 0.6622 | 0.6768 | 0.6910 | 0.7047 | 0.7179 | 0.7308 | 0.7433 | 0.7556 | 0.7793 | 0.8021 | 0.8243 | 0.8458 |
| 730 | h | 497.2 | 1200.5 | 1202.7 | 1212.6 | 1221.9 | 1230.8 | 1239.3 | 1247.5 | 1255.4 | 1263.0 | 1270.4 | 1277.7 | 1291.7 | 1305.1 | 1318.1 | 1330.7 |
| (507.80) | s | 0.6982 | 1.4251 | 1.4275 | 1.4376 | 1.4471 | 1.4560 | 1.4645 | 1.4725 | 1.4802 | 1.4876 | 1.4947 | 1.5016 | 1.5147 | 1.5270 | 1.5387 | 1.5499 |
| | v | 0.0207 | 0.6180 | 0.6191 | 0.6353 | 0.6506 | 0.6652 | 0.6794 | 0.6930 | 0.7062 | 0.7190 | 0.7314 | 0.7436 | 0.7671 | 0.7898 | 0.8118 | 0.8331 |
| 740 | h | 499.0 | 1200.2 | 1200.9 | 1210.9 | 1220.4 | 1229.4 | 1238.0 | 1246.3 | 1254.2 | 1261.9 | 1269.4 | 1276.7 | 1290.8 | 1304.3 | 1317.4 | 1330.0 |
| (509.34) | s | 0.7001 | 1.4237 | 1.4244 | 1.4347 | 1.4443 | 1.4534 | 1.4619 | 1.4700 | 1.4778 | 1.4853 | 1.4925 | 1.4994 | 1.5125 | 1.5249 | 1.5367 | 1.5480 |
| | v | 0.0207 | 0.6093 | | 0.6240 | 0.6393 | 0.6540 | 0.6680 | 0.6816 | 0.6947 | 0.7075 | 0.7198 | 0.7319 | 0.7553 | 0.7778 | 0.7996 | 0.8208 |
| 750 | h | 500.8 | 1200.0 | | 1209.3 | 1218.8 | 1227.9 | 1236.6 | 1245.0 | 1253.0 | 1260.8 | 1268.4 | 1275.7 | 1289.9 | 1303.5 | 1316.6 | 1329.4 |
| (510.86) | s | 0.7019 | 1.4223 | | 1.4318 | 1.4415 | 1.4507 | 1.4593 | 1.4676 | 1.4754 | 1.4830 | 1.4902 | 1.4972 | 1.5104 | 1.5229 | 1.5347 | 1.5460 |
| | v | 0.0208 | 0.5884 | | 0.5970 | 0.6124 | 0.6270 | 0.6409 | 0.6544 | 0.6674 | 0.6800 | 0.6922 | 0.7041 | 0.7271 | 0.7492 | 0.7705 | 0.7912 |
| 775 | h | 505.3 | 1199.3 | | 1205.0 | 1214.9 | 1224.3 | 1233.2 | 1241.8 | 1250.0 | 1258.0 | 1265.7 | 1273.2 | 1287.7 | 1301.5 | 1314.8 | 1327.7 |
| (514.59) | s | 0.7064 | 1.4188 | | 1.4245 | 1.4346 | 1.4441 | 1.4530 | 1.4614 | 1.4695 | 1.4772 | 1.4846 | 1.4917 | 1.5051 | 1.5178 | 1.5298 | 1.5412 |
| | v | 0.0209 | 0.5687 | | 0.5715 | 0.5869 | 0.6015 | 0.6154 | 0.6288 | 0.6416 | 0.6541 | 0.6662 | 0.6779 | 0.7006 | 0.7223 | 0.7433 | 0.7635 |
| 800 | h | 509.7 | 1198.6 | | 1200.5 | 1210.8 | 1220.5 | 1229.8 | 1238.6 | 1247.0 | 1255.2 | 1263.1 | 1270.7 | 1285.4 | 1299.4 | 1312.9 | 1325.9 |
| (518.23) | s | 0.7108 | 1.4153 | | 1.4172 | 1.4277 | 1.4375 | 1.4467 | 1.4553 | 1.4636 | 1.4715 | 1.4790 | 1.4863 | 1.5000 | 1.5129 | 1.5250 | 1.5366 |
| | v | 0.0210 | 0.5502 | | | 0.5628 | 0.5774 | 0.5913 | 0.6046 | 0.6174 | 0.6297 | 0.6417 | 0.6533 | 0.6757 | 0.6971 | 0.7176 | 0.7375 |
| 825 | h | 514.0 | 1197.9 | | | 1206.6 | 1216.7 | 1226.2 | 1235.3 | 1243.9 | 1252.3 | 1260.3 | 1268.1 | 1283.1 | 1297.3 | 1311.0 | 1324.2 |
| (521.79) | s | 0.7151 | 1.4119 | | | 1.4208 | 1.4309 | 1.4404 | 1.4493 | 1.4578 | 1.4658 | 1.4735 | 1.4809 | 1.4949 | 1.5080 | 1.5203 | 1.5320 |
| | v | 0.0211 | 0.5327 | | | 0.5400 | 0.5546 | 0.5685 | 0.5818 | 0.5945 | 0.6067 | 0.6186 | 0.6301 | 0.6521 | 0.6732 | 0.6934 | 0.7130 |
| 850 | h | 518.3 | 1197.1 | | | 1202.3 | 1212.7 | 1222.5 | 1231.9 | 1240.8 | 1249.3 | 1257.5 | 1265.5 | 1280.7 | 1295.2 | 1309.0 | 1322.4 |
| (525.26) | s | 0.7194 | 1.4085 | | | 1.4138 | 1.4243 | 1.4341 | 1.4433 | 1.4520 | 1.4602 | 1.4681 | 1.4756 | 1.4899 | 1.5032 | 1.5157 | 1.5275 |
| | v | 0.0211 | 0.5162 | | | 0.5183 | 0.5330 | 0.5469 | 0.5601 | 0.5728 | 0.5850 | 0.5967 | 0.6081 | 0.6299 | 0.6507 | 0.6706 | 0.6898 |
| 875 | h | 522.5 | 1196.3 | | | 1197.8 | 1208.6 | 1218.8 | 1228.4 | 1237.6 | 1246.3 | 1254.7 | 1262.8 | 1278.4 | 1293.1 | 1307.1 | 1320.6 |
| (528.66) | s | 0.7235 | 1.4052 | | | 1.4068 | 1.4177 | 1.4278 | 1.4373 | 1.4462 | 1.4547 | 1.4627 | 1.4704 | 1.4849 | 1.4984 | 1.5111 | 1.5231 |
| | v | 0.0212 | 0.5006 | | | | 0.5124 | 0.5264 | 0.5396 | 0.5522 | 0.5644 | 0.5760 | 0.5873 | 0.6089 | 0.6294 | 0.6491 | 0.6680 |
| 900 | h | 526.6 | 1195.4 | | | | 1204.4 | 1215.0 | 1224.9 | 1234.3 | 1243.2 | 1251.8 | 1260.1 | 1275.9 | 1290.9 | 1305.1 | 1318.8 |
| (531.98) | s | 0.7275 | 1.4020 | | | | 1.4111 | 1.4216 | 1.4313 | 1.4405 | 1.4492 | 1.4574 | 1.4653 | 1.4800 | 1.4938 | 1.5066 | 1.5187 |
| | v | 0.0213 | 0.4858 | | | | 0.4928 | 0.5068 | 0.5201 | 0.5327 | 0.5448 | 0.5564 | 0.5676 | 0.5890 | 0.6093 | 0.6286 | 0.6473 |
| 925 | h | 530.6 | 1194.6 | | | | 1200.1 | 1211.0 | 1221.2 | 1230.9 | 1240.1 | 1248.9 | 1257.4 | 1273.5 | 1288.7 | 1303.1 | 1317.0 |
| (535.24) | s | 0.7315 | 1.3989 | | | | 1.4044 | 1.4153 | 1.4253 | 1.4348 | 1.4437 | 1.4521 | 1.4602 | 1.4752 | 1.4892 | 1.5022 | 1.5144 |
| | v | 0.0214 | 0.4717 | | | | 0.4740 | 0.4882 | 0.5015 | 0.5141 | 0.5262 | 0.5377 | 0.5489 | 0.5701 | 0.5901 | 0.6092 | 0.6276 |
| 950 | h | 534.6 | 1193.7 | | | | 1195.5 | 1206.9 | 1217.5 | 1227.4 | 1236.9 | 1245.9 | 1254.6 | 1271.0 | 1286.4 | 1301.1 | 1315.2 |
| (538.42) | s | 0.7354 | 1.3958 | | | | 1.3976 | 1.4089 | 1.4194 | 1.4291 | 1.4382 | 1.4469 | 1.4551 | 1.4704 | 1.4846 | 1.4978 | 1.5103 |
| | v | 0.0215 | 0.4583 | | | | | 0.4703 | 0.4837 | 0.4964 | 0.5084 | 0.5199 | 0.5310 | 0.5521 | 0.5720 | 0.5908 | 0.6089 |
| 975 | h | 538.5 | 1192.8 | | | | | 1202.7 | 1213.6 | 1223.9 | 1233.6 | 1242.9 | 1251.7 | 1268.5 | 1284.2 | 1299.1 | 1313.3 |
| (541.55) | s | 0.7393 | 1.3927 | | | | | 1.4026 | 1.4134 | 1.4234 | 1.4328 | 1.4416 | 1.4500 | 1.4657 | 1.4801 | 1.4935 | 1.5061 |
| | v | 0.0216 | 0.4456 | | | | | 0.4533 | 0.4668 | 0.4795 | 0.4915 | 0.5030 | 0.5140 | 0.5350 | 0.5546 | 0.5733 | 0.5912 |
| 1000 | h | 542.4 | 1191.8 | | | | | 1198.3 | 1209.7 | 1220.3 | 1230.3 | 1239.8 | 1248.8 | 1265.9 | 1281.9 | 1297.0 | 1311.4 |
| (544.61) | s | 0.7430 | 1.3897 | | | | | 1.3961 | 1.4073 | 1.4177 | 1.4273 | 1.4364 | 1.4450 | 1.4610 | 1.4757 | 1.4893 | 1.5021 |
| | v | 0.0218 | 0.4218 | | | | | | 0.4349 | 0.4477 | 0.4598 | 0.4713 | 0.4823 | 0.5030 | 0.5224 | 0.5407 | 0.5582 |
| 1050 | h | 550.0 | 1189.9 | | | | | | 1201.5 | 1212.8 | 1223.4 | 1233.4 | 1242.9 | 1260.7 | 1277.2 | 1292.8 | 1307.6 |
| (550.57) | s | 0.7504 | 1.3838 | | | | | | 1.3952 | 1.4062 | 1.4165 | 1.4260 | 1.4350 | 1.4517 | 1.4669 | 1.4809 | 1.4940 |
| | v | 0.0220 | 0.4001 | | | | | | 0.4053 | 0.4184 | 0.4307 | 0.4422 | 0.4532 | 0.4738 | 0.4929 | 0.5110 | 0.5281 |
| 1100 | h | 557.4 | 1187.8 | | | | | | 1192.6 | 1204.8 | 1216.1 | 1226.7 | 1236.7 | 1255.3 | 1272.4 | 1288.5 | 1303.7 |
| (556.31) | s | 0.7575 | 1.3780 | | | | | | 1.3827 | 1.3946 | 1.4055 | 1.4156 | 1.4251 | 1.4425 | 1.4583 | 1.4728 | 1.4862 |
| | v | 0.0221 | 0.3802 | | | | | | | 0.3912 | 0.4037 | 0.4154 | 0.4264 | 0.4470 | 0.4659 | 0.4837 | 0.5005 |
| 1150 | h | 564.0 | 1185.6 | | | | | | | 1196.3 | 1208.5 | 1219.7 | 1230.2 | 1249.6 | 1267.5 | 1284.1 | 1299.8 |
| (561.86) | s | 0.7644 | 1.3723 | | | | | | | 1.3828 | 1.3944 | 1.4052 | 1.4152 | 1.4334 | 1.4497 | 1.4647 | 1.4786 |
| | v | 0.0223 | 0.3619 | | | | | | | 0.3657 | 0.3786 | 0.3905 | 0.4016 | 0.4222 | 0.4410 | 0.4586 | 0.4752 |
| 1200 | h | 571.7 | 1183.4 | | | | | | | 1187.2 | 1200.4 | 1212.4 | 1223.5 | 1243.9 | 1262.4 | 1279.6 | 1295.7 |
| (567.22) | s | 0.7711 | 1.3667 | | | | | | | 1.3705 | 1.3832 | 1.3947 | 1.4052 | 1.4243 | 1.4413 | 1.4568 | 1.4710 |

Temperature—Degrees Fahrenheit

| Abs. Press. Lb./Sq. In. (Sat. Temp.) | | 700° | 720° | 740° | 760° | 780° | 800° | 820° | 840° | 860° | 880° | 900° | 1000° | 1100° | 1200° | 1400° | 1600° |
|---|---|---|---|---|---|---|---|---|---|---|---|---|---|---|---|---|---|
| | v | 0.8801 | 0.9008 | 0.9213 | 0.9414 | 0.9613 | 0.9809 | 1.0003 | 1.0194 | 1.0384 | 1.0573 | 1.0760 | 1.1679 | 1.2577 | 1.3461 | 1.5201 | 1.6918 |
| 720 | h | 1343.7 | 1355.7 | 1367.6 | 1379.3 | 1390.8 | 1402.3 | 1413.6 | 1424.9 | 1436.1 | 1447.2 | 1458.3 | 1513.3 | 1568.1 | 1623.1 | 1734.4 | 1848.5 |
| (506.25) | s | 1.5626 | 1.5729 | 1.5828 | 1.5925 | 1.6019 | 1.6110 | 1.6200 | 1.6287 | 1.6372 | 1.6456 | 1.6538 | 1.6929 | 1.7292 | 1.7634 | 1.8267 | 1.8850 |
| | v | 0.8668 | 0.8874 | 0.9076 | 0.9275 | 0.9472 | 0.9666 | 0.9857 | 1.0047 | 1.0235 | 1.0421 | 1.0606 | 1.1514 | 1.2400 | 1.3273 | 1.4990 | 1.6685 |
| 730 | h | 1343.1 | 1355.2 | 1367.1 | 1378.8 | 1390.4 | 1401.8 | 1413.2 | 1424.5 | 1435.7 | 1446.8 | 1457.9 | 1513.0 | 1567.9 | 1622.9 | 1734.3 | 1848.4 |
| (507.80) | s | 1.5606 | 1.5710 | 1.5810 | 1.5907 | 1.6001 | 1.6092 | 1.6182 | 1.6269 | 1.6355 | 1.6439 | 1.6521 | 1.6912 | 1.7275 | 1.7617 | 1.8251 | 1.8834 |
| | v | 0.8539 | 0.8743 | 0.8943 | 0.9140 | 0.9334 | 0.9526 | 0.9716 | 0.9903 | 1.0089 | 1.0273 | 1.0456 | 1.1352 | 1.2228 | 1.3090 | 1.4785 | 1.6457 |
| 740 | h | 1342.4 | 1354.6 | 1366.5 | 1378.3 | 1389.9 | 1401.4 | 1412.7 | 1424.0 | 1435.3 | 1446.4 | 1457.6 | 1512.7 | 1567.6 | 1622.7 | 1734.1 | 1848.3 |
| (509.34) | s | 1.5587 | 1.5691 | 1.5791 | 1.5888 | 1.5983 | 1.6075 | 1.6164 | 1.6252 | 1.6338 | 1.6422 | 1.6504 | 1.6896 | 1.7259 | 1.7602 | 1.8236 | 1.8818 |
| | v | 0.8414 | 0.8615 | 0.8813 | 0.9008 | 0.9201 | 0.9391 | 0.9578 | 0.9763 | 0.9947 | 1.0129 | 1.0310 | 1.1196 | 1.2061 | 1.2912 | 1.4586 | 1.6236 |
| 750 | h | 1341.8 | 1354.0 | 1366.0 | 1377.7 | 1389.4 | 1400.9 | 1412.3 | 1423.6 | 1434.9 | 1446.1 | 1457.2 | 1512.4 | 1567.4 | 1622.4 | 1734.0 | 1848.1 |
| (510.86) | s | 1.5568 | 1.5672 | 1.5773 | 1.5870 | 1.5965 | 1.6057 | 1.6147 | 1.6235 | 1.6321 | 1.6405 | 1.6487 | 1.6879 | 1.7243 | 1.7586 | 1.8220 | 1.8803 |
| | v | 0.8114 | 0.8311 | 0.8505 | 0.8695 | 0.8882 | 0.9067 | 0.9249 | 0.9429 | 0.9608 | 0.9785 | 0.9961 | 1.0821 | 1.1661 | 1.2486 | 1.4109 | 1.5709 |
| 775 | h | 1340.2 | 1352.5 | 1364.6 | 1376.5 | 1388.2 | 1399.8 | 1411.2 | 1422.6 | 1433.9 | 1445.1 | 1456.3 | 1511.7 | 1566.8 | 1621.9 | 1733.6 | 1847.8 |
| (514.59) | s | 1.5522 | 1.5627 | 1.5728 | 1.5826 | 1.5922 | 1.6014 | 1.6105 | 1.6193 | 1.6279 | 1.6364 | 1.6446 | 1.6840 | 1.7204 | 1.7547 | 1.8182 | 1.8766 |
| | v | 0.7833 | 0.8026 | 0.8215 | 0.8400 | 0.8583 | 0.8763 | 0.8941 | 0.9116 | 0.9290 | 0.9462 | 0.9633 | 1.0470 | 1.1286 | 1.2088 | 1.3662 | 1.5214 |
| 800 | h | 1338.6 | 1351.0 | 1363.2 | 1375.2 | 1387.0 | 1398.6 | 1410.1 | 1421.6 | 1432.9 | 1444.2 | 1455.4 | 1511.0 | 1566.2 | 1621.4 | 1733.2 | 1847.5 |
| (518.23) | s | 1.5476 | 1.5582 | 1.5684 | 1.5783 | 1.5879 | 1.5972 | 1.6063 | 1.6152 | 1.6238 | 1.6323 | 1.6407 | 1.6801 | 1.7166 | 1.7510 | 1.8146 | 1.8729 |
| | v | 0.7569 | 0.7757 | 0.7942 | 0.8123 | 0.8302 | 0.8478 | 0.8651 | 0.8822 | 0.8991 | 0.9159 | 0.9326 | 1.0141 | 1.0934 | 1.1713 | 1.3243 | 1.4749 |
| 825 | h | 1337.0 | 1349.5 | 1361.8 | 1373.8 | 1385.7 | 1397.4 | 1409.0 | 1420.5 | 1431.9 | 1443.3 | 1454.5 | 1510.3 | 1565.6 | 1620.9 | 1732.8 | 1847.2 |
| (521.79) | s | 1.5431 | 1.5538 | 1.5642 | 1.5741 | 1.5838 | 1.5931 | 1.6023 | 1.6112 | 1.6199 | 1.6284 | 1.6368 | 1.6763 | 1.7130 | 1.7474 | 1.8110 | 1.8694 |
| | v | 0.7320 | 0.7505 | 0.7685 | 0.7863 | 0.8037 | 0.8209 | 0.8378 | 0.8545 | 0.8711 | 0.8874 | 0.9037 | 0.9830 | 1.0603 | 1.1360 | 1.2848 | 1.4311 |
| 850 | h | 1335.4 | 1348.0 | 1360.4 | 1372.5 | 1384.5 | 1396.3 | 1407.9 | 1419.5 | 1430.9 | 1442.3 | 1453.6 | 1509.5 | 1564.9 | 1620.4 | 1732.4 | 1846.9 |
| (525.26) | s | 1.5388 | 1.5496 | 1.5600 | 1.5700 | 1.5797 | 1.5892 | 1.5983 | 1.6073 | 1.6160 | 1.6246 | 1.6330 | 1.6727 | 1.7094 | 1.7438 | 1.8075 | 1.8660 |
| | v | 0.7085 | 0.7266 | 0.7443 | 0.7617 | 0.7788 | 0.7955 | 0.8121 | 0.8284 | 0.8446 | 0.8605 | 0.8764 | 0.9538 | 1.0290 | 1.1028 | 1.2475 | 1.3899 |
| 875 | h | 1333.8 | 1346.5 | 1359.0 | 1371.2 | 1383.2 | 1395.1 | 1406.8 | 1418.4 | 1429.9 | 1441.4 | 1452.7 | 1508.8 | 1564.3 | 1619.9 | 1732.0 | 1846.6 |
| (528.66) | s | 1.5345 | 1.5454 | 1.5559 | 1.5660 | 1.5758 | 1.5853 | 1.5945 | 1.6035 | 1.6123 | 1.6209 | 1.6293 | 1.6691 | 1.7059 | 1.7404 | 1.8042 | 1.8627 |
| | v | 0.6863 | 0.7041 | 0.7215 | 0.7385 | 0.7552 | 0.7716 | 0.7878 | 0.8038 | 0.8195 | 0.8351 | 0.8506 | 0.9262 | 0.9995 | 1.0714 | 1.2124 | 1.3509 |
| 900 | h | 1332.1 | 1345.0 | 1357.5 | 1369.9 | 1382.0 | 1393.9 | 1405.7 | 1417.4 | 1428.9 | 1440.4 | 1451.8 | 1508.1 | 1563.7 | 1619.3 | 1731.6 | 1846.3 |
| (531.98) | s | 1.5303 | 1.5413 | 1.5519 | 1.5620 | 1.5719 | 1.5814 | 1.5907 | 1.5998 | 1.6086 | 1.6172 | 1.6257 | 1.6656 | 1.7025 | 1.7371 | 1.8009 | 1.8595 |
| | v | 0.6653 | 0.6828 | 0.6998 | 0.7165 | 0.7329 | 0.7490 | 0.7648 | 0.7804 | 0.7958 | 0.8111 | 0.8262 | 0.9000 | 0.9716 | 1.0417 | 1.1791 | 1.3141 |
| 925 | h | 1330.4 | 1343.4 | 1356.1 | 1368.5 | 1380.7 | 1392.7 | 1404.6 | 1416.3 | 1427.9 | 1439.5 | 1450.9 | 1507.3 | 1563.1 | 1618.8 | 1731.2 | 1845.9 |
| (535.24) | s | 1.5261 | 1.5372 | 1.5479 | 1.5582 | 1.5681 | 1.5777 | 1.5871 | 1.5962 | 1.6050 | 1.6137 | 1.6222 | 1.6622 | 1.6992 | 1.7338 | 1.7977 | 1.8563 |
| | v | 0.6453 | 0.6625 | 0.6793 | 0.6957 | 0.7117 | 0.7275 | 0.7430 | 0.7583 | 0.7734 | 0.7883 | 0.8031 | 0.8753 | 0.9451 | 1.0136 | 1.1476 | 1.2792 |
| 950 | h | 1328.7 | 1341.9 | 1354.7 | 1367.2 | 1379.5 | 1391.6 | 1403.5 | 1415.3 | 1426.9 | 1438.5 | 1450.0 | 1506.6 | 1562.5 | 1618.3 | 1730.8 | 1845.6 |
| (538.42) | s | 1.5221 | 1.5333 | 1.5440 | 1.5544 | 1.5644 | 1.5741 | 1.5835 | 1.5926 | 1.6015 | 1.6102 | 1.6187 | 1.6589 | 1.6959 | 1.7306 | 1.7946 | 1.8532 |
| | v | 0.6264 | 0.6433 | 0.6598 | 0.6759 | 0.6916 | 0.7071 | 0.7223 | 0.7373 | 0.7521 | 0.7667 | 0.7812 | 0.8518 | 0.9200 | 0.9869 | 1.1177 | 1.2461 |
| 975 | h | 1327.0 | 1340.3 | 1353.2 | 1365.8 | 1378.2 | 1390.4 | 1402.3 | 1414.2 | 1425.9 | 1437.6 | 1449.1 | 1505.9 | 1561.9 | 1617.8 | 1730.4 | 1845.3 |
| (541.55) | s | 1.5180 | 1.5294 | 1.5403 | 1.5507 | 1.5607 | 1.5705 | 1.5799 | 1.5891 | 1.5981 | 1.6068 | 1.6154 | 1.6557 | 1.6928 | 1.7275 | 1.7916 | 1.8503 |
| | v | 0.6084 | 0.6251 | 0.6413 | 0.6571 | 0.6726 | 0.6878 | 0.7027 | 0.7173 | 0.7318 | 0.7462 | 0.7604 | 0.8294 | 0.8962 | 0.9615 | 1.0893 | 1.2146 |
| 1000 | h | 1325.3 | 1338.7 | 1351.7 | 1364.4 | 1376.9 | 1389.2 | 1401.2 | 1413.1 | 1424.9 | 1436.6 | 1448.2 | 1505.1 | 1561.3 | 1617.3 | 1730.0 | 1845.0 |
| (544.61) | s | 1.5141 | 1.5256 | 1.5365 | 1.5470 | 1.5572 | 1.5670 | 1.5765 | 1.5857 | 1.5947 | 1.6035 | 1.6121 | 1.6525 | 1.6897 | 1.7245 | 1.7886 | 1.8474 |
| | v | 0.5750 | 0.5911 | 0.6069 | 0.6222 | 0.6371 | 0.6518 | 0.6662 | 0.6803 | 0.6943 | 0.7081 | 0.7217 | 0.7880 | 0.8519 | 0.9144 | 1.0365 | 1.1562 |
| 1050 | h | 1321.8 | 1335.5 | 1348.8 | 1361.7 | 1374.3 | 1386.7 | 1398.9 | 1411.0 | 1422.9 | 1434.7 | 1446.3 | 1503.7 | 1560.0 | 1616.2 | 1729.2 | 1844.4 |
| (550.57) | s | 1.5064 | 1.5181 | 1.5292 | 1.5399 | 1.5502 | 1.5601 | 1.5698 | 1.5791 | 1.5882 | 1.5970 | 1.6057 | 1.6464 | 1.6837 | 1.7186 | 1.7829 | 1.8417 |
| | v | 0.5445 | 0.5602 | 0.5755 | 0.5904 | 0.6049 | 0.6191 | 0.6329 | 0.6466 | 0.6601 | 0.6734 | 0.6866 | 0.7503 | 0.8117 | 0.8716 | 0.9885 | 1.1031 |
| 1100 | h | 1318.3 | 1332.2 | 1345.8 | 1358.9 | 1371.7 | 1384.3 | 1396.6 | 1408.8 | 1420.8 | 1432.7 | 1444.5 | 1502.2 | 1558.8 | 1615.2 | 1728.4 | 1843.8 |
| (556.31) | s | 1.4989 | 1.5108 | 1.5222 | 1.5330 | 1.5435 | 1.5535 | 1.5633 | 1.5727 | 1.5819 | 1.5908 | 1.5995 | 1.6405 | 1.6780 | 1.7130 | 1.7775 | 1.8363 |
| | v | 0.5166 | 0.5320 | 0.5469 | 0.5613 | 0.5754 | 0.5892 | 0.6026 | 0.6158 | 0.6289 | 0.6418 | 0.6544 | 0.7158 | 0.7749 | 0.8325 | 0.9447 | 1.0547 |
| 1150 | h | 1314.7 | 1328.9 | 1342.7 | 1356.1 | 1369.1 | 1381.8 | 1394.3 | 1406.6 | 1418.8 | 1430.7 | 1442.6 | 1500.7 | 1557.6 | 1614.1 | 1727.7 | 1843.1 |
| (561.86) | s | 1.4915 | 1.5037 | 1.5153 | 1.5263 | 1.5369 | 1.5471 | 1.5570 | 1.5665 | 1.5758 | 1.5848 | 1.5936 | 1.6348 | 1.6725 | 1.7076 | 1.7722 | 1.8312 |
| | v | 0.4909 | 0.5060 | 0.5206 | 0.5347 | 0.5484 | 0.5617 | 0.5748 | 0.5876 | 0.6003 | 0.6127 | 0.6250 | 0.6843 | 0.7412 | 0.7967 | 0.9046 | 1.0101 |
| 1200 | h | 1311.0 | 1325.6 | 1339.6 | 1353.2 | 1366.4 | 1379.3 | 1392.0 | 1404.4 | 1416.7 | 1428.8 | 1440.7 | 1499.2 | 1556.4 | 1613.1 | 1726.9 | 1842.5 |
| (567.22) | s | 1.4843 | 1.4968 | 1.5086 | 1.5198 | 1.5306 | 1.5409 | 1.5509 | 1.5605 | 1.5699 | 1.5790 | 1.5879 | 1.6293 | 1.6672 | 1.7025 | 1.7672 | 1.8263 |

## Table 3. Superheated Vapor

| Abs. Press. Lb./Sq. In. (Sat. Temp.) | | Sat. Liquid | Sat. Vapor | 580° | 590° | 600° | 610° | 620° | 630° | 640° | 650° | 660° | 670° | 680° | 690° | 700° | 720° |
|---|---|---|---|---|---|---|---|---|---|---|---|---|---|---|---|---|---|
| | | | | Temperature—Degrees Fahrenheit | | | | | | | | | | | | | |
| 1250 | v | 0.0225 | 0.3450 | 0.3551 | 0.3672 | 0.3785 | 0.3891 | 0.3992 | 0.4088 | 0.4180 | 0.4269 | 0.4354 | 0.4437 | 0.4518 | 0.4597 | 0.4673 | 0.482 |
| (572.42) | h | 578.6 | 1181.0 | 1191.7 | 1204.6 | 1216.5 | 1227.5 | 1237.9 | 1247.8 | 1257.2 | 1266.3 | 1275.0 | 1283.4 | 1291.5 | 1299.5 | 1307.2 | 1322 |
| | s | 0.7776 | 1.3612 | 1.3716 | 1.3839 | 1.3951 | 1.4055 | 1.4152 | 1.4243 | 1.4329 | 1.4411 | 1.4489 | 1.4564 | 1.4636 | 1.4706 | 1.4773 | 1.49 |
| 1300 | v | 0.0227 | 0.3293 | 0.3328 | 0.3454 | 0.3569 | 0.3677 | 0.3778 | 0.3874 | 0.3966 | 0.4054 | 0.4139 | 0.4222 | 0.4301 | 0.4379 | 0.4454 | 0.46 |
| (577.46) | h | 585.4 | 1178.6 | 1182.4 | 1196.4 | 1209.0 | 1220.7 | 1231.6 | 1242.0 | 1251.8 | 1261.2 | 1270.2 | 1278.9 | 1287.3 | 1295.4 | 1303.4 | 1318 |
| | s | 0.7840 | 1.3559 | 1.3596 | 1.3730 | 1.3850 | 1.3959 | 1.4061 | 1.4156 | 1.4246 | 1.4331 | 1.4412 | 1.4489 | 1.4563 | 1.4634 | 1.4703 | 1.48 |
| 1350 | v | 0.0229 | 0.3148 | | 0.3247 | 0.3366 | 0.3476 | 0.3578 | 0.3675 | 0.3767 | 0.3855 | 0.3939 | 0.4021 | 0.4100 | 0.4176 | 0.4251 | 0.43 |
| (582.35) | h | 592.1 | 1176.1 | | 1187.6 | 1201.2 | 1213.6 | 1225.1 | 1236.0 | 1246.2 | 1256.0 | 1265.3 | 1274.3 | 1282.9 | 1291.3 | 1299.5 | 1315 |
| | s | 0.7902 | 1.3506 | | 1.3616 | 1.3746 | 1.3862 | 1.3969 | 1.4069 | 1.4163 | 1.4251 | 1.4335 | 1.4415 | 1.4491 | 1.4564 | 1.4635 | 1.47 |
| 1400 | v | 0.0231 | 0.3012 | | 0.3050 | 0.3174 | 0.3286 | 0.3390 | 0.3488 | 0.3580 | 0.3668 | 0.3753 | 0.3834 | 0.3912 | 0.3988 | 0.4062 | 0.42 |
| (587.10) | h | 598.7 | 1173.4 | | 1178.1 | 1193.0 | 1206.2 | 1218.4 | 1229.8 | 1240.4 | 1250.6 | 1260.3 | 1269.6 | 1278.5 | 1287.1 | 1295.5 | 1311 |
| | s | 0.7963 | 1.3454 | | 1.3498 | 1.3639 | 1.3764 | 1.3877 | 1.3982 | 1.4079 | 1.4171 | 1.4258 | 1.4341 | 1.4419 | 1.4495 | 1.4567 | 1.47 |
| 1450 | v | 0.0233 | 0.2884 | | | 0.2991 | 0.3107 | 0.3213 | 0.3312 | 0.3405 | 0.3493 | 0.3578 | 0.3658 | 0.3736 | 0.3812 | 0.3885 | 0.40 |
| (591.73) | h | 605.2 | 1170.7 | | | 1184.1 | 1198.4 | 1211.3 | 1223.3 | 1234.5 | 1245.0 | 1255.1 | 1264.7 | 1273.9 | 1282.8 | 1291.4 | 1307 |
| | s | 0.8023 | 1.3402 | | | 1.3528 | 1.3663 | 1.3784 | 1.3894 | 1.3996 | 1.4092 | 1.4182 | 1.4267 | 1.4348 | 1.4426 | 1.4500 | 1.46 |
| 1500 | v | 0.0235 | 0.2765 | | | 0.2815 | 0.2936 | 0.3045 | 0.3146 | 0.3240 | 0.3329 | 0.3413 | 0.3494 | 0.3572 | 0.3647 | 0.3719 | 0.38 |
| (596.23) | h | 611.6 | 1167.9 | | | 1174.5 | 1190.1 | 1203.9 | 1216.6 | 1228.3 | 1239.3 | 1249.8 | 1259.7 | 1269.3 | 1278.4 | 1287.2 | 1304 |
| | s | 0.8082 | 1.3351 | | | 1.3412 | 1.3559 | 1.3688 | 1.3805 | 1.3912 | 1.4012 | 1.4105 | 1.4194 | 1.4278 | 1.4358 | 1.4434 | 1.45 |
| 1600 | v | 0.0239 | 0.2548 | | | | 0.2614 | 0.2733 | 0.2839 | 0.2936 | 0.3027 | 0.3112 | 0.3193 | 0.3271 | 0.3345 | 0.3417 | 0.35 |
| (604.90) | h | 624.1 | 1162.1 | | | | 1171.5 | 1187.8 | 1202.2 | 1215.2 | 1227.3 | 1238.7 | 1249.4 | 1259.6 | 1269.3 | 1278.7 | 1296 |
| | s | 0.8196 | 1.3249 | | | | 1.3337 | 1.3489 | 1.3622 | 1.3741 | 1.3850 | 1.3952 | 1.4047 | 1.4137 | 1.4222 | 1.4303 | 1.44 |
| 1700 | v | 0.0243 | 0.2354 | | | | | 0.2443 | 0.2558 | 0.2661 | 0.2755 | 0.2842 | 0.2924 | 0.3002 | 0.3077 | 0.3148 | 0.32 |
| (613.15) | h | 636.3 | 1155.9 | | | | | 1169.2 | 1186.1 | 1201.0 | 1214.4 | 1226.8 | 1238.4 | 1249.3 | 1259.7 | 1269.7 | 1288 |
| | s | 0.8306 | 1.3149 | | | | | 1.3273 | 1.3429 | 1.3564 | 1.3686 | 1.3797 | 1.3900 | 1.3997 | 1.4087 | 1.4173 | 1.43 |
| 1800 | v | 0.0247 | 0.2179 | | | | | | 0.2296 | 0.2407 | 0.2506 | 0.2597 | 0.2681 | 0.2760 | 0.2835 | 0.2907 | 0.30 |
| (621.03) | h | 648.3 | 1149.4 | | | | | | 1167.7 | 1185.1 | 1200.3 | 1214.0 | 1226.7 | 1238.5 | 1249.7 | 1260.3 | 1280 |
| | s | 0.8412 | 1.3049 | | | | | | 1.3219 | 1.3377 | 1.3515 | 1.3638 | 1.3751 | 1.3855 | 1.3952 | 1.4044 | 1.42 |
| 1900 | v | 0.0252 | 0.2021 | | | | | | 0.2040 | 0.2168 | 0.2276 | 0.2371 | 0.2459 | 0.2540 | 0.2616 | 0.2688 | 0.28 |
| (628.58) | h | 660.1 | 1142.4 | | | | | | 1145.8 | 1167.0 | 1184.7 | 1200.2 | 1214.2 | 1227.1 | 1239.1 | 1250.4 | 1271 |
| | s | 0.8516 | 1.2949 | | | | | | 1.2981 | 1.3174 | 1.3335 | 1.3474 | 1.3598 | 1.3711 | 1.3816 | 1.3915 | 1.40 |
| 2000 | v | 0.0257 | 0.1878 | | | | | | | 0.1936 | 0.2058 | 0.2161 | 0.2253 | 0.2337 | 0.2415 | 0.2489 | 0.26 |
| (635.82) | h | 671.7 | 1135.1 | | | | | | | 1145.6 | 1167.0 | 1184.9 | 1200.6 | 1214.8 | 1227.8 | 1240.0 | 1262 |
| | s | 0.8619 | 1.2849 | | | | | | | 1.2945 | 1.3139 | 1.3300 | 1.3439 | 1.3564 | 1.3678 | 1.3783 | 1.39 |
| 2100 | v | 0.0262 | 0.1746 | | | | | | | | 0.1846 | 0.1962 | 0.2061 | 0.2150 | 0.2231 | 0.2306 | 0.24 |
| (642.77) | h | 683.3 | 1127.4 | | | | | | | | 1146.3 | 1167.7 | 1185.8 | 1201.6 | 1215.9 | 1229.0 | 1252 |
| | s | 0.8721 | 1.2748 | | | | | | | | 1.2919 | 1.3112 | 1.3272 | 1.3412 | 1.3536 | 1.3651 | 1.38 |
| 2200 | v | 0.0268 | 0.1625 | | | | | | | | 0.1633 | 0.1768 | 0.1879 | 0.1973 | 0.2058 | 0.2135 | 0.22 |
| (649.46) | h | 694.8 | 1119.2 | | | | | | | | 1121.0 | 1147.8 | 1169.1 | 1187.1 | 1203.0 | 1217.4 | 1243 |
| | s | 0.8820 | 1.2646 | | | | | | | | 1.2665 | 1.2903 | 1.3093 | 1.3252 | 1.3390 | 1.3515 | 1.37 |
| 2300 | v | 0.0274 | 0.1513 | | | | | | | | | 0.1575 | 0.1702 | 0.1806 | 0.1897 | 0.1978 | 0.21 |
| (655.91) | h | 706.5 | 1110.4 | | | | | | | | | 1123.8 | 1150.0 | 1171.1 | 1189.1 | 1204.9 | 1232 |
| | s | 0.8921 | 1.2541 | | | | | | | | | 1.2661 | 1.2895 | 1.3080 | 1.3237 | 1.3375 | 1.36 |
| 2400 | v | 0.0280 | 0.1407 | | | | | | | | | | 0.1526 | 0.1644 | 0.1742 | 0.1828 | 0.19 |
| (662.12) | h | 718.4 | 1101.1 | | | | | | | | | | 1128.2 | 1152.9 | 1173.7 | 1191.5 | 1221 |
| | s | 0.9023 | 1.2434 | | | | | | | | | | 1.2673 | 1.2893 | 1.3074 | 1.3228 | 1.34 |
| 2500 | v | 0.0287 | 0.1307 | | | | | | | | | | 0.1342 | 0.1484 | 0.1594 | 0.1686 | 0.18 |
| (668.13) | h | 730.6 | 1091.1 | | | | | | | | | | 1099.8 | 1132.3 | 1156.6 | 1176.8 | 1210 |
| | s | 0.9126 | 1.2322 | | | | | | | | | | 1.2400 | 1.2687 | 1.2899 | 1.3073 | 1.33 |
| 2600 | v | 0.0295 | 0.1213 | | | | | | | | | | | 0.1319 | 0.1447 | 0.1549 | 0.17 |
| (673.94) | h | 743.0 | 1080.2 | | | | | | | | | | | 1107.1 | 1137.3 | 1160.6 | 1197 |
| | s | 0.9232 | 1.2205 | | | | | | | | | | | 1.2443 | 1.2706 | 1.2908 | 1.32 |
| 2700 | v | 0.0305 | 0.1123 | | | | | | | | | | | 0.1137 | 0.1299 | 0.1415 | 0.15 |
| (679.55) | h | 756.2 | 1068.3 | | | | | | | | | | | 1072.8 | 1114.3 | 1142.5 | 1184 |
| | s | 0.9342 | 1.2082 | | | | | | | | | | | 1.2122 | 1.2484 | 1.2728 | 1.30 |

Temperature—Degrees Fahrenheit

| 740° | 760° | 780° | 800° | 820° | 840° | 860° | 880° | 900° | 950° | 1000° | 1050° | 1100° | 1200° | 1400° | 1600° | | Abs. Press. Lb./Sq. In. (Sat. Temp.) |
|---|---|---|---|---|---|---|---|---|---|---|---|---|---|---|---|---|---|
| 0.4963 | 0.5101 | 0.5235 | 0.5365 | 0.5492 | 0.5617 | 0.5740 | 0.5860 | 0.5979 | 0.6269 | 0.6552 | 0.6830 | 0.7102 | 0.7637 | 0.8677 | 0.9692 | v | |
| 1336.5 | 1350.3 | 1363.7 | 1376.8 | 1389.6 | 1402.2 | 1414.6 | 1426.8 | 1438.9 | 1468.5 | 1497.7 | 1526.5 | 1555.1 | 1612.0 | 1726.1 | 1841.9 | h | **1250** |
| 1.5021 | 1.5135 | 1.5244 | 1.5349 | 1.5450 | 1.5547 | 1.5642 | 1.5734 | 1.5823 | 1.6037 | 1.6241 | 1.6435 | 1.6621 | 1.6975 | 1.7624 | 1.8215 | s | (572.42) |
| 0.4739 | 0.4874 | 0.5004 | 0.5131 | 0.5255 | 0.5377 | 0.5496 | 0.5613 | 0.5728 | 0.6010 | 0.6284 | 0.6553 | 0.6816 | 0.7333 | 0.8336 | 0.9314 | v | |
| 1333.3 | 1347.3 | 1361.0 | 1374.3 | 1387.2 | 1400.0 | 1412.5 | 1424.8 | 1437.0 | 1466.9 | 1496.2 | 1525.1 | 1553.9 | 1611.0 | 1725.3 | 1841.3 | h | **1300** |
| 1.4957 | 1.5073 | 1.5184 | 1.5290 | 1.5392 | 1.5491 | 1.5587 | 1.5679 | 1.5769 | 1.5985 | 1.6190 | 1.6385 | 1.6572 | 1.6927 | 1.7577 | 1.8169 | s | (577.46) |
| 0.4531 | 0.4664 | 0.4791 | 0.4915 | 0.5036 | 0.5155 | 0.5271 | 0.5385 | 0.5496 | 0.5770 | 0.6036 | 0.6296 | 0.6551 | 0.7051 | 0.8020 | 0.8965 | v | |
| 1330.0 | 1344.3 | 1358.2 | 1371.7 | 1384.8 | 1397.7 | 1410.3 | 1422.8 | 1435.1 | 1465.2 | 1494.7 | 1523.8 | 1552.6 | 1609.9 | 1724.5 | 1840.6 | h | **1350** |
| 1.4894 | 1.5012 | 1.5125 | 1.5233 | 1.5336 | 1.5436 | 1.5533 | 1.5626 | 1.5717 | 1.5935 | 1.6140 | 1.6336 | 1.6524 | 1.6881 | 1.7532 | 1.8125 | s | (582.35) |
| 0.4338 | 0.4468 | 0.4593 | 0.4714 | 0.4833 | 0.4948 | 0.5061 | 0.5172 | 0.5281 | 0.5547 | 0.5805 | 0.6058 | 0.6305 | 0.6789 | 0.7727 | 0.8640 | v | |
| 1326.7 | 1341.3 | 1355.4 | 1369.1 | 1382.4 | 1395.4 | 1408.2 | 1420.8 | 1433.1 | 1463.5 | 1493.2 | 1522.4 | 1551.4 | 1608.9 | 1723.7 | 1840.0 | h | **1400** |
| 1.4832 | 1.4953 | 1.5068 | 1.5177 | 1.5282 | 1.5383 | 1.5480 | 1.5575 | 1.5666 | 1.5886 | 1.6093 | 1.6289 | 1.6478 | 1.6836 | 1.7489 | 1.8083 | s | (587.10) |
| 0.4157 | 0.4285 | 0.4408 | 0.4527 | 0.4643 | 0.4756 | 0.4866 | 0.4974 | 0.5080 | 0.5339 | 0.5591 | 0.5836 | 0.6076 | 0.6545 | 0.7454 | 0.8338 | v | |
| 1323.4 | 1338.3 | 1352.6 | 1366.5 | 1380.0 | 1393.1 | 1406.0 | 1418.7 | 1431.2 | 1461.8 | 1491.6 | 1521.0 | 1550.1 | 1607.8 | 1722.9 | 1839.4 | h | **1450** |
| 1.4772 | 1.4895 | 1.5011 | 1.5122 | 1.5228 | 1.5330 | 1.5429 | 1.5524 | 1.5617 | 1.5838 | 1.6046 | 1.6244 | 1.6434 | 1.6792 | 1.7447 | 1.8042 | s | (591.73) |
| 0.3989 | 0.4114 | 0.4235 | 0.4352 | 0.4465 | 0.4576 | 0.4684 | 0.4789 | 0.4893 | 0.5146 | 0.5390 | 0.5629 | 0.5862 | 0.6318 | 0.7199 | 0.8056 | v | |
| 1320.0 | 1335.2 | 1349.7 | 1363.8 | 1377.5 | 1390.8 | 1403.9 | 1416.7 | 1429.3 | 1460.1 | 1490.1 | 1519.6 | 1548.9 | 1606.8 | 1722.1 | 1838.8 | h | **1500** |
| 1.4712 | 1.4837 | 1.4956 | 1.5068 | 1.5176 | 1.5279 | 1.5379 | 1.5475 | 1.5569 | 1.5791 | 1.6001 | 1.6200 | 1.6390 | 1.6750 | 1.7406 | 1.8002 | s | (596.23) |
| 0.3682 | 0.3804 | 0.3921 | 0.4034 | 0.4144 | 0.4250 | 0.4353 | 0.4454 | 0.4553 | 0.4794 | 0.5027 | 0.5253 | 0.5474 | 0.5906 | 0.6738 | 0.7545 | v | |
| 1313.0 | 1328.8 | 1343.9 | 1358.4 | 1372.5 | 1386.1 | 1399.5 | 1412.5 | 1425.3 | 1456.6 | 1487.0 | 1516.9 | 1546.4 | 1604.6 | 1720.5 | 1837.5 | h | **1600** |
| 1.4595 | 1.4725 | 1.4848 | 1.4964 | 1.5075 | 1.5181 | 1.5283 | 1.5381 | 1.5476 | 1.5702 | 1.5914 | 1.6115 | 1.6307 | 1.6669 | 1.7328 | 1.7926 | s | (604.90) |
| 0.3410 | 0.3529 | 0.3643 | 0.3753 | 0.3859 | 0.3961 | 0.4061 | 0.4158 | 0.4253 | 0.4484 | 0.4706 | 0.4922 | 0.5132 | 0.5542 | 0.6330 | 0.7094 | v | |
| 1305.8 | 1322.3 | 1337.9 | 1352.9 | 1367.3 | 1381.3 | 1395.0 | 1408.3 | 1421.4 | 1453.2 | 1484.0 | 1514.1 | 1543.8 | 1602.5 | 1718.9 | 1836.2 | h | **1700** |
| 1.4480 | 1.4616 | 1.4743 | 1.4863 | 1.4977 | 1.5086 | 1.5190 | 1.5290 | 1.5387 | 1.5616 | 1.5831 | 1.6034 | 1.6228 | 1.6592 | 1.7255 | 1.7854 | s | (613.15) |
| 0.3166 | 0.3284 | 0.3395 | 0.3502 | 0.3605 | 0.3705 | 0.3801 | 0.3895 | 0.3986 | 0.4208 | 0.4421 | 0.4627 | 0.4828 | 0.5218 | 0.5968 | 0.6693 | v | |
| 1298.4 | 1315.5 | 1331.8 | 1347.2 | 1362.1 | 1376.5 | 1390.4 | 1404.1 | 1417.4 | 1449.7 | 1480.8 | 1511.3 | 1541.3 | 1600.4 | 1717.3 | 1835.0 | h | **1800** |
| 1.4367 | 1.4509 | 1.4641 | 1.4765 | 1.4882 | 1.4994 | 1.5100 | 1.5203 | 1.5301 | 1.5534 | 1.5752 | 1.5957 | 1.6153 | 1.6520 | 1.7185 | 1.7786 | s | (621.03) |
| 0.2947 | 0.3063 | 0.3173 | 0.3277 | 0.3378 | 0.3474 | 0.3568 | 0.3659 | 0.3747 | 0.3961 | 0.4165 | 0.4363 | 0.4556 | 0.4929 | 0.5644 | 0.6334 | v | |
| 1290.6 | 1308.6 | 1325.4 | 1341.5 | 1356.8 | 1371.5 | 1385.8 | 1399.7 | 1413.3 | 1446.1 | 1477.7 | 1508.5 | 1538.8 | 1598.2 | 1715.7 | 1833.7 | h | **1900** |
| 1.4256 | 1.4404 | 1.4541 | 1.4669 | 1.4790 | 1.4904 | 1.5014 | 1.5118 | 1.5219 | 1.5456 | 1.5676 | 1.5883 | 1.6081 | 1.6450 | 1.7118 | 1.7721 | s | (628.58) |
| 0.2748 | 0.2863 | 0.2972 | 0.3074 | 0.3172 | 0.3267 | 0.3358 | 0.3446 | 0.3532 | 0.3738 | 0.3935 | 0.4126 | 0.4311 | 0.4668 | 0.5352 | 0.6011 | v | |
| 1282.6 | 1301.4 | 1319.0 | 1335.5 | 1351.3 | 1366.5 | 1381.2 | 1395.3 | 1409.2 | 1442.5 | 1474.5 | 1505.7 | 1536.2 | 1596.[illegible] | 1714.1 | 1832.5 | h | **2000** |
| 1.4145 | 1.4300 | 1.4443 | 1.4576 | 1.4700 | 1.4818 | 1.4930 | 1.5036 | 1.5139 | 1.5380 | 1.5603 | 1.5813 | 1.6012 | 1.6384 | 1.7055 | 1.7660 | s | (635.82) |
| 0.2567 | 0.2682 | 0.2789 | 0.2890 | 0.2986 | 0.3078 | 0.3167 | 0.3253 | 0.3337 | 0.3537 | 0.3727 | 0.3911 | 0.4089 | 0.4433 | 0.5088 | 0.5719 | v | |
| 1274.3 | 1294.0 | 1312.3 | 1329.5 | 1345.8 | 1361.4 | 1376.4 | 1390.9 | 1405.0 | 1438.9 | 1471.4 | 1502.9 | 1533.6 | 1593.9 | 1712.5 | 1831.2 | h | **2100** |
| 1.4035 | 1.4197 | 1.4346 | 1.4484 | 1.4612 | 1.4733 | 1.4848 | 1.4957 | 1.5062 | 1.5306 | 1.5533 | 1.5745 | 1.5945 | 1.6320 | 1.6995 | 1.7601 | s | (642.77) |
| 0.2400 | 0.2514 | 0.2621 | 0.2721 | 0.2816 | 0.2907 | 0.2994 | 0.3078 | 0.3159 | 0.3354 | 0.3538 | 0.3715 | 0.3887 | 0.4218 | 0.4849 | 0.5453 | v | |
| 1265.7 | 1286.3 | 1305.4 | 1323.3 | 1340.1 | 1356.2 | 1371.5 | 1386.4 | 1400.8 | 1435.3 | 1468.2 | 1500.0 | 1531.1 | 1591.8 | 1710.9 | 1830.0 | h | **2200** |
| 1.3925 | 1.4095 | 1.4250 | 1.4393 | 1.4526 | 1.4650 | 1.4768 | 1.4880 | 1.4986 | 1.5235 | 1.5465 | 1.5679 | 1.5881 | 1.6259 | 1.6937 | 1.7545 | s | (649.46) |
| 0.2247 | 0.2362 | 0.2468 | 0.2567 | 0.2661 | 0.2750 | 0.2835 | 0.2917 | 0.2997 | 0.3186 | 0.3365 | 0.3537 | 0.3703 | 0.4023 | 0.4630 | 0.5211 | v | |
| 1256.7 | 1278.4 | 1298.4 | 1316.9 | 1334.3 | 1350.8 | 1366.6 | 1381.8 | 1396.5 | 1431.6 | 1464.9 | 1497.1 | 1528.5 | 1589.6 | 1709.3 | 1828.7 | h | **2300** |
| 1.3814 | 1.3994 | 1.4156 | 1.4304 | 1.4441 | 1.4569 | 1.4690 | 1.4804 | 1.4913 | 1.5166 | 1.5399 | 1.5615 | 1.5820 | 1.6200 | 1.6881 | 1.7491 | s | (655.91) |
| 0.2105 | 0.2221 | 0.2327 | 0.2425 | 0.2518 | 0.2606 | 0.2689 | 0.2770 | 0.2848 | 0.3033 | 0.3207 | 0.3373 | 0.3534 | 0.3843 | 0.4429 | 0.4988 | v | |
| 1247.3 | 1270.2 | 1291.1 | 1310.3 | 1328.4 | 1345.4 | 1361.6 | 1377.2 | 1392.2 | 1427.9 | 1461.7 | 1494.2 | 1525.9 | 1587.4 | 1707.7 | 1827.4 | h | **2400** |
| 1.3702 | 1.3892 | 1.4061 | 1.4215 | 1.4357 | 1.4489 | 1.4613 | 1.4730 | 1.4842 | 1.5099 | 1.5335 | 1.5554 | 1.5760 | 1.6143 | 1.6827 | 1.7439 | s | (662.12) |
| 0.1973 | 0.2090 | 0.2196 | 0.2294 | 0.2386 | 0.2473 | 0.2555 | 0.2634 | 0.2710 | 0.2891 | 0.3061 | 0.3222 | 0.3379 | 0.3678 | 0.4244 | 0.4784 | v | |
| 1237.6 | 1261.8 | 1283.6 | 1303.6 | 1322.3 | 1339.9 | 1356.5 | 1372.5 | 1387.8 | 1424.2 | 1458.4 | 1491.3 | 1523.2 | 1585.3 | 1706.1 | 1826.2 | h | **2500** |
| 1.3589 | 1.3789 | 1.3967 | 1.4127 | 1.4274 | 1.4411 | 1.4538 | 1.4658 | 1.4772 | 1.5034 | 1.5273 | 1.5495 | 1.5702 | 1.6088 | 1.6775 | 1.7389 | s | (668.13) |
| 0.1849 | 0.1967 | 0.2074 | 0.2172 | 0.2263 | 0.2349 | 0.2431 | 0.2509 | 0.2584 | 0.2760 | 0.2926 | 0.3083 | 0.3236 | 0.3526 | 0.4074 | 0.4595 | v | |
| 1227.3 | 1252.9 | 1275.8 | 1296.8 | 1316.1 | 1334.2 | 1351.4 | 1367.7 | 1383.4 | 1420.4 | 1455.1 | 1488.4 | 1520.6 | 1583.1 | 1704.4 | 1824.9 | h | **2600** |
| 1.3474 | 1.3686 | 1.3873 | 1.4040 | 1.4192 | 1.4333 | 1.4464 | 1.4587 | 1.4703 | 1.4970 | 1.5212 | 1.5437 | 1.5646 | 1.6035 | 1.6725 | 1.7341 | s | (673.94) |
| 0.1732 | 0.1853 | 0.1960 | 0.2059 | 0.2150 | 0.2235 | 0.2315 | 0.2392 | 0.2466 | 0.2639 | 0.2801 | 0.2955 | 0.3103 | 0.3385 | 0.3916 | 0.4420 | v | |
| 1216.5 | 1243.8 | 1267.9 | 1289.7 | 1309.7 | 1328.5 | 1346.1 | 1362.9 | 1378.9 | 1416.6 | 1451.8 | 1485.5 | 1518.0 | 1580.9 | 1702.8 | 1823.7 | h | **2700** |
| 1.3357 | 1.3582 | 1.3778 | 1.3953 | 1.4111 | 1.4256 | 1.4391 | 1.4517 | 1.4635 | 1.4908 | 1.5154 | 1.5380 | 1.5592 | 1.5983 | 1.6677 | 1.7294 | s | (679.55) |

# Table 3. Superheated Vapor

| Abs. Press. Lb./Sq. In. (Sat. Temp.) | | Sat. Liquid | Sat. Vapor | Temperature—Degrees Fahrenheit 690° | 700° | 710° | 720° | 730° | 740° | 750° | 760° | 770° | 780° | 790° | 800° | 820° |
|---|---|---|---|---|---|---|---|---|---|---|---|---|---|---|---|---|
| | v | 0.0315 | 0.1035 | 0.1141 | 0.1281 | 0.1387 | 0.1475 | 0.1553 | 0.1622 | 0.1686 | 0.1745 | 0.1801 | 0.1854 | 0.1904 | 0.1953 | 0.2043 |
| **2800** | h | 770.1 | 1054.8 | 1085.2 | 1121.4 | 1148.0 | 1169.7 | 1188.5 | 1205.1 | 1220.3 | 1234.2 | 1247.3 | 1259.6 | 1271.3 | 1282.4 | 1303.2 |
| (684.99) | s | 0.9459 | 1.1946 | 1.2210 | 1.2524 | 1.2752 | 1.2938 | 1.3095 | 1.3236 | 1.3362 | 1.3477 | 1.3583 | 1.3683 | 1.3777 | 1.3865 | 1.4030 |
| | v | 0.0329 | 0.0947 | | 0.1143 | 0.1266 | 0.1362 | 0.1444 | 0.1517 | 0.1583 | 0.1644 | 0.1701 | 0.1754 | 0.1805 | 0.1853 | 0.1944 |
| **2900** | h | 785.4 | 1039.0 | | 1095.9 | 1128.7 | 1153.9 | 1174.8 | 1193.0 | 1209.3 | 1224.3 | 1238.1 | 1251.1 | 1263.3 | 1274.9 | 1296.6 |
| (690.26) | s | 0.9587 | 1.1792 | | 1.2286 | 1.2568 | 1.2782 | 1.2959 | 1.3111 | 1.3246 | 1.3369 | 1.3482 | 1.3587 | 1.3685 | 1.3778 | 1.3949 |
| | v | 0.0346 | 0.0858 | | 0.0984 | 0.1142 | 0.1251 | 0.1339 | 0.1416 | 0.1485 | 0.1548 | 0.1606 | 0.1660 | 0.1711 | 0.1760 | 0.1851 |
| **3000** | h | 802.5 | 1020.3 | | 1060.7 | 1106.5 | 1136.1 | 1159.9 | 1180.1 | 1197.8 | 1213.8 | 1228.5 | 1242.2 | 1255.1 | 1267.2 | 1289.7 |
| (695.36) | s | 0.9731 | 1.1615 | | 1.1966 | 1.2359 | 1.2611 | 1.2812 | 1.2981 | 1.3128 | 1.3259 | 1.3380 | 1.3490 | 1.3594 | 1.3690 | 1.3868 |
| | v | 0.0371 | 0.0753 | | | 0.1012 | 0.1140 | 0.1238 | 0.1320 | 0.1391 | 0.1456 | 0.1516 | 0.1571 | 0.1623 | 0.1672 | 0.1763 |
| **3100** | h | 825.0 | 993.1 | | | 1079.6 | 1115.7 | 1143.5 | 1166.2 | 1185.6 | 1202.9 | 1218.6 | 1233.0 | 1246.6 | 1259.3 | 1282.7 |
| (700.31) | s | 0.9919 | 1.1368 | | | 1.2112 | 1.2419 | 1.2654 | 1.2844 | 1.3005 | 1.3147 | 1.3275 | 1.3392 | 1.3501 | 1.3602 | 1.3787 |
| | v | 0.0444 | 0.0580 | | | 0.0852 | 0.1027 | 0.1138 | 0.1226 | 0.1301 | 0.1369 | 0.1430 | 0.1486 | 0.1539 | 0.1589 | 0.1681 |
| **3200** | h | 872.4 | 934.4 | | | 1037.9 | 1092.0 | 1125.2 | 1151.1 | 1172.6 | 1191.4 | 1208.2 | 1223.5 | 1237.8 | 1251.1 | 1275.6 |
| (705.11) | s | 1.0320 | 1.0852 | | | 1.1740 | 1.2201 | 1.2481 | 1.2698 | 1.2877 | 1.3031 | 1.3168 | 1.3292 | 1.3407 | 1.3513 | 1.3706 |
| | v | 0.0503 | 0.0503 | | | 0.0843 | 0.1020 | 0.1132 | 0.1220 | 0.1295 | 0.1363 | 0.1424 | 0.1480 | 0.1533 | 0.1583 | 0.1676 |
| **3206.2** | h | 902.7 | 902.7 | | | 1035.8 | 1090.7 | 1124.0 | 1150.2 | 1171.7 | 1190.6 | 1207.5 | 1222.9 | 1237.2 | 1250.5 | 1275.1 |
| (705.40) | s | 1.0580 | 1.0580 | | | 1.1723 | 1.2190 | 1.2471 | 1.2690 | 1.2869 | 1.3024 | 1.3162 | 1.3287 | 1.3402 | 1.3508 | 1.3701 |
| | v | | | 0.0293 | 0.0320 | 0.0545 | 0.0905 | 0.1037 | 0.1134 | 0.1215 | 0.1285 | 0.1348 | 0.1406 | 0.1459 | 0.1510 | 0.1602 |
| **3300** | h | | | 760.3 | 792.6 | 926.7 | 1062.6 | 1104.8 | 1134.6 | 1158.7 | 1179.2 | 1197.4 | 1213.6 | 1228.6 | 1242.6 | 1268.2 |
| | s | | | 0.9348 | 0.9628 | 1.0780 | 1.1936 | 1.2292 | 1.2542 | 1.2742 | 1.2911 | 1.3059 | 1.3191 | 1.3311 | 1.3423 | 1.3625 |
| | v | | | 0.0290 | 0.0312 | 0.0363 | 0.0773 | 0.0935 | 0.1042 | 0.1129 | 0.1204 | 0.1270 | 0.1329 | 0.1384 | 0.1435 | 0.1529 |
| **3400** | h | | | 757.4 | 785.9 | 832.8 | 1024.1 | 1081.3 | 1116.1 | 1143.6 | 1166.2 | 1185.9 | 1203.3 | 1219.2 | 1233.9 | 1260.6 |
| | s | | | 0.9320 | 0.9566 | 0.9969 | 1.1598 | 1.2080 | 1.2372 | 1.2600 | 1.2786 | 1.2946 | 1.3088 | 1.3215 | 1.3333 | 1.3543 |
| | v | | | 0.0287 | 0.0306 | 0.0342 | 0.0557 | 0.0823 | 0.0952 | 0.1047 | 0.1126 | 0.1195 | 0.1256 | 0.1312 | 0.1364 | 0.1458 |
| **3500** | h | | | 754.7 | 780.5 | 818.5 | 945.0 | 1052.4 | 1095.6 | 1127.3 | 1152.4 | 1173.7 | 1192.6 | 1209.4 | 1224.9 | 1252.9 |
| | s | | | 0.9291 | 0.9515 | 0.9841 | 1.0918 | 1.1825 | 1.2186 | 1.2449 | 1.2656 | 1.2829 | 1.2983 | 1.3118 | 1.3241 | 1.3462 |
| | v | | | 0.0285 | 0.0301 | 0.0329 | 0.0410 | 0.0701 | 0.0860 | 0.0965 | 0.1049 | 0.1121 | 0.1185 | 0.1243 | 0.1296 | 0.1392 |
| **3600** | h | | | 752.0 | 776.1 | 808.2 | 872.9 | 1013.4 | 1071.9 | 1109.5 | 1137.6 | 1160.8 | 1181.3 | 1199.2 | 1215.6 | 1244.9 |
| | s | | | 0.9264 | 0.9472 | 0.9748 | 1.0298 | 1.1484 | 1.1974 | 1.2286 | 1.2518 | 1.2707 | 1.2873 | 1.3017 | 1.3148 | 1.3379 |
| | v | | | 0.0282 | 0.0297 | 0.0320 | 0.0371 | 0.0557 | 0.0764 | 0.0885 | 0.0975 | 0.1051 | 0.1117 | 0.1177 | 0.1231 | 0.1328 |
| **3700** | h | | | 749.6 | 772.4 | 800.8 | 847.6 | 959.4 | 1043.7 | 1089.7 | 1121.6 | 1147.2 | 1169.4 | 1188.5 | 1205.9 | 1236.8 |
| | s | | | 0.9238 | 0.9436 | 0.9679 | 1.0078 | 1.1021 | 1.1727 | 1.2109 | 1.2372 | 1.2581 | 1.2760 | 1.2914 | 1.3053 | 1.3296 |
| | v | | | 0.0279 | 0.0293 | 0.0313 | 0.0349 | 0.0450 | 0.0664 | 0.0804 | 0.0903 | 0.0983 | 0.1052 | 0.1113 | 0.1169 | 0.1268 |
| **3800** | h | | | 747.4 | 769.2 | 795.5 | 833.8 | 908.2 | 1009.0 | 1067.1 | 1104.3 | 1132.9 | 1156.9 | 1177.4 | 1195.9 | 1228.4 |
| | s | | | 0.9214 | 0.9403 | 0.9629 | 0.9955 | 1.0583 | 1.1426 | 1.1908 | 1.2216 | 1.2449 | 1.2643 | 1.2808 | 1.2955 | 1.3211 |
| | v | | | 0.0277 | 0.0290 | 0.0308 | 0.0337 | 0.0400 | 0.0558 | 0.0721 | 0.0832 | 0.0917 | 0.0989 | 0.1052 | 0.1109 | 0.1210 |
| **3900** | h | | | 745.1 | 766.3 | 791.2 | 824.3 | 879.1 | 966.7 | 1040.9 | 1085.5 | 1117.5 | 1143.7 | 1165.9 | 1185.5 | 1219.7 |
| | s | | | 0.9190 | 0.9373 | 0.9588 | 0.9869 | 1.0331 | 1.1065 | 1.1681 | 1.2048 | 1.2310 | 1.2522 | 1.2700 | 1.2856 | 1.3126 |
| | v | | | 0.0275 | 0.0287 | 0.0303 | 0.0328 | 0.0373 | 0.0478 | 0.0638 | 0.0761 | 0.0852 | 0.0927 | 0.0993 | 0.1052 | 0.1155 |
| **4000** | h | | | 743.1 | 763.8 | 787.3 | 817.5 | 860.4 | 930.7 | 1010.6 | 1064.2 | 1101.1 | 1129.8 | 1153.9 | 1174.8 | 1210.7 |
| | s | | | 0.9168 | 0.9347 | 0.9549 | 0.9806 | 1.0168 | 1.0757 | 1.1420 | 1.1862 | 1.2163 | 1.2397 | 1.2591 | 1.2757 | 1.3040 |
| | v | | | 0.0267 | 0.0276 | 0.0286 | 0.0301 | 0.0320 | 0.0350 | 0.0395 | 0.0467 | 0.0561 | 0.0652 | 0.0730 | 0.0798 | 0.0911 |
| **4500** | h | | | 735.4 | 753.5 | 773.1 | 795.2 | 821.4 | 852.9 | 894.8 | 947.7 | 1000.8 | 1047.5 | 1084.3 | 1113.9 | 1162.5 |
| | s | | | 0.9078 | 0.9235 | 0.9403 | 0.9593 | 0.9814 | 1.0078 | 1.0426 | 1.0861 | 1.1294 | 1.1672 | 1.1968 | 1.2204 | 1.2586 |
| | v | | | 0.0260 | 0.0268 | 0.0276 | 0.0287 | 0.0300 | 0.0316 | 0.0337 | 0.0368 | 0.0410 | 0.0465 | 0.0527 | 0.0593 | 0.0715 |
| **5000** | h | | | 729.6 | 746.4 | 764.5 | 783.9 | 804.8 | 828.4 | 856.1 | 887.8 | 925.0 | 966.2 | 1007.6 | 1047.1 | 1108.7 |
| | s | | | 0.9006 | 0.9152 | 0.9307 | 0.9472 | 0.9649 | 0.9847 | 1.0076 | 1.0337 | 1.0641 | 1.0975 | 1.1307 | 1.1622 | 1.2107 |
| | v | | | 0.0255 | 0.0262 | 0.0269 | 0.0278 | 0.0288 | 0.0299 | 0.0313 | 0.0331 | 0.0354 | 0.0384 | 0.0420 | 0.0463 | 0.0560 |
| **5500** | h | | | 725.7 | 741.3 | 758.0 | 775.6 | 794.3 | 814.2 | 836.5 | 860.8 | 887.4 | 917.5 | 950.4 | 985.0 | 1052.8 |
| | s | | | 0.8955 | 0.9090 | 0.9233 | 0.9382 | 0.9540 | 0.9708 | 0.9892 | 1.0092 | 1.0309 | 1.0553 | 1.0817 | 1.1093 | 1.1627 |

| 840° | 860° | 880° | 900° | 920° | 940° | 960° | 980° | 1000° | 1050° | 1100° | 1150° | 1200° | 1400° | 1600° | | Abs. Press. Lb./Sq. In. (Sat. Temp.) |
|---|---|---|---|---|---|---|---|---|---|---|---|---|---|---|---|---|
| | | | | | | | | | | | Temperature—Degrees Fahrenheit | | | | | |
| 0.2128 | 0.2208 | 0.2284 | 0.2356 | 0.2426 | 0.2494 | 0.2559 | 0.2623 | 0.2685 | 0.2835 | 0.2979 | 0.3118 | 0.3254 | 0.3769 | 0.4258 | v | |
| 1322.6 | 1340.8 | 1357.9 | 1374.3 | 1390.1 | 1405.3 | 1420.1 | 1434.4 | 1448.5 | 1482.5 | 1515.4 | 1547.3 | 1578.7 | 1701.2 | 1822.4 | h | **2800** |
| 1.4180 | 1.4318 | 1.4448 | 1.4569 | 1.4684 | 1.4794 | 1.4898 | 1.4999 | 1.5096 | 1.5325 | 1.5539 | 1.5741 | 1.5933 | 1.6630 | 1.7249 | s | (684.99) |
| 0.2028 | 0.2108 | 0.2183 | 0.2254 | 0.2323 | 0.2390 | 0.2454 | 0.2516 | 0.2577 | 0.2723 | 0.2864 | 0.3000 | 0.3132 | 0.3632 | 0.4107 | v | |
| 1316.6 | 1335.3 | 1352.9 | 1369.7 | 1385.8 | 1401.3 | 1416.3 | 1430.9 | 1445.1 | 1479.6 | 1512.7 | 1544.9 | 1576.5 | 1699.6 | 1821.1 | h | **2900** |
| 1.4104 | 1.4247 | 1.4379 | 1.4504 | 1.4621 | 1.4733 | 1.4839 | 1.4941 | 1.5040 | 1.5272 | 1.5487 | 1.5691 | 1.5884 | 1.6584 | 1.7205 | s | (690.26) |
| 0.1935 | 0.2014 | 0.2088 | 0.2159 | 0.2227 | 0.2292 | 0.2355 | 0.2416 | 0.2476 | 0.2620 | 0.2757 | 0.2890 | 0.3018 | 0.3505 | 0.3966 | v | |
| 1310.4 | 1329.7 | 1347.8 | 1365.0 | 1381.4 | 1397.2 | 1412.5 | 1427.3 | 1441.8 | 1476.6 | 1510.0 | 1542.5 | 1574.3 | 1698.0 | 1819.9 | h | **3000** |
| 1.4029 | 1.4176 | 1.4312 | 1.4439 | 1.4559 | 1.4673 | 1.4781 | 1.4885 | 1.4984 | 1.5219 | 1.5437 | 1.5642 | 1.5837 | 1.6540 | 1.7163 | s | (695.36) |
| 0.1847 | 0.1926 | 0.1999 | 0.2070 | 0.2137 | 0.2201 | 0.2263 | 0.2323 | 0.2382 | 0.2523 | 0.2657 | 0.2786 | 0.2911 | 0.3386 | 0.3834 | v | |
| 1304.2 | 1324.1 | 1342.7 | 1360.3 | 1377.1 | 1393.2 | 1408.7 | 1423.7 | 1438.4 | 1473.6 | 1507.4 | 1540.1 | 1572.1 | 1696.3 | 1818.6 | h | **3100** |
| 1.3953 | 1.4105 | 1.4245 | 1.4376 | 1.4498 | 1.4614 | 1.4724 | 1.4830 | 1.4930 | 1.5168 | 1.5388 | 1.5594 | 1.5790 | 1.6497 | 1.7122 | s | (700.31) |
| 0.1765 | 0.1843 | 0.1916 | 0.1986 | 0.2052 | 0.2115 | 0.2176 | 0.2236 | 0.2293 | 0.2431 | 0.2563 | 0.2689 | 0.2811 | 0.3274 | 0.3710 | v | |
| 1297.8 | 1318.3 | 1337.5 | 1355.5 | 1372.6 | 1389.0 | 1404.8 | 1420.1 | 1434.9 | 1470.6 | 1504.7 | 1537.7 | 1569.9 | 1694.7 | 1817.3 | h | **3200** |
| 1.3879 | 1.4035 | 1.4179 | 1.4313 | 1.4438 | 1.4556 | 1.4668 | 1.4775 | 1.4877 | 1.5118 | 1.5340 | 1.5548 | 1.5745 | 1.6455 | 1.7082 | s | (705.11) |
| 0.1760 | 0.1838 | 0.1911 | 0.1981 | 0.2047 | 0.2110 | 0.2171 | 0.2231 | 0.2288 | 0.2426 | 0.2557 | 0.2683 | 0.2806 | 0.3267 | 0.3703 | v | |
| 1297.4 | 1317.9 | 1337.2 | 1355.2 | 1372.3 | 1388.7 | 1404.6 | 1419.9 | 1434.7 | 1470.4 | 1504.5 | 1537.5 | 1569.8 | 1694.6 | 1817.2 | h | **3206.2** |
| 1.3874 | 1.4031 | 1.4175 | 1.4309 | 1.4434 | 1.4552 | 1.4665 | 1.4772 | 1.4874 | 1.5115 | 1.5337 | 1.5545 | 1.5742 | 1.6452 | 1.7080 | s | (705.40) |
| 0.1687 | 0.1765 | 0.1838 | 0.1907 | 0.1972 | 0.2035 | 0.2095 | 0.2153 | 0.2210 | 0.2346 | 0.2474 | 0.2598 | 0.2718 | 0.3169 | 0.3594 | v | |
| 1291.3 | 1312.5 | 1332.1 | 1350.6 | 1368.1 | 1384.8 | 1400.9 | 1416.5 | 1431.5 | 1467.6 | 1502.0 | 1535.2 | 1567.7 | 1693.1 | 1816.1 | h | **3300** |
| 1.3804 | 1.3966 | 1.4113 | 1.4251 | 1.4378 | 1.4499 | 1.4613 | 1.4722 | 1.4825 | 1.5068 | 1.5292 | 1.5502 | 1.5701 | 1.6414 | 1.7043 | s | |
| 0.1613 | 0.1691 | 0.1764 | 0.1832 | 0.1897 | 0.1959 | 0.2018 | 0.2076 | 0.2132 | 0.2265 | 0.2391 | 0.2512 | 0.2629 | 0.3070 | 0.3484 | v | |
| 1284.6 | 1306.5 | 1326.7 | 1345.7 | 1363.5 | 1380.6 | 1397.0 | 1412.8 | 1428.0 | 1464.6 | 1499.3 | 1532.8 | 1565.5 | 1691.5 | 1814.8 | h | **3400** |
| 1.3729 | 1.3896 | 1.4048 | 1.4189 | 1.4319 | 1.4442 | 1.4558 | 1.4669 | 1.4774 | 1.5020 | 1.5246 | 1.5458 | 1.5658 | 1.6375 | 1.7005 | s | |
| 0.1543 | 0.1621 | 0.1694 | 0.1762 | 0.1826 | 0.1887 | 0.1946 | 0.2003 | 0.2058 | 0.2189 | 0.2313 | 0.2431 | 0.2546 | 0.2977 | 0.3381 | v | |
| 1277.8 | 1300.4 | 1321.2 | 1340.7 | 1358.9 | 1376.4 | 1393.0 | 1409.1 | 1424.5 | 1461.5 | 1496.6 | 1530.4 | 1563.3 | 1689.8 | 1813.6 | h | **3500** |
| 1.3654 | 1.3827 | 1.3983 | 1.4127 | 1.4261 | 1.4386 | 1.4505 | 1.4617 | 1.4723 | 1.4972 | 1.5201 | 1.5414 | 1.5615 | 1.6336 | 1.6968 | s | |
| 0.1477 | 0.1556 | 0.1627 | 0.1695 | 0.1758 | 0.1819 | 0.1878 | 0.1934 | 0.1988 | 0.2117 | 0.2239 | 0.2355 | 0.2467 | 0.2889 | 0.3284 | v | |
| 1270.8 | 1294.2 | 1315.6 | 1335.6 | 1354.3 | 1372.1 | 1389.0 | 1405.3 | 1421.0 | 1458.5 | 1493.9 | 1527.9 | 1561.0 | 1688.2 | 1812.3 | h | **3600** |
| 1.3579 | 1.3757 | 1.3919 | 1.4067 | 1.4203 | 1.4331 | 1.4452 | 1.4566 | 1.4674 | 1.4926 | 1.5157 | 1.5372 | 1.5574 | 1.6298 | 1.6932 | s | |
| 0.1414 | 0.1493 | 0.1565 | 0.1632 | 0.1695 | 0.1755 | 0.1813 | 0.1868 | 0.1922 | 0.2049 | 0.2168 | 0.2283 | 0.2393 | 0.2806 | 0.3192 | v | |
| 1263.6 | 1287.7 | 1309.9 | 1330.4 | 1349.6 | 1367.7 | 1385.0 | 1401.6 | 1417.5 | 1455.4 | 1491.2 | 1525.5 | 1558.8 | 1686.6 | 1811.0 | h | **3700** |
| 1.3504 | 1.3686 | 1.3854 | 1.4006 | 1.4147 | 1.4277 | 1.4400 | 1.4516 | 1.4625 | 1.4880 | 1.5113 | 1.5330 | 1.5534 | 1.6261 | 1.6897 | s | |
| 0.1354 | 0.1434 | 0.1506 | 0.1572 | 0.1635 | 0.1695 | 0.1752 | 0.1806 | 0.1859 | 0.1984 | 0.2102 | 0.2214 | 0.2322 | 0.2727 | 0.3104 | v | |
| 1256.3 | 1281.1 | 1304.0 | 1325.1 | 1344.8 | 1363.3 | 1381.0 | 1397.8 | 1413.9 | 1452.3 | 1488.4 | 1523.0 | 1556.6 | 1684.9 | 1809.7 | h | **3800** |
| 1.3428 | 1.3616 | 1.3789 | 1.3946 | 1.4089 | 1.4223 | 1.4348 | 1.4465 | 1.4577 | 1.4835 | 1.5070 | 1.5288 | 1.5494 | 1.6225 | 1.6862 | s | |
| 0.1297 | 0.1377 | 0.1450 | 0.1516 | 0.1578 | 0.1637 | 0.1693 | 0.1748 | 0.1800 | 0.1923 | 0.2039 | 0.2149 | 0.2255 | 0.2652 | 0.3022 | v | |
| 1248.9 | 1274.4 | 1298.0 | 1319.9 | 1339.9 | 1358.9 | 1376.9 | 1394.0 | 1410.4 | 1449.2 | 1485.7 | 1520.6 | 1554.4 | 1683.3 | 1808.5 | h | **3900** |
| 1.3352 | 1.3547 | 1.3724 | 1.3887 | 1.4033 | 1.4169 | 1.4297 | 1.4416 | 1.4529 | 1.4790 | 1.5028 | 1.5248 | 1.5455 | 1.6189 | 1.6828 | s | |
| 0.1243 | 0.1323 | 0.1396 | 0.1462 | 0.1524 | 0.1582 | 0.1638 | 0.1692 | 0.1743 | 0.1865 | 0.1979 | 0.2088 | 0.2192 | 0.2581 | 0.2943 | v | |
| 1241.1 | 1267.8 | 1292.1 | 1314.4 | 1335.0 | 1354.4 | 1372.7 | 1390.1 | 1406.8 | 1446.1 | 1482.9 | 1518.1 | 1552.1 | 1681.7 | 1807.2 | h | **4000** |
| 1.3275 | 1.3479 | 1.3662 | 1.3827 | 1.3978 | 1.4117 | 1.4247 | 1.4369 | 1.4482 | 1.4747 | 1.4987 | 1.5209 | 1.5417 | 1.6154 | 1.6795 | s | |
| 0.1004 | 0.1086 | 0.1160 | 0.1226 | 0.1287 | 0.1345 | 0.1399 | 0.1451 | 0.1500 | 0.1613 | 0.1720 | 0.1820 | 0.1917 | 0.2273 | 0.2602 | v | |
| 1200.2 | 1232.6 | 1260.9 | 1286.5 | 1309.7 | 1331.3 | 1351.4 | 1370.4 | 1388.4 | 1430.4 | 1469.1 | 1505.7 | 1540.8 | 1673.5 | 1800.9 | h | **4500** |
| 1.2879 | 1.3127 | 1.3339 | 1.3529 | 1.3699 | 1.3854 | 1.3997 | 1.4129 | 1.4253 | 1.4536 | 1.4789 | 1.5019 | 1.5235 | 1.5990 | 1.6640 | s | |
| 0.0811 | 0.0894 | 0.0968 | 0.1036 | 0.1097 | 0.1154 | 0.1206 | 0.1256 | 0.1303 | 0.1413 | 0.1513 | 0.1606 | 0.1696 | 0.2027 | 0.2329 | v | |
| 1155.7 | 1194.0 | 1227.0 | 1256.5 | 1282.8 | 1306.9 | 1329.2 | 1350.0 | 1369.5 | 1414.3 | 1455.0 | 1493.2 | 1529.5 | 1665.3 | 1794.5 | h | **5000** |
| 1.2471 | 1.2764 | 1.3012 | 1.3231 | 1.3424 | 1.3596 | 1.3755 | 1.3900 | 1.4034 | 1.4336 | 1.4602 | 1.4844 | 1.5066 | 1.5839 | 1.6499 | s | |
| 0.0657 | 0.0739 | 0.0813 | 0.0880 | 0.0940 | 0.0997 | 0.1048 | 0.1097 | 0.1143 | 0.1248 | 0.1343 | 0.1432 | 0.1516 | 0.1825 | 0.2106 | v | |
| 1108.4 | 1152.6 | 1190.6 | 1224.1 | 1253.8 | 1280.6 | 1305.3 | 1328.1 | 1349.3 | 1397.6 | 1440.8 | 1480.6 | 1518.2 | 1657.0 | 1788.1 | h | **5500** |
| 1.2058 | 1.2396 | 1.2682 | 1.2930 | 1.3147 | 1.3340 | 1.3515 | 1.3675 | 1.3821 | 1.4146 | 1.4427 | 1.4679 | 1.4908 | 1.5699 | 1.6369 | s | |

# Table 4. Compressed Liquid

| Abs. Press. Lb./Sq. In. (Sat. Temp.) | | Temperature—Degrees Fahrenheit 32° | 100° | 200° | 300° | 400° | 500° | 600° | 620° | 640° | 660° | 680° | 690° | 700° | 705.4° |
|---|---|---|---|---|---|---|---|---|---|---|---|---|---|---|---|
| Saturated Liquid | p | 0.08854 | 0.9492 | 11.526 | 67.013 | 247.31 | 680.8 | 1542.9 | 1786.6 | 2059.7 | 2365.4 | 2708.1 | 2895.1 | 3093.7 | 3206.2 |
| | $v_f$ | 0.016022 | 0.016132 | 0.016634 | 0.017449 | 0.018639 | 0.020432 | 0.023629 | 0.02466 | 0.02598 | 0.02777 | 0.03054 | 0.03277 | 0.03692 | 0.05030 |
| | $h_f$ | 0 | 67.97 | 167.99 | 269.59 | 374.97 | 487.82 | 617.0 | 646.7 | 678.6 | 714.2 | 757.3 | 784.4 | 823.3 | 902.7 |
| | $s_f$ | 0 | 0.12948 | 0.29382 | 0.43694 | 0.56638 | 0.68871 | 0.8131 | 0.8398 | 0.8679 | 0.8987 | 0.9351 | 0.9578 | 0.9905 | 1.0580 |
| 200 (381.79) | $(v-v_f)\cdot 10^5$ | −1.1 | −1.1 | −1.1 | −1.1 | | | | | | | | | | |
| | $(h-h_f)$ | +0.61 | +0.54 | +0.41 | +0.23 | | | | | | | | | | |
| | $(s-s_f)\cdot 10^3$ | +0.03 | −0.05 | −0.21 | −0.21 | | | | | | | | | | |
| 400 (444.59) | $(v-v_f)\cdot 10^5$ | −2.3 | −2.1 | −2.2 | −2.8 | −2.1 | | | | | | | | | |
| | $(h-h_f)$ | +1.21 | +1.09 | +0.88 | +0.61 | +0.16 | | | | | | | | | |
| | $(s-s_f)\cdot 10^3$ | +0.04 | −0.16 | −0.47 | −0.56 | −0.40 | | | | | | | | | |
| 600 (486.21) | $(v-v_f)\cdot 10^5$ | −3.5 | −3.2 | −3.4 | −4.3 | −4.4 | | | | | | | | | |
| | $(h-h_f)$ | +1.80 | +1.67 | +1.31 | +0.97 | +0.39 | | | | | | | | | |
| | $(s-s_f)\cdot 10^3$ | +0.07 | −0.27 | −0.74 | −0.94 | −0.96 | | | | | | | | | |
| 800 (518.23) | $(v-v_f)\cdot 10^5$ | −4.6 | −4.0 | −4.4 | −5.6 | −6.5 | −1.7 | | | | | | | | |
| | $(h-h_f)$ | +2.39 | +2.17 | +1.78 | +1.35 | +0.61 | −0.05 | | | | | | | | |
| | $(s-s_f)\cdot 10^3$ | +0.10 | −0.40 | −0.97 | −1.27 | −1.48 | −0.53 | | | | | | | | |
| 1000 (544.61) | $(v-v_f)\cdot 10^5$ | −5.7 | −5.1 | −5.4 | −6.9 | −8.7 | −6.4 | | | | | | | | |
| | $(h-h_f)$ | +2.99 | +2.70 | +2.21 | +1.75 | +0.84 | −0.14 | | | | | | | | |
| | $(s-s_f)\cdot 10^3$ | +0.15 | −0.53 | −1.20 | −1.64 | −2.00 | −1.41 | | | | | | | | |
| 1500 (596.23) | $(v-v_f)\cdot 10^5$ | −8.4 | −7.5 | −8.1 | −10.4 | −14.1 | −17.3 | | | | | | | | |
| | $(h-h_f)$ | +4.48 | +3.99 | +3.36 | +2.70 | +1.44 | −0.29 | | | | | | | | |
| | $(s-s_f)\cdot 10^3$ | +0.20 | −0.86 | −1.79 | −2.53 | −3.32 | −3.56 | | | | | | | | |
| 2000 (635.82) | $(v-v_f)\cdot 10^5$ | −11.0 | −9.9 | −10.8 | −13.8 | −19.5 | −27.8 | −32.6 | −24 | | | | | | |
| | $(h-h_f)$ | +5.97 | +5.31 | +4.51 | +3.64 | +2.03 | −0.38 | −2.5 | −1.8 | | | | | | |
| | $(s-s_f)\cdot 10^3$ | +0.22 | −1.18 | −2.39 | −3.42 | −4.57 | −5.58 | −4.3 | −2.6 | | | | | | |
| 2500 (668.13) | $(v-v_f)\cdot 10^5$ | −13.7 | −12.3 | −13.4 | −17.2 | −24.8 | −37.7 | −61.9 | −65 | −67 | −48 | | | | |
| | $(h-h_f)$ | +7.49 | +6.58 | +5.63 | +4.55 | +2.66 | −0.41 | −4.9 | −5.6 | −5.4 | −3.1 | | | | |
| | $(s-s_f)\cdot 10^3$ | +0.25 | −1.48 | −2.97 | −4.25 | −5.79 | −7.54 | −8.5 | −8.2 | −6.9 | −3.4 | | | | |
| 3000 (695.36) | $(v-v_f)\cdot 10^5$ | −16.3 | −14.7 | −16.0 | −20.7 | −30.0 | −47.1 | −87.9 | −101 | −122 | −146 | −172 | −166 | | |
| | $(h-h_f)$ | +9.00 | +7.88 | +6.76 | +5.49 | +3.33 | −0.41 | −6.9 | −8.7 | −10.3 | −11.8 | −12.2 | −8.9 | | |
| | $(s-s_f)\cdot 10^3$ | +0.28 | −1.79 | −3.56 | −5.12 | −7.03 | −9.42 | −12.4 | −13.0 | −13.4 | −13.3 | −12.0 | −8.2 | | |
| 3206.2 (705.40) | $(v-v_f)\cdot 10^5$ | −17.5 | −15.7 | −17.1 | −22.2 | −32.1 | −51.0 | −98.0 | −114 | −139 | −177 | −240 | −299 | −354 | 0 |
| | $(h-h_f)$ | +9.61 | +8.45 | +7.25 | +5.90 | +3.62 | −0.40 | −7.6 | −9.8 | −12.1 | −14.6 | −17.6 | −19.4 | −21.6 | 0 |
| | $(s-s_f)\cdot 10^3$ | +0.29 | −1.93 | −3.80 | −5.50 | −7.54 | −10.19 | −14.0 | −15.0 | −16.0 | −16.8 | −17.8 | −18.4 | −19.2 | 0 |
| 3500 | $(v-v_f)\cdot 10^5$ | −19.0 | −16.9 | −18.5 | −24.2 | −35.0 | −56.1 | −111.1 | −133 | −166 | −215 | −312 | −407 | −634 | −1815 |
| | $(h-h_f)$ | +10.44 | +9.17 | +7.90 | +6.44 | +4.01 | −0.34 | −8.6 | −11.2 | −14.3 | −17.8 | −24.2 | −29.7 | −42.8 | −104.1 |
| | $(s-s_f)\cdot 10^3$ | +0.30 | −2.08 | −4.14 | −5.97 | −8.21 | −11.24 | −16.0 | −17.4 | −19.1 | −20.9 | −24.9 | −28.7 | −39.0 | −91.0 |

| | | | | | | | | | | | | | | | |
|---|---|---|---|---|---|---|---|---|---|---|---|---|---|---|---|
| **4000** | $(v-v_f)\cdot 10^5$ | −21.5 | −19.2 | −21.0 | −27.5 | −40.0 | −64.5 | −132.2 | −160 | −202 | −270 | −400 | −528 | −821 | −2079 |
| | $(h-h_f)$ | +11.88 | +10.49 | +9.03 | +7.41 | +4.71 | −0.16 | −10.0 | −13.3 | −17.4 | −22.5 | −32.4 | −41.2 | −59.5 | −126.6 |
| | $(s-s_f)\cdot 10^3$ | +0.29 | −2.42 | −4.74 | −6.77 | −9.40 | −13.03 | −19.3 | −21.4 | −24.0 | −27.2 | −34.3 | −41.0 | −55.8 | −112.7 |
| **4500** | $(v-v_f)\cdot 10^5$ | −24.1 | −21.4 | −23.5 | −30.7 | −44.9 | −72.5 | −151.5 | −184 | −234 | −313 | −464 | −611 | −937 | −2219 |
| | $(h-h_f)$ | +13.32 | +11.80 | +10.15 | +8.40 | +5.40 | +0.02 | −11.1 | −15.1 | −19.9 | −26.4 | −38.4 | −49.0 | −69.8 | −138.9 |
| | $(s-s_f)\cdot 10^3$ | +0.26 | −2.74 | −5.33 | −7.60 | −10.58 | −14.80 | −22.4 | −25.1 | −28.3 | −32.7 | −41.8 | −50.0 | −67.0 | −125.6 |
| **5000** | $(v-v_f)\cdot 10^5$ | −26.7 | −23.6 | −26.0 | −34.0 | −49.6 | −80.5 | −169.3 | −206 | −262 | −350 | −518 | −677 | −1017 | −2309 |
| | $(h-h_f)$ | +14.75 | +13.08 | +11.30 | +9.36 | +6.08 | +0.25 | −12.1 | −16.7 | −22.0 | −29.5 | −43.0 | −54.7 | −76.9 | −146.7 |
| | $(s-s_f)\cdot 10^3$ | +0.22 | −3.07 | −5.92 | −8.40 | −11.74 | −16.47 | −25.3 | −28.5 | −32.2 | −37.5 | −47.9 | −57.2 | −75.3 | −134.6 |
| **5500** | $(v-v_f)\cdot 10^5$ | −29.2 | −25.7 | −28.4 | −37.2 | −54.2 | −88.3 | −186.1 | −227 | −288 | −384 | −562 | −727 | −1076 | −2375 |
| | $(h-h_f)$ | +16.18 | +14.41 | +12.47 | +10.36 | +6.78 | +0.52 | −13.1 | −18.0 | −23.7 | −31.8 | −46.4 | −58.7 | −82.0 | −152.6 |
| | $(s-s_f)\cdot 10^3$ | +0.20 | −3.39 | −6.47 | −9.20 | −12.86 | −18.10 | −28.0 | −31.5 | −35.6 | −41.5 | −52.7 | −62.4 | −81.5 | −141.4 |
| **6000** | $(v-v_f)\cdot 10^5$ | −31.7 | −27.8 | −30.8 | −40.5 | −58.7 | −96.1 | −202.9 | | | | | | | |
| | $(h-h_f)$ | +17.60 | +15.72 | +13.62 | +11.39 | +7.50 | +0.77 | −14.0 | | | | | | | |
| | $(s-s_f)\cdot 10^3$ | +0.10 | −3.72 | −7.06 | −10.00 | −13.96 | −19.57 | −30.6 | | | | | | | |

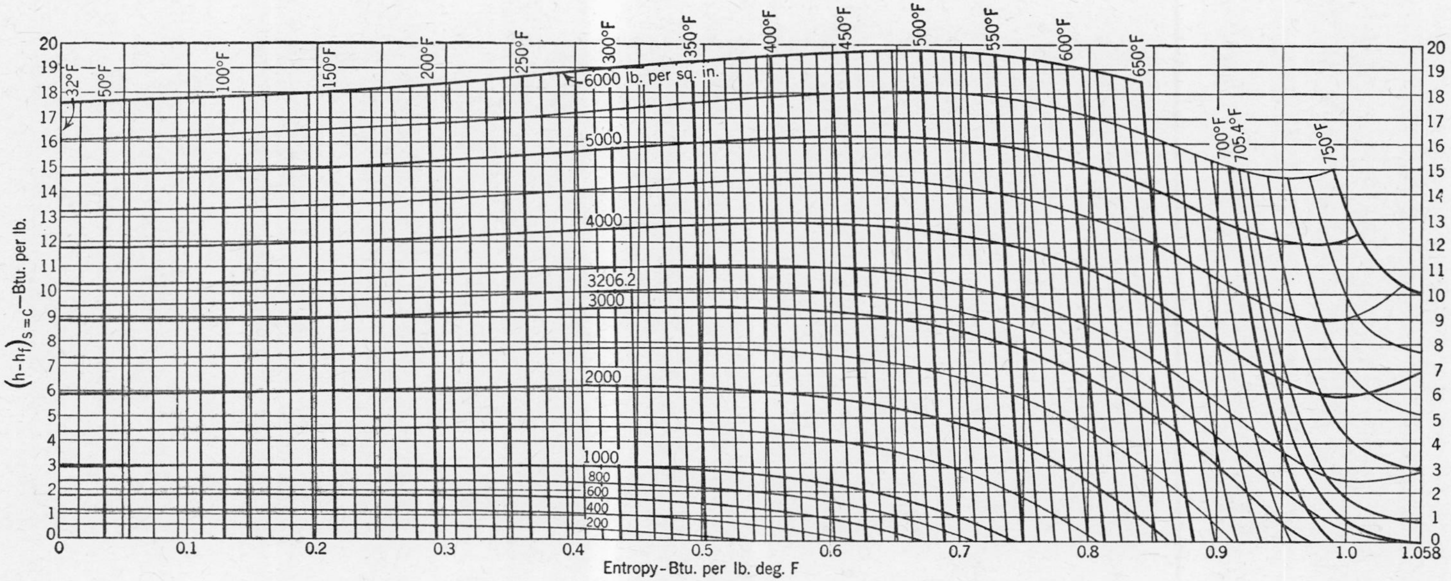

FIGURE 3. Enthalpy changes for constant entropy.

Departure of enthalpy from the value for saturated liquid at the same ENTROPY. The work of isentropic steady flow compression is the difference between two values of the ordinate at the same entropy.

# Table 5. Saturation: Solid-Vapor

| Temp. Fahr. | Abs. Press. Lb./Sq. In. | Specific Volume | | Enthalpy | | | Entropy | | | Internal Energy | | | Temp. Fahr. |
|---|---|---|---|---|---|---|---|---|---|---|---|---|---|
| | | Sat. Solid | Sat. Vapor | Sat. Solid | Subl. | Sat. Vapor | Sat. Solid | Subl. | Sat. Vapor | Sat. Solid | Subl. | Sat. Vapor | |
| t | p | $v_i$ | $v_g \times 10^{-3}$ | $h_i$ | $h_{ig}$ | $h_g$ | $s_i$ | $s_{ig}$ | $s_g$ | $u_i$ | $u_{ig}$ | $u_g$ | t |
| **32°** | 0.0885 | 0.01747 | 3.306 | −143.35 | 1219.1 | 1075.8 | −0.2916 | 2.4793 | 2.1877 | −143.35 | 1164.9 | 1021.6 | **32°** |
| **30** | 0.0808 | 0.01747 | 3.609 | −144.35 | 1219.3 | 1074.9 | −0.2936 | 2.4897 | 2.1961 | −144.35 | 1165.4 | 1021.0 | **30** |
| **25** | 0.0640 | 0.01746 | 4.508 | −146.84 | 1219.6 | 1072.8 | −0.2987 | 2.5159 | 2.2172 | −146.84 | 1166.2 | 1019.4 | **25** |
| **20** | 0.0505 | 0.01745 | 5.658 | −149.31 | 1219.9 | 1070.6 | −0.3038 | 2.5425 | 2.2387 | −149.31 | 1167.0 | 1017.7 | **20** |
| **15°** | 0.0396 | 0.01745 | 7.14 | −151.75 | 1220.2 | 1068.4 | −0.3089 | 2.5698 | 2.2609 | −151.75 | 1167.9 | 1016.1 | **15°** |
| **10** | 0.0309 | 0.01744 | 9.05 | −154.17 | 1220.4 | 1066.2 | −0.3141 | 2.5977 | 2.2836 | −154.17 | 1168.6 | 1014.4 | **10** |
| **5** | 0.0240 | 0.01743 | 11.53 | −156.56 | 1220.6 | 1064.0 | −0.3192 | 2.6260 | 2.3068 | −156.56 | 1169.4 | 1012.8 | **5** |
| **0** | 0.0185 | 0.01742 | 14.77 | −158.93 | 1220.7 | 1061.8 | −0.3241 | 2.6546 | 2.3305 | −158.93 | 1170.0 | 1011.1 | **0** |
| **−5** | 0.0142 | 0.01742 | 19.04 | −161.27 | 1220.9 | 1059.6 | −0.3294 | 2.6842 | 2.3548 | −161.27 | 1170.8 | 1009.5 | **−5** |
| **−10** | 0.0108 | 0.01741 | 24.67 | −163.59 | 1221.0 | 1057.4 | −0.3346 | 2.7143 | 2.3797 | −163.59 | 1171.4 | 1007.8 | **−10** |
| **−15°** | 0.0082 | 0.01740 | 32.1 | −165.89 | 1221.1 | 1055.2 | −0.3397 | 2.7449 | 2.4052 | −165.89 | 1172.1 | 1006.2 | **−15°** |
| **−20** | 0.0062 | 0.01739 | 42.2 | −168.16 | 1221.2 | 1053.0 | −0.3448 | 2.7764 | 2.4316 | −168.16 | 1172.7 | 1004.5 | **−20** |
| **−25** | 0.0046 | 0.01739 | 55.8 | −170.40 | 1221.2 | 1050.8 | −0.3500 | 2.8085 | 2.4585 | −170.40 | 1173.3 | 1002.9 | **−25** |
| **−30** | 0.0035 | 0.01738 | 74.1 | −172.63 | 1221.2 | 1048.6 | −0.3551 | 2.8411 | 2.4860 | −172.63 | 1173.8 | 1001.2 | **−30** |
| **−35** | 0.0025 | 0.01737 | 99.3 | −174.82 | 1221.2 | 1046.4 | −0.3602 | 2.8745 | 2.5143 | −174.82 | 1174.4 | 999.6 | **−35** |
| **−40** | 0.0019 | 0.01737 | 133.9 | −177.00 | 1221.2 | 1044.2 | −0.3654 | 2.9087 | 2.5433 | −177.00 | 1174.9 | 997.9 | **−40** |

# Table 6. Viscosity $\times 10^7$ (lb. sec. ft.$^{-2}$)

**Temperature—Degrees Fahrenheit**

| | Sat. Vapor | 32° | 200° | 400° | 600° | 800° | 1000° | 1200° |
|---|---|---|---|---|---|---|---|---|
| Sat. Liquid | | 375 | 64 | 29 | 18 | | | |
| Sat. Vapor | | 2.0 | 2.7 | 4.0 | 7.2 | | | |
| **p Lb./Sq. In.** | | | | | | | | |
| **0** | | 2.0 | 2.7 | 3.5 | 4.3 | 5.1 | 5.8 | 6.4 |
| **500** | 4.7 | | | | 5.1 | 5.7 | 6.3 | 6.9 |
| **1000** | 6.0 | | | | 6.0 | 6.4 | 6.9 | 7.4 |
| **1500** | 7.1 | | | | 7.1 | 7.1 | 7.5 | 7.9 |
| **2000** | 8.1 | | | | | 7.9 | 8.1 | 8.4 |
| **2500** | 9.3 | | | | | 8.7 | 8.7 | 9.0 |
| **3000** | 10.0 | | | | | 9.5 | 9.3 | 9.5 |
| **3500** | | | | | | 10.3 | 9.8 | 10.0 |

# Table 7. Heat Conductivity $\times 10^3$ (Btu. hr.$^{-1}$ ft.$^{-2}$ °F$^{-1}$ ft.)

**Temperature—Degrees Fahrenheit**

| | Sat. Vapor | 32° | 200° | 400° | 600° | 800° | 1000° |
|---|---|---|---|---|---|---|---|
| Sat. Liquid | | 319 | 392 | 382 | 293 | | |
| Sat. Vapor | | 9.2 | 13.5 | 21.1 | 38.5 | | |
| **p Lb./Sq. In.** | | | | | | | |
| **0** | | 9.2 | 13.3 | 18.4 | 23.8 | 29.2 | 34.7 |
| **250** | 21.1 | | | | 24.7 | 29.6 | 34.9 |
| **500** | 25.1 | | | | 26.0 | 30.2 | 35.2 |
| **1000** | 31.6 | | | | 30.1 | 31.4 | 35.7 |
| **1500** | 37.9 | | | | 37.6 | 33.2 | 36.4 |
| **1750** | 40.8 | | | | | 34.3 | 36.8 |
| **2000** | 44.5 | | | | | 35.5 | 37.2 |

# Table 8. Conversion Factors

## Basic Definitions and Ratios:

Length: 2.54000 cm. = 1 in.
Mass: 453.5924 grams = 1 pound
Pressure: (*a*) One standard atmosphere = pressure of 76 cm. mercury column (density 13.5951 g. per cu. cm., gravity 980.665 cm. per sec. per sec.)
(*b*) One bar = $10^6$ dynes per sq. cm.

Energy: (*a*) 1000 international steam-table calories (1000 IT. cal.) = $\frac{1}{860}$ international kilowatt-hour
(*b*) 1 international electrical watt = 1.0003 absolute watts
(*c*) 1 British thermal unit = 251.996 IT. cal

Temperature: The Fahrenheit temperature scale used in the tables is defined as $t_F = 1.8t + 32$, where $t$ is the temperature on the International (Centigrade) Scale.
The Absolute Fahrenheit Scale is $T_F = t_F + 459.69 = 1.8\,(t + 273.16)$

## Table Quantities

### Specific Volume

1 ft.$^3$/lb. = 62.428 cm.$^3$/g. 1 cm.$^3$/g. = 0.0160185 ft.$^3$/lb.

### Energy (Specific)

| | Abs. joules / g. | Kg. m. / g. | Ft. lb. / lb. | Int. joules / g. | Int. w. hr. / g. | IT. cal. / g. | Btu. / lb. | Lb. ft.$^3$ / in.$^2$ lb. | Atm. cm$^3$ / g. | Hp. hr. / lb. |
|---|---|---|---|---|---|---|---|---|---|---|
| $10^3$ Abs. joules / g. | 1000* | 101.972 | 334553 | 999.7 | 0.277694 | 238.817 | 429.87 | 2323.28 | 9869.23 | 0.168966 |
| 10 Kg. m. / g. | 98.0665* | 10* | 32808 | 98.0371 | 0.0272325 | 23.420 | 42.156 | 227.836 | 967.841 | 0.0165699 |
| $10^6$ Ft. lb. / lb. | 2989.1 | 304.8* | 1000000* | 2988.2 | 0.8300 | 713.83 | 1284.91 | 6944.44 | 29500 | 0.50505 |
| $10^3$ Int. joules / g. | 1000.3 | 102.002 | 334651 | 1000* | 0.27778 | 238.89 | 430.00 | 2324.0 | 9872.2 | 0.16902 |
| $10^{-5}$ Int. kw. hr. / g. | 36.011 | 3.6721 | 12047 | 36* | 0.01* | 8.6* | 15.4800 | 83.663 | 355.40 | 0.006085 |
| 10 IT. cal. / g. | 41.873 | 4.2699 | 14009 | 41.8605 | 0.0116279 | 10* | 18* | 97.28 | 413.255 | 0.0070751 |
| $10^2$ Btu. / lb. | 232.63 | 23.722 | 77826 | 232.56 | 0.064600 | 55.5556 | 100* | 540.46 | 2295.8 | 0.039306 |
| $10^4$ Lb. ft.$^3$ / in.$^2$ lb. | 4304.3 | 438.91 | 1440000* | 4303.0 | 1.1953 | 1027.9 | 1850.27 | 10000* | 42480 | 0.72727 |
| $10^4$ Atm. cm.$^3$ / g. | 1013.25 | 103.323 | 338985 | 1012.95 | 0.281374 | 241.98 | 435.57 | 2354.1 | 10000* | 0.17120 |
| $10^{-3}$ Hp. hr. / lb. | 5.9184 | 0.60350 | 1980* | 5.9166 | 0.00164349 | 1.4134 | 2.5441 | 13.750 | 58.409 | 0.001* |

### Entropy (Specific) and Specific Heat

1 Btu./°F lb. = 1 IT. cal./°C g.

Other conversion factors can be obtained from the Energy (Specific) table by means of the factors
1 °F$^{-1}$ = 1.8 °C$^{-1}$, and 1 °C$^{-1}$ = 0.55556 °F$^{-1}$

## Pressure

| Units | Atm. | kg./cm.$^2$ | lb./in.$^2$ | bar | mm. Hg. (0°C) | in. Hg. (32°F) | ft. $H_2O$ (60°F) |
|---|---|---|---|---|---|---|---|
| 1 Atmosphere | 1* | 1.033228 | 14.6959 | 1.013250 | 760* | 29.921 | 33.934 |
| 1 kg./cm.$^2$ | 0.967841 | 1* | 14.2233 | 0.980665* | 735.559 | 28.959 | 32.843 |
| 10 lb./in.$^2$ | 0.68046 | 0.70307 | 10* | 0.689476 | 517.149 | 20.360 | 23.091 |
| 1 bar | 0.986923 | 1.019716 | 14.5038 | 1* | 750.062 | 29.530 | 33.490 |
| 1 meter Hg. (0°C) | 1.31579 | 1.35951 | 19.3368 | 1.333224 | 1000* | 39.370 | 44.65 |
| 10 in. Hg. (32°F) | 0.33421 | 0.34532 | 4.9115 | 0.33864 | 254* | 10* | 11.341 |
| 100 ft. $H_2O$ (60°F) | 2.9469 | 3.0448 | 43.308 | 2.9859 | 2239.6 | 88.175 | 100* |

## Energy

| Units | Abs. joules | Kg. m. | Ft. lb. | Int. joules | Int. w. hr. | IT. cal. | Btu. | Lb./in.$^2$ × ft.$^3$ | Atm. × cm.$^3$ | Hp. hr. |
|---|---|---|---|---|---|---|---|---|---|---|
| $10^4$ Abs. joules | 10000* | 1019.72 | 7375.62 | 9997 | 2.77694 | 2388.17 | 9.4770 | 51.2196 | 98692.3 | 0.00372506 |
| $10^4$ Kg. m. | 98066.5 | 10000* | 72330 | 98037.1 | 27.2325 | 23420 | 92.938 | 502.293 | 9678.41 | 0.0365304 |
| $10^4$ Ft. lb. | 13558.2 | 1382.55 | 10000* | 13554.1 | 3.7650 | 3237.9 | 12.8491 | 69.4444 | 133809 | 0.0050505 |
| $10^4$ Int. joules | 10003 | 1020.02 | 7377.8 | 10000* | 2.7778 | 2388.9 | 9.4799 | 51.235 | 98722 | 0.0037262 |
| $10^{-3}$ Int. Kw. hr. | 3601.1 | 367.21 | 2656.0 | 3600* | 1* | 860* | 3.41275 | 18.4446 | 35540 | 0.0013414 |
| $10^4$ IT. cal. | 41873 | 4269.9 | 30884 | 41860.5 | 11.6279 | 10000* | 39.683 | 214.47 | 413255 | 0.0155980 |
| 10 Btu. | 10551.8 | 1075.99 | 7782.6 | 10548.7 | 2.93019 | 2519.96 | 10* | 54.046 | 104138 | 0.0039306 |
| $10^2$ Lb./in.$^2$ × ft.$^3$ | 19523.8 | 1990.87 | 14400* | 19517.9 | 5.4216 | 4662.6 | 18.5027 | 100* | 192685 | 0.0072727 |
| $10^5$ Atm. × cm.$^3$ | 10132.5 | 1033.23 | 7473.35 | 10129.5 | 2.81374 | 2419.8 | 9.6026 | 51.898 | 100000* | 0.0037744 |
| $10^{-3}$ Hp. hr. | 2684.5 | 273.75 | 1980* | 2683.7 | 0.74548 | 641.1 | 2.5441 | 13.750 | 26494 | 0.001* |

* Exact value, by definition.

## TABLE OF EQUIVALENT TEMPERATURES

| Cent. | Fahr. | Cent. | Fahr. | Cent. | Fahr. | Cent. | Fahr. | Cent. | Fahr. | Cent. | Fahr. | Cent. | Fahr. | Cent. | Fahr. |
|---|---|---|---|---|---|---|---|---|---|---|---|---|---|---|---|
| −50° | −58° | 75° | 167° | 200° | 392° | 325° | 617° | 450° | 842° | 575° | 1067° | 700° | 1292° | 825° | 1517° |
| −45 | −49 | 80 | 176 | 205 | 401 | 330 | 626 | 455 | 851 | 580 | 1076 | 705 | 1301 | 830 | 1526 |
| −40 | −40 | 85 | 185 | 210 | 410 | 335 | 635 | 460 | 860 | 585 | 1085 | 710 | 1310 | 835 | 1535 |
| −35 | −31 | 90 | 194 | 215 | 419 | 340 | 644 | 465 | 869 | 590 | 1094 | 715 | 1319 | 840 | 1544 |
| −30 | −22 | 95 | 203 | 220 | 428 | 345 | 653 | 470 | 878 | 595 | 1103 | 720 | 1328 | 845 | 1553 |
| −25° | −13° | 100° | 212° | 225° | 437° | 350° | 662° | 475° | 887° | 600° | 1112° | 725° | 1337° | 850° | 1562° |
| −20 | − 4 | 105 | 221 | 230 | 446 | 355 | 671 | 480 | 896 | 605 | 1121 | 730 | 1346 | 855 | 1571 |
| −15 | + 5 | 110 | 230 | 235 | 455 | 360 | 680 | 485 | 905 | 610 | 1130 | 735 | 1355 | 860 | 1580 |
| −10 | 14 | 115 | 239 | 240 | 464 | 365 | 689 | 490 | 914 | 615 | 1139 | 740 | 1364 | 865 | 1589 |
| − 5 | 23 | 120 | 248 | 245 | 473 | 370 | 698 | 495 | 923 | 620 | 1148 | 745 | 1373 | 870 | 1598 |
| 0° | 32° | 125° | 257° | 250° | 482° | 375° | 707° | 500° | 932° | 625° | 1157° | 750° | 1382° | 875° | 1607° |
| 5 | 41 | 130 | 266 | 255 | 491 | 380 | 716 | 505 | 941 | 630 | 1166 | 755 | 1391 | 880 | 1616 |
| 10 | 50 | 135 | 275 | 260 | 500 | 385 | 725 | 510 | 950 | 635 | 1175 | 760 | 1400 | 885 | 1625 |
| 15 | 59 | 140 | 284 | 265 | 509 | 390 | 734 | 515 | 959 | 640 | 1184 | 765 | 1409 | 890 | 1634 |
| 20 | 68 | 145 | 293 | 270 | 518 | 395 | 743 | 520 | 968 | 645 | 1193 | 770 | 1418 | 895 | 1643 |
| 25° | 77° | 150° | 302° | 275° | 527° | 400° | 752° | 525° | 977° | 650° | 1202° | 775° | 1427° | 900° | 1652° |
| 30 | 86 | 155 | 311 | 280 | 536 | 405 | 761 | 530 | 986 | 655 | 1211 | 780 | 1436 | 905 | 1661 |
| 35 | 95 | 160 | 320 | 285 | 545 | 410 | 770 | 535 | 995 | 660 | 1220 | 785 | 1445 | 910 | 1670 |
| 40 | 104 | 165 | 329 | 290 | 554 | 415 | 779 | 540 | 1004 | 665 | 1229 | 790 | 1454 | 915 | 1679 |
| 45 | 113 | 170 | 338 | 295 | 563 | 420 | 788 | 545 | 1013 | 670 | 1238 | 795 | 1463 | 920 | 1688 |
| 50° | 122° | 175° | 347° | 300° | 572° | 425° | 797° | 550° | 1022° | 675° | 1247° | 800° | 1472° | 925° | 1697° |
| 55 | 131 | 180 | 356 | 305 | 581 | 430 | 806 | 555 | 1031 | 680 | 1256 | 805 | 1481 | 930 | 1706 |
| 60 | 140 | 185 | 365 | 310 | 590 | 435 | 815 | 560 | 1040 | 685 | 1265 | 810 | 1490 | 935 | 1715 |
| 65 | 149 | 190 | 374 | 315 | 599 | 440 | 824 | 565 | 1049 | 690 | 1274 | 815 | 1499 | 940 | 1724 |
| 70 | 158 | 195 | 383 | 320 | 608 | 445 | 833 | 570 | 1058 | 695 | 1283 | 820 | 1508 | 945 | 1733 |

## TABLE OF VALUES FOR INTERPOLATION IN ABOVE

| | | |
|---|---|---|
| 1°C = 1.8°F | 4°C = 7.2°F | 7°C = 12.6°F |
| 2 = 3.6 | 5 = 9.0 | 8 = 14.4 |
| 3 = 5.4 | 6 = 10.8 | 9 = 16.2 |

All decimals are exact.

| | | |
|---|---|---|
| 1°F = 0.55°C | 4°F = 2.22°C | 7°F = 3.88°C |
| 2 = 1.11 | 5 = 2.77 | 8 = 4.44 |
| 3 = 1.66 | 6 = 3.33 | 9 = 5.00 |

All decimals are repeating decimals.

# Table 9. Thermometer Calibration Formulas*

Sulfur boiling point: $t_p = 444.600 + 0.0908028\,(p - 760) - 0.000047573\,(p - 760)^2 + 0.00000004361\,(p - 760)^3$

Mercury boiling point: $t_p = 356.580 + 0.0730951\,(p - 760) - 0.000039866\,(p - 760)^2 + 0.00000003191\,(p - 760)^3$

Water boiling point: $t_p = 100.000 + 0.0368578\,(p - 760) - 0.000020159\,(p - 760)^2 + 0.00000001621\,(p - 760)^3$

Water freezing point: $t = -\,0.0000129\,(p - 760) - 0.00000072\,H$ [Air Saturated]

Temperatures are on the international temperature scale degrees Centigrade.
$p$ is absolute pressure in mm. of mercury (density 13.5951 g./cm.$^3$, gravity 980.665 cm./sec.$^2$).
The formulations are valid for 660 mm. $< p <$ 860 mm.
$H$ = head of water in mm.

$k = \frac{C_P}{C_V}$

$S_2 - S_1 = C_{V0} \ln \frac{T_2}{T_1} + R \ln \frac{V_2}{V_1}$ NON FLOW $Tds = dU + PdV$

$= C_{P0} \ln \frac{T_2}{T_1} - R \ln \frac{P_2}{P_1}$ FLOW PROCESS $Tds = dH - VdP$

$C_V = \frac{R}{K-1}$

$C_P = \frac{KR}{K-1}$

FOR REV ADIABATIC PROCESS

$Tds = dU + PdV = C_{V0}\,dT + PdV = 0$

FOR PROCESS
ISOBARIC N=0
ISOTHERM N=1
ISENTROPIC N=K
ISOMETRIC N=∞

* J. A. Beattie, B. E. Blaisdell, J. Kaminsky — Proc. Am. Acad. Arts & Sci. **71**, 327 (1937). J. A. Beattie, T. C. Huang, M. Benedict - Ibid. **72**, 137 (1938).

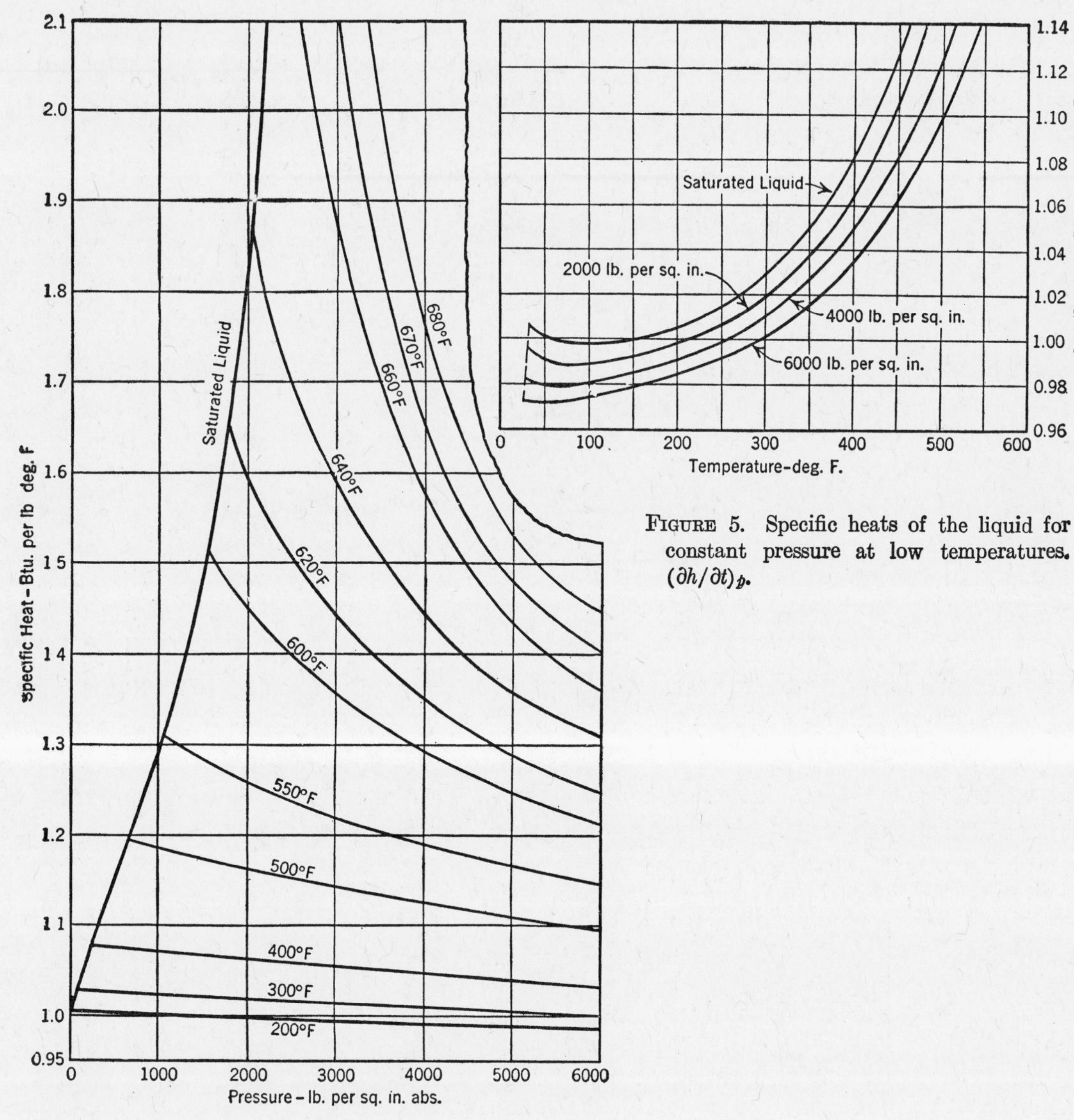


FIGURE 4. Specific heats of the liquid for constant pressure. $(\partial h/\partial t)_p$.

FIGURE 5. Specific heats of the liquid for constant pressure at low temperatures. $(\partial h/\partial t)_p$.

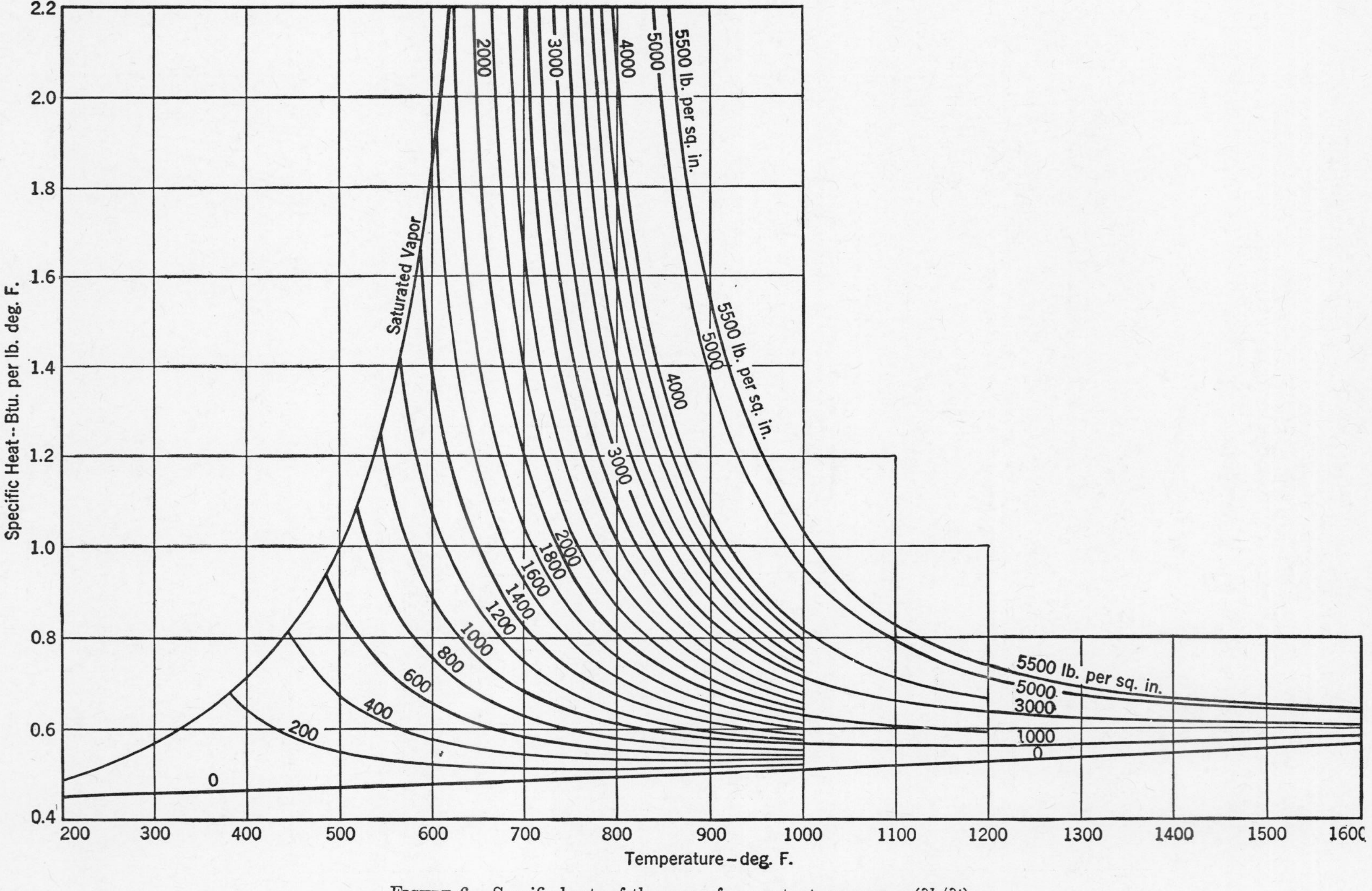


FIGURE 6. Specific heats of the vapor for constant pressure. $(\partial h/\partial t)_p$.

Numbers on curves give pressures in lb. per sq. in. abs.

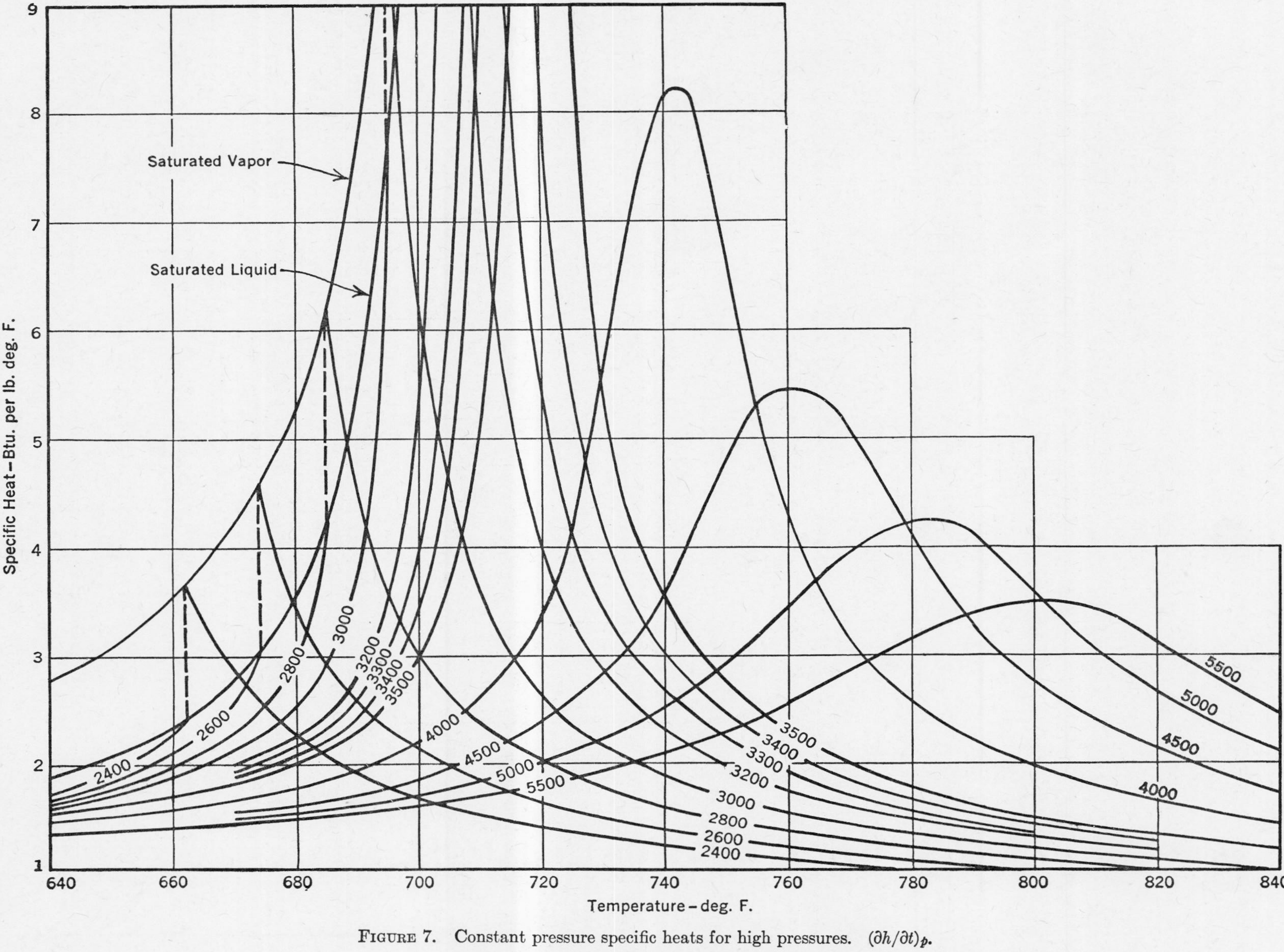


FIGURE 7. Constant pressure specific heats for high pressures. $(\partial h/\partial t)_p$.

Numbers on curves give pressures in lb. per. sq. in. abs.

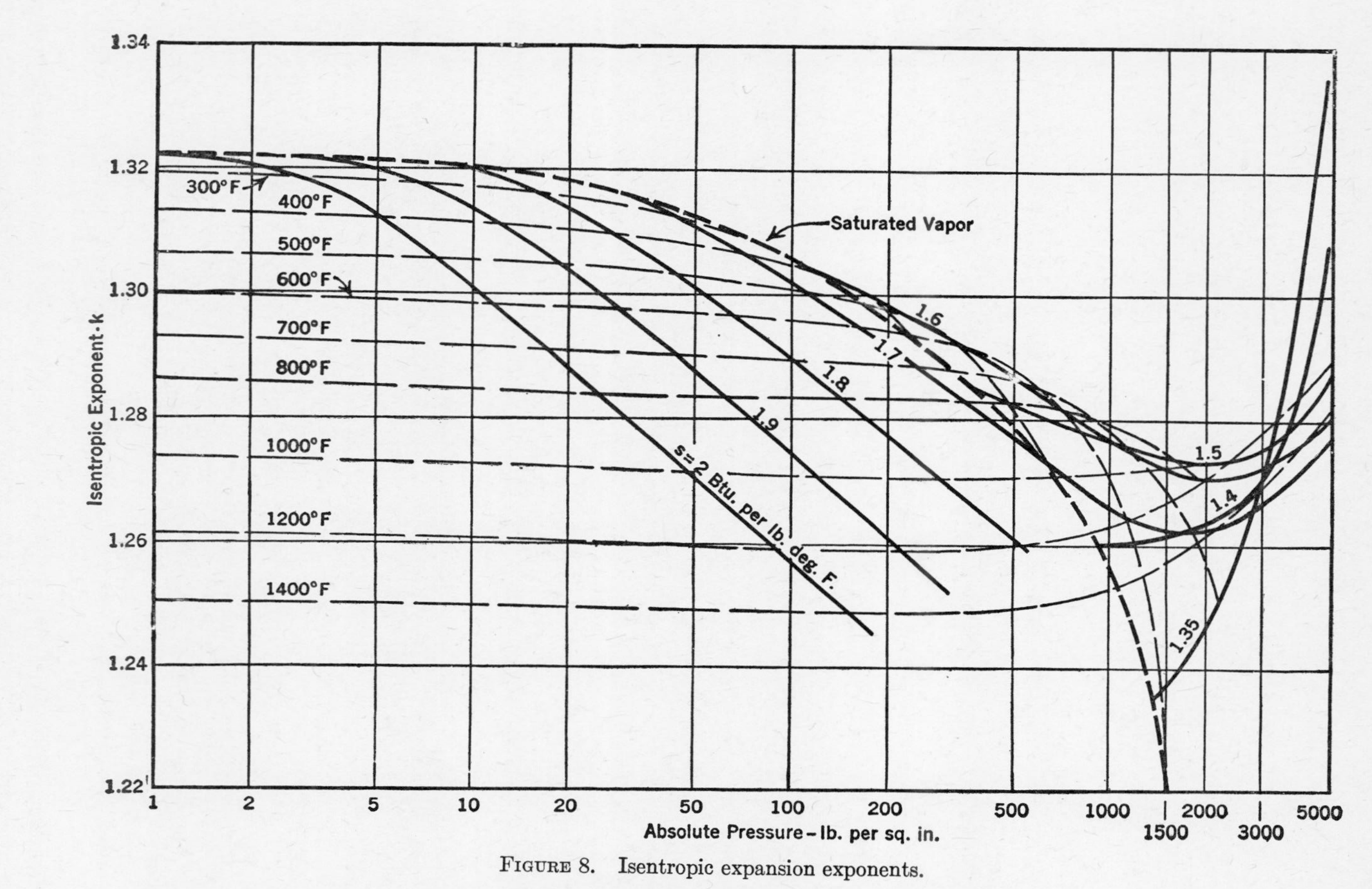


FIGURE 8. Isentropic expansion exponents.

$$k = -\left[\frac{\partial\,(\log p)}{\partial\,(\log v)}\right]_s = -\frac{v}{p}\left(\frac{\partial p}{\partial v}\right)_s = -\frac{v}{p}\left(\frac{\partial p}{\partial v}\right)_T \frac{C_p}{C_v}$$

**For small changes in pressure (or volume) along an isentropic, $pv^k$ = constant.**

# ACKNOWLEDGMENTS

Pages 84-87 are reprinted from "Four Place Tables," by Edward V. Huntington, and are used by permission of Houghton Mifflin Company.

Pages 88-89 are reprinted from "Steam Tables and Diagrams," by Lionel S. Marks and Harvey N. Davis, by permission of Longmans, Green and Company.

# Table 10. Logarithms to the Base 10

| | 0 | 1 | 2 | 3 | 4 | 5 | 6 | 7 | 8 | 9 | 10 |
|---|---|---|---|---|---|---|---|---|---|---|---|
| **1.00** | 0.0000 | 0004 | 0009 | 0013 | 0017 | 0022 | 0026 | 0030 | 0035 | 0039 | 0043 |
| 1.01 | 0043 | 0048 | 0052 | 0056 | 0060 | 0065 | 0069 | 0073 | 0077 | 0082 | 0086 |
| 1.02 | 0086 | 0090 | 0095 | 0099 | 0103 | 0107 | 0111 | 0116 | 0120 | 0124 | 0128 |
| 1.03 | 0128 | 0133 | 0137 | 0141 | 0145 | 0149 | 0154 | 0158 | 0162 | 0166 | 0170 |
| 1.04 | 0170 | 0175 | 0179 | 0183 | 0187 | 0191 | 0195 | 0199 | 0204 | 0208 | 0212 |
| 1.05 | 0212 | 0216 | 0220 | 0224 | 0228 | 0233 | 0237 | 0241 | 0245 | 0249 | 0253 |
| 1.06 | 0253 | 0257 | 0261 | 0265 | 0269 | 0273 | 0278 | 0282 | 0286 | 0290 | 0294 |
| 1.07 | 0294 | 0298 | 0302 | 0306 | 0310 | 0314 | 0318 | 0322 | 0326 | 0330 | 0334 |
| 1.08 | 0334 | 0338 | 0342 | 0346 | 0350 | 0354 | 0358 | 0362 | 0366 | 0370 | 0374 |
| 1.09 | 0374 | 0378 | 0382 | 0386 | 0390 | 0394 | 0398 | 0402 | 0406 | 0410 | 0414 |
| **1.10** | 0.0414 | 0418 | 0422 | 0426 | 0430 | 0434 | 0438 | 0441 | 0445 | 0449 | 0453 |
| 1.11 | 0453 | 0457 | 0461 | 0465 | 0469 | 0473 | 0477 | 0481 | 0484 | 0488 | 0492 |
| 1.12 | 0492 | 0496 | 0500 | 0504 | 0508 | 0512 | 0515 | 0519 | 0523 | 0527 | 0531 |
| 1.13 | 0531 | 0535 | 0538 | 0542 | 0546 | 0550 | 0554 | 0558 | 0561 | 0565 | 0569 |
| 1.14 | 0569 | 0573 | 0577 | 0580 | 0584 | 0588 | 0592 | 0596 | 0599 | 0603 | 0607 |
| 1.15 | 0607 | 0611 | 0615 | 0618 | 0622 | 0626 | 0630 | 0633 | 0637 | 0641 | 0645 |
| 1.16 | 0645 | 0648 | 0652 | 0656 | 0660 | 0663 | 0667 | 0671 | 0674 | 0678 | 0682 |
| 1.17 | 0682 | 0686 | 0689 | 0693 | 0697 | 0700 | 0704 | 0708 | 0711 | 0715 | 0719 |
| 1.18 | 0719 | 0722 | 0726 | 0730 | 0734 | 0737 | 0741 | 0745 | 0748 | 0752 | 0755 |
| 1.19 | 0755 | 0759 | 0763 | 0766 | 0770 | 0774 | 0777 | 0781 | 0785 | 0788 | 0792 |
| **1.20** | 0.0792 | 0795 | 0799 | 0803 | 0806 | 0810 | 0813 | 0817 | 0821 | 0824 | 0828 |
| 1.21 | 0828 | 0831 | 0835 | 0839 | 0842 | 0846 | 0849 | 0853 | 0856 | 0860 | 0864 |
| 1.22 | 0864 | 0867 | 0871 | 0874 | 0878 | 0881 | 0885 | 0888 | 0892 | 0896 | 0899 |
| 1.23 | 0899 | 0903 | 0906 | 0910 | 0913 | 0917 | 0920 | 0924 | 0927 | 0931 | 0934 |
| 1.24 | 0934 | 0938 | 0941 | 0945 | 0948 | 0952 | 0955 | 0959 | 0962 | 0966 | 0969 |
| 1.25 | 0969 | 0973 | 0976 | 0980 | 0983 | 0986 | 0990 | 0993 | 0997 | 1000 | 1004 |
| 1.26 | 1004 | 1007 | 1011 | 1014 | 1017 | 1021 | 1024 | 1028 | 1031 | 1035 | 1038 |
| 1.27 | 1038 | 1041 | 1045 | 1048 | 1052 | 1055 | 1059 | 1062 | 1065 | 1069 | 1072 |
| 1.28 | 1072 | 1075 | 1079 | 1082 | 1086 | 1089 | 1092 | 1096 | 1099 | 1103 | 1106 |
| 1.29 | 1106 | 1109 | 1113 | 1116 | 1119 | 1123 | 1126 | 1129 | 1133 | 1136 | 1139 |
| **1.30** | 0.1139 | 1143 | 1146 | 1149 | 1153 | 1156 | 1159 | 1163 | 1166 | 1169 | 1173 |
| 1.31 | 1173 | 1176 | 1179 | 1183 | 1186 | 1189 | 1193 | 1196 | 1199 | 1202 | 1206 |
| 1.32 | 1206 | 1209 | 1212 | 1216 | 1219 | 1222 | 1225 | 1229 | 1232 | 1235 | 1239 |
| 1.33 | 1239 | 1242 | 1245 | 1248 | 1252 | 1255 | 1258 | 1261 | 1265 | 1268 | 1271 |
| 1.34 | 1271 | 1274 | 1278 | 1281 | 1284 | 1287 | 1290 | 1294 | 1297 | 1300 | 1303 |
| 1.35 | 1303 | 1307 | 1310 | 1313 | 1316 | 1319 | 1323 | 1326 | 1329 | 1332 | 1335 |
| 1.36 | 1335 | 1339 | 1342 | 1345 | 1348 | 1351 | 1355 | 1358 | 1361 | 1364 | 1367 |
| 1.37 | 1367 | 1370 | 1374 | 1377 | 1380 | 1383 | 1386 | 1389 | 1392 | 1396 | 1399 |
| 1.38 | 1399 | 1402 | 1405 | 1408 | 1411 | 1414 | 1418 | 1421 | 1424 | 1427 | 1430 |
| 1.39 | 1430 | 1433 | 1436 | 1440 | 1443 | 1446 | 1449 | 1452 | 1455 | 1458 | 1461 |
| **1.40** | 0.1461 | 1464 | 1467 | 1471 | 1474 | 1477 | 1480 | 1483 | 1486 | 1489 | 1492 |
| 1.41 | 1492 | 1495 | 1498 | 1501 | 1504 | 1508 | 1511 | 1514 | 1517 | 1520 | 1523 |
| 1.42 | 1523 | 1526 | 1529 | 1532 | 1535 | 1538 | 1541 | 1544 | 1547 | 1550 | 1553 |
| 1.43 | 1553 | 1556 | 1559 | 1562 | 1565 | 1569 | 1572 | 1575 | 1578 | 1581 | 1584 |
| 1.44 | 1584 | 1587 | 1590 | 1593 | 1596 | 1599 | 1602 | 1605 | 1608 | 1611 | 1614 |
| 1.45 | 1614 | 1617 | 1620 | 1623 | 1626 | 1629 | 1632 | 1635 | 1638 | 1641 | 1644 |
| 1.46 | 1644 | 1647 | 1649 | 1652 | 1655 | 1658 | 1661 | 1664 | 1667 | 1670 | 1673 |
| 1.47 | 1673 | 1676 | 1679 | 1682 | 1685 | 1688 | 1691 | 1694 | 1697 | 1700 | 1703 |
| 1.48 | 1703 | 1706 | 1708 | 1711 | 1714 | 1717 | 1720 | 1723 | 1726 | 1729 | 1732 |
| 1.49 | 1732 | 1735 | 1738 | 1741 | 1744 | 1746 | 1749 | 1752 | 1755 | 1758 | 1761 |

## Table 10. Logarithms to the Base 10

| | 0 | 1 | 2 | 3 | 4 | 5 | 6 | 7 | 8 | 9 | 10 |
|---|---|---|---|---|---|---|---|---|---|---|---|
| **1.50** | 0.1761 | 1764 | 1767 | 1770 | 1772 | 1775 | 1778 | 1781 | 1784 | 1787 | 1790 |
| 1.51 | 1790 | 1793 | 1796 | 1798 | 1801 | 1804 | 1807 | 1810 | 1813 | 1816 | 1818 |
| 1.52 | 1818 | 1821 | 1824 | 1827 | 1830 | 1833 | 1836 | 1838 | 1841 | 1844 | 1847 |
| 1.53 | 1847 | 1850 | 1853 | 1855 | 1858 | 1861 | 1864 | 1867 | 1870 | 1872 | 1875 |
| 1.54 | 1875 | 1878 | 1881 | 1884 | 1886 | 1889 | 1892 | 1895 | 1898 | 1901 | 1903 |
| 1.55 | 1903 | 1906 | 1909 | 1912 | 1915 | 1917 | 1920 | 1923 | 1926 | 1928 | 1931 |
| 1.56 | 1931 | 1934 | 1937 | 1940 | 1942 | 1945 | 1948 | 1951 | 1953 | 1956 | 1959 |
| 1.57 | 1959 | 1962 | 1965 | 1967 | 1970 | 1973 | 1976 | 1978 | 1981 | 1984 | 1987 |
| 1.58 | 1987 | 1989 | 1992 | 1995 | 1998 | 2000 | 2003 | 2006 | 2009 | 2011 | 2014 |
| 1.59 | 2014 | 2017 | 2019 | 2022 | 2025 | 2028 | 2030 | 2033 | 2036 | 2038 | 2041 |
| **1.60** | 0.2041 | 2044 | 2047 | 2049 | 2052 | 2055 | 2057 | 2060 | 2063 | 2066 | 2068 |
| 1.61 | 2068 | 2071 | 2074 | 2076 | 2079 | 2082 | 2084 | 2087 | 2090 | 2092 | 2095 |
| 1.62 | 2095 | 2098 | 2101 | 2103 | 2106 | 2109 | 2111 | 2114 | 2117 | 2119 | 2122 |
| 1.63 | 2122 | 2125 | 2127 | 2130 | 2133 | 2135 | 2138 | 2140 | 2143 | 2146 | 2148 |
| 1.64 | 2148 | 2151 | 2154 | 2156 | 2159 | 2162 | 2164 | 2167 | 2170 | 2172 | 2175 |
| 1.65 | 2175 | 2177 | 2180 | 2183 | 2185 | 2188 | 2191 | 2193 | 2196 | 2198 | 2201 |
| 1.66 | 2201 | 2204 | 2206 | 2209 | 2212 | 2214 | 2217 | 2219 | 2222 | 2225 | 2227 |
| 1.67 | 2227 | 2230 | 2232 | 2235 | 2238 | 2240 | 2243 | 2245 | 2248 | 2251 | 2253 |
| 1.68 | 2253 | 2256 | 2258 | 2261 | 2263 | 2266 | 2269 | 2271 | 2274 | 2276 | 2279 |
| 1.69 | 2279 | 2281 | 2284 | 2287 | 2289 | 2292 | 2294 | 2297 | 2299 | 2302 | 2304 |
| **1.70** | 0.2304 | 2307 | 2310 | 2312 | 2315 | 2317 | 2320 | 2322 | 2325 | 2327 | 2330 |
| 1.71 | 2330 | 2333 | 2335 | 2338 | 2340 | 2343 | 2345 | 2348 | 2350 | 2353 | 2355 |
| 1.72 | 2355 | 2358 | 2360 | 2363 | 2365 | 2368 | 2370 | 2373 | 2375 | 2378 | 2380 |
| 1.73 | 2380 | 2383 | 2385 | 2388 | 2390 | 2393 | 2395 | 2398 | 2400 | 2403 | 2405 |
| 1.74 | 2405 | 2408 | 2410 | 2413 | 2415 | 2418 | 2420 | 2423 | 2425 | 2428 | 2430 |
| 1.75 | 2430 | 2433 | 2435 | 2438 | 2440 | 2443 | 2445 | 2448 | 2450 | 2453 | 2455 |
| 1.76 | 2455 | 2458 | 2460 | 2463 | 2465 | 2467 | 2470 | 2472 | 2475 | 2477 | 2480 |
| 1.77 | 2480 | 2482 | 2485 | 2487 | 2490 | 2492 | 2494 | 2497 | 2499 | 2502 | 2504 |
| 1.78 | 2504 | 2507 | 2509 | 2512 | 2514 | 2516 | 2519 | 2521 | 2524 | 2526 | 2529 |
| 1.79 | 2529 | 2531 | 2533 | 2536 | 2538 | 2541 | 2543 | 2545 | 2548 | 2550 | 2553 |
| **1.80** | 0.2553 | 2555 | 2558 | 2560 | 2562 | 2565 | 2567 | 2570 | 2572 | 2574 | 2577 |
| 1.81 | 2577 | 2579 | 2582 | 2584 | 2586 | 2589 | 2591 | 2594 | 2596 | 2598 | 2601 |
| 1.82 | 2601 | 2603 | 2605 | 2608 | 2610 | 2613 | 2615 | 2617 | 2620 | 2622 | 2625 |
| 1.83 | 2625 | 2627 | 2629 | 2632 | 2634 | 2636 | 2639 | 2641 | 2643 | 2646 | 2648 |
| 1.84 | 2648 | 2651 | 2653 | 2655 | 2658 | 2660 | 2662 | 2665 | 2667 | 2669 | 2672 |
| 1.85 | 2672 | 2674 | 2676 | 2679 | 2681 | 2683 | 2686 | 2688 | 2690 | 2693 | 2695 |
| 1.86 | 2695 | 2697 | 2700 | 2702 | 2704 | 2707 | 2709 | 2711 | 2714 | 2716 | 2718 |
| 1.87 | 2718 | 2721 | 2723 | 2725 | 2728 | 2730 | 2732 | 2735 | 2737 | 2739 | 2742 |
| 1.88 | 2742 | 2744 | 2746 | 2749 | 2751 | 2753 | 2755 | 2758 | 2760 | 2762 | 2765 |
| 1.89 | 2765 | 2767 | 2769 | 2772 | 2774 | 2776 | 2778 | 2781 | 2783 | 2785 | 2788 |
| **1.90** | 0.2788 | 2790 | 2792 | 2794 | 2797 | 2799 | 2801 | 2804 | 2806 | 2808 | 2810 |
| 1.91 | 2810 | 2813 | 2815 | 2817 | 2819 | 2822 | 2824 | 2826 | 2828 | 2831 | 2833 |
| 1.92 | 2833 | 2835 | 2838 | 2840 | 2842 | 2844 | 2847 | 2849 | 2851 | 2853 | 2856 |
| 1.93 | 2856 | 2858 | 2860 | 2862 | 2865 | 2867 | 2869 | 2871 | 2874 | 2876 | 2878 |
| 1.94 | 2878 | 2880 | 2882 | 2885 | 2887 | 2889 | 2891 | 2894 | 2896 | 2898 | 2900 |
| 1.95 | 2900 | 2903 | 2905 | 2907 | 2909 | 2911 | 2914 | 2916 | 2918 | 2920 | 2923 |
| 1.96 | 2923 | 2925 | 2927 | 2929 | 2931 | 2934 | 2936 | 2938 | 2940 | 2942 | 2945 |
| 1.97 | 2945 | 2947 | 2949 | 2951 | 2953 | 2956 | 2958 | 2960 | 2962 | 2964 | 2967 |
| 1.98 | 2967 | 2969 | 2971 | 2973 | 2975 | 2978 | 2980 | 2982 | 2984 | 2986 | 2989 |
| 1.99 | 2989 | 2991 | 2993 | 2995 | 2997 | 2999 | 3002 | 3004 | 3006 | 3008 | 3010 |

# Table 10. Logarithms to the Base 10

These two pages give the common logarithms of numbers between 1 and 10, correct to four places. Moving the decimal point $n$ places to the right (or left) in the number is equivalent to adding $n$ (or $-n$) to the logarithm. Thus, $\log 0.017453 = 0.2419 - 2$ [$= \bar{2}.2419$].

To facilitate interpolation, the tenths of the tabular differences are given at the end of each line, so that the differences themselves need not be considered. In using these aids, first find the nearest tabular entry, and then add (to move to the right) or subtract (to move to the left), as the case may require.

| | 0 | 1 | 2 | 3 | 4 | 5 | 6 | 7 | 8 | 9 | 10 | Tenths of the Tabular Difference 1 | 2 | 3 | 4 | 5 |
|---|---|---|---|---|---|---|---|---|---|---|---|---|---|---|---|---|
| **1.0** | 0.0000 | 0043 | 0086 | 0128 | 0170 | 0212 | 0253 | 0294 | 0334 | 0374 | 0414 | | | | | |
| 1.1 | 0414 | 0453 | 0492 | 0531 | 0569 | 0607 | 0645 | 0682 | 0719 | 0755 | 0792 | | | | | |
| 1.2 | 0792 | 0828 | 0864 | 0899 | 0934 | 0969 | 1004 | 1038 | 1072 | 1106 | 1139 | | | | | |
| 1.3 | 1139 | 1173 | 1206 | 1239 | 1271 | 1303 | 1335 | 1367 | 1399 | 1430 | 1461 | | | | | |
| 1.4 | 1461 | 1492 | 1523 | 1553 | 1584 | 1614 | 1644 | 1673 | 1703 | 1732 | 1761 | | | | | |
| 1.5 | 1761 | 1790 | 1818 | 1847 | 1875 | 1903 | 1931 | 1959 | 1987 | 2014 | 2041 | | | | | |
| 1.6 | 2041 | 2068 | 2095 | 2122 | 2148 | 2175 | 2201 | 2227 | 2253 | 2279 | 2304 | | | | | |
| 1.7 | 2304 | 2330 | 2355 | 2380 | 2405 | 2430 | 2455 | 2480 | 2504 | 2529 | 2553 | | | | | |
| 1.8 | 2553 | 2577 | 2601 | 2625 | 2648 | 2672 | 2695 | 2718 | 2742 | 2765 | 2788 | | | | | |
| 1.9 | 2788 | 2810 | 2833 | 2856 | 2878 | 2900 | 2923 | 2945 | 2967 | 2989 | 3010 | | | | | |
| **2.0** | 0.3010 | 3032 | 3054 | 3075 | 3096 | 3118 | 3139 | 3160 | 3181 | 3201 | 3222 | 2 | 4 | 6 | 8 | 11 |
| 2.1 | 3222 | 3243 | 3263 | 3284 | 3304 | 3324 | 3345 | 3365 | 3385 | 3404 | 3424 | 2 | 4 | 6 | 8 | 10 |
| 2.2 | 3424 | 3444 | 3464 | 3483 | 3502 | 3522 | 3541 | 3560 | 3579 | 3598 | 3617 | 2 | 4 | 6 | 8 | 10 |
| 2.3 | 3617 | 3636 | 3655 | 3674 | 3692 | 3711 | 3729 | 3747 | 3766 | 3784 | 3802 | 2 | 4 | 5 | 7 | 9 |
| 2.4 | 3802 | 3820 | 3838 | 3856 | 3874 | 3892 | 3909 | 3927 | 3945 | 3962 | 3979 | 2 | 4 | 5 | 7 | 9 |
| 2.5 | 3979 | 3997 | 4014 | 4031 | 4048 | 4065 | 4082 | 4099 | 4116 | 4133 | 4150 | 2 | 3 | 5 | 7 | 9 |
| 2.6 | 4150 | 4166 | 4183 | 4200 | 4216 | 4232 | 4249 | 4265 | 4281 | 4298 | 4314 | 2 | 3 | 5 | 7 | 8 |
| 2.7 | 4314 | 4330 | 4346 | 4362 | 4378 | 4393 | 4409 | 4425 | 4440 | 4456 | 4472 | 2 | 3 | 5 | 6 | 8 |
| 2.8 | 4472 | 4487 | 4502 | 4518 | 4533 | 4548 | 4564 | 4579 | 4594 | 4609 | 4624 | 2 | 3 | 5 | 6 | 8 |
| 2.9 | 4624 | 4639 | 4654 | 4669 | 4683 | 4698 | 4713 | 4728 | 4742 | 4757 | 4771 | 1 | 3 | 4 | 6 | 7 |
| **3.0** | 0.4771 | 4786 | 4800 | 4814 | 4829 | 4843 | 4857 | 4871 | 4886 | 4900 | 4914 | 1 | 3 | 4 | 6 | 7 |
| 3.1 | 4914 | 4928 | 4942 | 4955 | 4969 | 4983 | 4997 | 5011 | 5024 | 5038 | 5051 | 1 | 3 | 4 | 6 | 7 |
| 3.2 | 5051 | 5065 | 5079 | 5092 | 5105 | 5119 | 5132 | 5145 | 5159 | 5172 | 5185 | 1 | 3 | 4 | 5 | 7 |
| 3.3 | 5185 | 5198 | 5211 | 5224 | 5237 | 5250 | 5263 | 5276 | 5289 | 5302 | 5315 | 1 | 3 | 4 | 5 | 6 |
| 3.4 | 5315 | 5328 | 5340 | 5353 | 5366 | 5378 | 5391 | 5403 | 5416 | 5428 | 5441 | 1 | 3 | 4 | 5 | 6 |
| 3.5 | 5441 | 5453 | 5465 | 5478 | 5490 | 5502 | 5514 | 5527 | 5539 | 5551 | 5563 | 1 | 2 | 4 | 5 | 6 |
| 3.6 | 5563 | 5575 | 5587 | 5599 | 5611 | 5623 | 5635 | 5647 | 5658 | 5670 | 5682 | 1 | 2 | 4 | 5 | 6 |
| 3.7 | 5682 | 5694 | 5705 | 5717 | 5729 | 5740 | 5752 | 5763 | 5775 | 5786 | 5798 | 1 | 2 | 3 | 5 | 6 |
| 3.8 | 5798 | 5809 | 5821 | 5832 | 5843 | 5855 | 5866 | 5877 | 5888 | 5899 | 5911 | 1 | 2 | 3 | 5 | 6 |
| 3.9 | 5911 | 5922 | 5933 | 5944 | 5955 | 5966 | 5977 | 5988 | 5999 | 6010 | 6021 | 1 | 2 | 3 | 4 | 6 |
| **4.0** | 0.6021 | 6031 | 6042 | 6053 | 6064 | 6075 | 6085 | 6096 | 6107 | 6117 | 6128 | 1 | 2 | 3 | 4 | 5 |
| 4.1 | 6128 | 6138 | 6149 | 6160 | 6170 | 6180 | 6191 | 6201 | 6212 | 6222 | 6232 | 1 | 2 | 3 | 4 | 5 |
| 4.2 | 6232 | 6243 | 6253 | 6263 | 6274 | 6284 | 6294 | 6304 | 6314 | 6325 | 6335 | 1 | 2 | 3 | 4 | 5 |
| 4.3 | 6335 | 6345 | 6355 | 6365 | 6375 | 6385 | 6395 | 6405 | 6415 | 6425 | 6435 | 1 | 2 | 3 | 4 | 5 |
| 4.4 | 6435 | 6444 | 6454 | 6464 | 6474 | 6484 | 6493 | 6503 | 6513 | 6522 | 6532 | 1 | 2 | 3 | 4 | 5 |
| 4.5 | 6532 | 6542 | 6551 | 6561 | 6571 | 6580 | 6590 | 6599 | 6609 | 6618 | 6628 | 1 | 2 | 3 | 4 | 5 |
| 4.6 | 6628 | 6637 | 6646 | 6656 | 6665 | 6675 | 6684 | 6693 | 6702 | 6712 | 6721 | 1 | 2 | 3 | 4 | 5 |
| 4.7 | 6721 | 6730 | 6739 | 6749 | 6758 | 6767 | 6776 | 6785 | 6794 | 6803 | 6812 | 1 | 2 | 3 | 4 | 5 |
| 4.8 | 6812 | 6821 | 6830 | 6839 | 6848 | 6857 | 6866 | 6875 | 6884 | 6893 | 6902 | 1 | 2 | 3 | 4 | 4 |
| 4.9 | 6902 | 6911 | 6920 | 6928 | 6937 | 6946 | 6955 | 6964 | 6972 | 6981 | 6990 | 1 | 2 | 3 | 4 | 4 |

**To avoid Interpolation in the first ten lines, use the special table on the preceding page.**

# Table 10. Logarithms to the Base 10

| | 0 | 1 | 2 | 3 | 4 | 5 | 6 | 7 | 8 | 9 | 10 | Tenths of the Tabular Difference | | | | |
|---|---|---|---|---|---|---|---|---|---|---|---|---|---|---|---|---|
| | | | | | | | | | | | | 1 | 2 | 3 | 4 | 5 |
| **5.0** | 0.6990 | 6998 | 7007 | 7016 | 7024 | 7033 | 7042 | 7050 | 7059 | 7067 | 7076 | 1 | 2 | 3 | 3 | 4 |
| 5.1 | 7076 | 7084 | 7093 | 7101 | 7110 | 7118 | 7126 | 7135 | 7143 | 7152 | 7160 | 1 | 2 | 3 | 3 | 4 |
| 5.2 | 7160 | 7168 | 7177 | 7185 | 7193 | 7202 | 7210 | 7218 | 7226 | 7235 | 7243 | 1 | 2 | 2 | 3 | 4 |
| 5.3 | 7243 | 7251 | 7259 | 7267 | 7275 | 7284 | 7292 | 7300 | 7308 | 7316 | 7324 | 1 | 2 | 2 | 3 | 4 |
| 5.4 | 7324 | 7332 | 7340 | 7348 | 7356 | 7364 | 7372 | 7380 | 7388 | 7396 | 7404 | 1 | 2 | 2 | 3 | 4 |
| 5.5 | 7404 | 7412 | 7419 | 7427 | 7435 | 7443 | 7451 | 7459 | 7466 | 7474 | 7482 | 1 | 2 | 2 | 3 | 4 |
| 5.6 | 7482 | 7490 | 7497 | 7505 | 7513 | 7520 | 7528 | 7536 | 7543 | 7551 | 7559 | 1 | 2 | 2 | 3 | 4 |
| 5.7 | 7559 | 7566 | 7574 | 7582 | 7589 | 7597 | 7604 | 7612 | 7619 | 7627 | 7634 | 1 | 2 | 2 | 3 | 4 |
| 5.8 | 7634 | 7642 | 7649 | 7657 | 7664 | 7672 | 7679 | 7686 | 7694 | 7701 | 7709 | 1 | 1 | 2 | 3 | 4 |
| 5.9 | 7709 | 7716 | 7723 | 7731 | 7738 | 7745 | 7752 | 7760 | 7767 | 7774 | 7782 | 1 | 1 | 2 | 3 | 4 |
| **6.0** | 0.7782 | 7789 | 7796 | 7803 | 7810 | 7818 | 7825 | 7832 | 7839 | 7846 | 7853 | 1 | 1 | 2 | 3 | 4 |
| 6.1 | 7853 | 7860 | 7868 | 7875 | 7882 | 7889 | 7896 | 7903 | 7910 | 7917 | 7924 | 1 | 1 | 2 | 3 | 4 |
| 6.2 | 7924 | 7931 | 7938 | 7945 | 7952 | 7959 | 7966 | 7973 | 7980 | 7987 | 7993 | 1 | 1 | 2 | 3 | 3 |
| 6.3 | 7993 | 8000 | 8007 | 8014 | 8021 | 8028 | 8035 | 8041 | 8048 | 8055 | 8062 | 1 | 1 | 2 | 3 | 3 |
| 6.4 | 8062 | 8069 | 8075 | 8082 | 8089 | 8096 | 8102 | 8109 | 8116 | 8122 | 8129 | 1 | 1 | 2 | 3 | 3 |
| 6.5 | 8129 | 8136 | 8142 | 8149 | 8156 | 8162 | 8169 | 8176 | 8182 | 8189 | 8195 | 1 | 1 | 2 | 3 | 3 |
| 6.6 | 8195 | 8202 | 8209 | 8215 | 8222 | 8228 | 8235 | 8241 | 8248 | 8254 | 8261 | 1 | 1 | 2 | 3 | 3 |
| 6.7 | 8261 | 8267 | 8274 | 8280 | 8287 | 8293 | 8299 | 8306 | 8312 | 8319 | 8325 | 1 | 1 | 2 | 3 | 3 |
| 6.8 | 8325 | 8331 | 8338 | 8344 | 8351 | 8357 | 8363 | 8370 | 8376 | 8382 | 8388 | 1 | 1 | 2 | 3 | 3 |
| 6.9 | 8388 | 8395 | 8401 | 8407 | 8414 | 8420 | 8426 | 8432 | 8439 | 8445 | 8451 | 1 | 1 | 2 | 3 | 3 |
| **7.0** | 0.8451 | 8457 | 8463 | 8470 | 8476 | 8482 | 8488 | 8494 | 8500 | 8506 | 8513 | 1 | 1 | 2 | 2 | 3 |
| 7.1 | 8513 | 8519 | 8525 | 8531 | 8537 | 8543 | 8549 | 8555 | 8561 | 8567 | 8573 | 1 | 1 | 2 | 2 | 3 |
| 7.2 | 8573 | 8579 | 8585 | 8591 | 8597 | 8603 | 8609 | 8615 | 8621 | 8627 | 8633 | 1 | 1 | 2 | 2 | 3 |
| 7.3 | 8633 | 8639 | 8645 | 8651 | 8657 | 8663 | 8669 | 8675 | 8681 | 8686 | 8692 | 1 | 1 | 2 | 2 | 3 |
| 7.4 | 8692 | 8698 | 8704 | 8710 | 8716 | 8722 | 8727 | 8733 | 8739 | 8745 | 8751 | 1 | 1 | 2 | 2 | 3 |
| 7.5 | 8751 | 8756 | 8762 | 8768 | 8774 | 8779 | 8785 | 8791 | 8797 | 8802 | 8808 | 1 | 1 | 2 | 2 | 3 |
| 7.6 | 8808 | 8814 | 8820 | 8825 | 8831 | 8837 | 8842 | 8848 | 8854 | 8859 | 8865 | 1 | 1 | 2 | 2 | 3 |
| 7.7 | 8865 | 8871 | 8876 | 8882 | 8887 | 8893 | 8899 | 8904 | 8910 | 8915 | 8921 | 1 | 1 | 2 | 2 | 3 |
| 7.8 | 8921 | 8927 | 8932 | 8938 | 8943 | 8949 | 8954 | 8960 | 8965 | 8971 | 8976 | 1 | 1 | 2 | 2 | 3 |
| 7.9 | 8976 | 8982 | 8987 | 8993 | 8998 | 9004 | 9009 | 9015 | 9020 | 9025 | 9031 | 1 | 1 | 2 | 2 | 3 |
| **8.0** | 0.9031 | 9036 | 9042 | 9047 | 9053 | 9058 | 9063 | 9069 | 9074 | 9079 | 9085 | 1 | 1 | 2 | 2 | 3 |
| 8.1 | 9085 | 9090 | 9096 | 9101 | 9106 | 9112 | 9117 | 9122 | 9128 | 9133 | 9138 | 1 | 1 | 2 | 2 | 3 |
| 8.2 | 9138 | 9143 | 9149 | 9154 | 9159 | 9165 | 9170 | 9175 | 9180 | 9186 | 9191 | 1 | 1 | 2 | 2 | 3 |
| 8.3 | 9191 | 9196 | 9201 | 9206 | 9212 | 9217 | 9222 | 9227 | 9232 | 9238 | 9243 | 1 | 1 | 2 | 2 | 3 |
| 8.4 | 9243 | 9248 | 9253 | 9258 | 9263 | 9269 | 9274 | 9279 | 9284 | 9289 | 9294 | 1 | 1 | 2 | 2 | 3 |
| 8.5 | 9294 | 9299 | 9304 | 9309 | 9315 | 9320 | 9325 | 9330 | 9335 | 9340 | 9345 | 1 | 1 | 2 | 2 | 3 |
| 8.6 | 9345 | 9350 | 9355 | 9360 | 9365 | 9370 | 9375 | 9380 | 9385 | 9390 | 9395 | 1 | 1 | 2 | 2 | 3 |
| 8.7 | 9395 | 9400 | 9405 | 9410 | 9415 | 9420 | 9425 | 9430 | 9435 | 9440 | 9445 | 0 | 1 | 1 | 2 | 2 |
| 8.8 | 9445 | 9450 | 9455 | 9460 | 9465 | 9469 | 9474 | 9479 | 9484 | 9489 | 9494 | 0 | 1 | 1 | 2 | 2 |
| 8.9 | 9494 | 9499 | 9504 | 9509 | 9513 | 9518 | 9523 | 9528 | 9533 | 9538 | 9542 | 0 | 1 | 1 | 2 | 2 |
| **9.0** | 0.9542 | 9547 | 9552 | 9557 | 9562 | 9566 | 9571 | 9576 | 9581 | 9586 | 9590 | 0 | 1 | 1 | 2 | 2 |
| 9.1 | 9590 | 9595 | 9600 | 9605 | 9609 | 9614 | 9619 | 9624 | 9628 | 9633 | 9638 | 0 | 1 | 1 | 2 | 2 |
| 9.2 | 9638 | 9643 | 9647 | 9652 | 9657 | 9661 | 9666 | 9671 | 9675 | 9680 | 9685 | 0 | 1 | 1 | 2 | 2 |
| 9.3 | 9685 | 9689 | 9694 | 9699 | 9703 | 9708 | 9713 | 9717 | 9722 | 9727 | 9731 | 0 | 1 | 1 | 2 | 2 |
| 9.4 | 9731 | 9736 | 9741 | 9745 | 9750 | 9754 | 9759 | 9763 | 9768 | 9773 | 9777 | 0 | 1 | 1 | 2 | 2 |
| 9.5 | 9777 | 9782 | 9786 | 9791 | 9795 | 9800 | 9805 | 9809 | 9814 | 9818 | 9823 | 0 | 1 | 1 | 2 | 2 |
| 9.6 | 9823 | 9827 | 9832 | 9836 | 9841 | 9845 | 9850 | 9854 | 9859 | 9863 | 9868 | 0 | 1 | 1 | 2 | 2 |
| 9.7 | 9868 | 9872 | 9877 | 9881 | 9886 | 9890 | 9894 | 9899 | 9903 | 9908 | 9912 | 0 | 1 | 1 | 2 | 2 |
| 9.8 | 9912 | 9917 | 9921 | 9926 | 9930 | 9934 | 9939 | 9943 | 9948 | 9952 | 9956 | 0 | 1 | 1 | 2 | 2 |
| 9.9 | 9956 | 9961 | 9965 | 9969 | 9974 | 9978 | 9983 | 9987 | 9991 | 9996 | | 0 | 1 | 1 | 2 | 2 |

# Table 11. Logarithms to the Base e

These two pages give the natural (hyperbolic, or Napierian) logarithms of numbers between 1 and 10, correct to four places. Moving the decimal point $n$ places to the right (or left) in the number is equivalent to adding $n$ times 2.3026 (or $n$ times $\bar{3}$.6974) to the logarithm.

| | | | |
|---|---|---|---|
| 1 | **2.3026** | 1 | **0.6974–3** |
| 2 | 4.6052 | 2 | 0.3948–5 |
| 3 | 6.9078 | 3 | 0.0922–7 |
| 4 | 9.2103 | 4 | 0.7897–10 |
| 5 | 11.5129 | 5 | 0.4871–12 |
| 6 | 13.8155 | 6 | 0.1845–14 |
| 7 | 16.1181 | 7 | 0.8819–17 |
| 8 | 18.4207 | 8 | 0.5793–19 |
| 9 | 20.7233 | 9 | 0.2767–21 |

| | 0 | 1 | 2 | 3 | 4 | 5 | 6 | 7 | 8 | 9 | 10 | Tenths of the Tabular Difference 1 | 2 | 3 | 4 | 5 |
|---|---|---|---|---|---|---|---|---|---|---|---|---|---|---|---|---|
| **1.0** | 0.0000 | 0100 | 0198 | 0296 | 0392 | 0488 | 0583 | 0677 | 0770 | 0862 | 0.0953 | 10 | 19 | 29 | 38 | 48 |
| 1.1 | 0953 | 1044 | 1133 | 1222 | 1310 | 1398 | 1484 | 1570 | 1655 | 1740 | 1823 | 9 | 17 | 26 | 35 | 44 |
| 1.2 | 1823 | 1906 | 1989 | 2070 | 2151 | 2231 | 2311 | 2390 | 2469 | 2546 | 2624 | 8 | 16 | 24 | 32 | 40 |
| 1.3 | 2624 | 2700 | 2776 | 2852 | 2927 | 3001 | 3075 | 3148 | 3221 | 3293 | 3365 | 7 | 15 | 22 | 30 | 37 |
| 1.4 | 3365 | 3436 | 3507 | 3577 | 3646 | 3716 | 3784 | 3853 | 3920 | 3988 | 4055 | 7 | 14 | 21 | 28 | 34 |
| 1.5 | 4055 | 4121 | 4187 | 4253 | 4318 | 4383 | 4447 | 4511 | 4574 | 4637 | 4700 | 6 | 13 | 19 | 26 | 32 |
| 1.6 | 4700 | 4762 | 4824 | 4886 | 4947 | 5008 | 5068 | 5128 | 5188 | 5247 | 5306 | 6 | 12 | 18 | 24 | 30 |
| 1.7 | 5306 | 5365 | 5423 | 5481 | 5539 | 5596 | 5653 | 5710 | 5766 | 5822 | 5878 | 6 | 11 | 17 | 23 | 29 |
| 1.8 | 5878 | 5933 | 5988 | 6043 | 6098 | 6152 | 6206 | 6259 | 6313 | 6366 | 6419 | 6 | 11 | 16 | 22 | 27 |
| 1.9 | 6419 | 6471 | 6523 | 6575 | 6627 | 6678 | 6729 | 6780 | 6831 | 6881 | 0.6931 | 5 | 10 | 15 | 21 | 26 |
| **2.0** | 0.6931 | 6981 | 7031 | 7080 | 7129 | 7178 | 7227 | 7275 | 7324 | 7372 | 7419 | 5 | 10 | 15 | 20 | 24 |
| 2.1 | 7419 | 7467 | 7514 | 7561 | 7608 | 7655 | 7701 | 7747 | 7793 | 7839 | 7885 | 5 | 9 | 14 | 19 | 23 |
| 2.2 | 7885 | 7930 | 7975 | 8020 | 8065 | 8109 | 8154 | 8198 | 8242 | 8286 | 8329 | 4 | 9 | 13 | 18 | 22 |
| 2.3 | 8329 | 8372 | 8416 | 8459 | 8502 | 8544 | 8587 | 8629 | 8671 | 8713 | 8755 | 4 | 9 | 13 | 17 | 21 |
| 2.4 | 8755 | 8796 | 8838 | 8879 | 8920 | 8961 | 9002 | 9042 | 9083 | 9123 | 9163 | 4 | 8 | 12 | 16 | 20 |
| 2.5 | 9163 | 9203 | 9243 | 9282 | 9322 | 9361 | 9400 | 9439 | 9478 | 9517 | 9555 | 4 | 8 | 12 | 16 | 20 |
| 2.6 | 9555 | 9594 | 9632 | 9670 | 9708 | 9746 | 9783 | 9821 | 9858 | 9895 | 0.9933 | 4 | 8 | 11 | 15 | 19 |
| 2.7 | 0.9933 | 9969 | 0006 | 0043 | 0080 | 0116 | 0152 | 0188 | 0225 | 0260 | 1.0296 | 4 | 7 | 11 | 15 | 18 |
| 2.8 | 1.0296 | 0332 | 0367 | 0403 | 0438 | 0473 | 0508 | 0543 | 0578 | 0613 | 0647 | 4 | 7 | 11 | 14 | 18 |
| 2.9 | 0647 | 0682 | 0716 | 0750 | 0784 | 0818 | 0852 | 0886 | 0919 | 0953 | 1.0986 | 3 | 7 | 10 | 14 | 17 |
| **3.0** | 1.0986 | 1019 | 1053 | 1086 | 1119 | 1151 | 1184 | 1217 | 1249 | 1282 | 1314 | 3 | 7 | 10 | 13 | 16 |
| 3.1 | 1314 | 1346 | 1378 | 1410 | 1442 | 1474 | 1506 | 1537 | 1569 | 1600 | 1632 | 3 | 6 | 10 | 13 | 16 |
| 3.2 | 1632 | 1663 | 1694 | 1725 | 1756 | 1787 | 1817 | 1848 | 1878 | 1909 | 1939 | 3 | 6 | 9 | 12 | 15 |
| 3.3 | 1939 | 1969 | 2000 | 2030 | 2060 | 2090 | 2119 | 2149 | 2179 | 2208 | 2238 | 3 | 6 | 9 | 12 | 15 |
| 3.4 | 2238 | 2267 | 2296 | 2326 | 2355 | 2384 | 2413 | 2442 | 2470 | 2499 | 2528 | 3 | 6 | 9 | 12 | 14 |
| 3.5 | 2528 | 2556 | 2585 | 2613 | 2641 | 2669 | 2698 | 2726 | 2754 | 2782 | 2809 | 3 | 6 | 8 | 11 | 14 |
| 3.6 | 2809 | 2837 | 2865 | 2892 | 2920 | 2947 | 2975 | 3002 | 3029 | 3056 | 3083 | 3 | 5 | 8 | 11 | 14 |
| 3.7 | 3083 | 3110 | 3137 | 3164 | 3191 | 3218 | 3244 | 3271 | 3297 | 3324 | 3350 | 3 | 5 | 8 | 11 | 13 |
| 3.8 | 3350 | 3376 | 3403 | 3429 | 3455 | 3481 | 3507 | 3533 | 3558 | 3584 | 3610 | 3 | 5 | 8 | 10 | 13 |
| 3.9 | 3610 | 3635 | 3661 | 3686 | 3712 | 3737 | 3762 | 3788 | 3813 | 3838 | 1.3863 | 3 | 5 | 8 | 10 | 13 |
| **4.0** | 1.3863 | 3888 | 3913 | 3938 | 3962 | 3987 | 4012 | 4036 | 4061 | 4085 | 4110 | 2 | 5 | 7 | 10 | 12 |
| 4.1 | 4110 | 4134 | 4159 | 4183 | 4207 | 4231 | 4255 | 4279 | 4303 | 4327 | 4351 | 2 | 5 | 7 | 10 | 12 |
| 4.2 | 4351 | 4375 | 4398 | 4422 | 4446 | 4469 | 4493 | 4516 | 4540 | 4563 | 4586 | 2 | 5 | 7 | 9 | 12 |
| 4.3 | 4586 | 4609 | 4633 | 4656 | 4679 | 4702 | 4725 | 4748 | 4770 | 4793 | 4816 | 2 | 5 | 7 | 9 | 11 |
| 4.4 | 4816 | 4839 | 4861 | 4884 | 4907 | 4929 | 4951 | 4974 | 4996 | 5019 | 5041 | 2 | 4 | 7 | 9 | 11 |
| 4.5 | 5041 | 5063 | 5085 | 5107 | 5129 | 5151 | 5173 | 5195 | 5217 | 5239 | 5261 | 2 | 4 | 7 | 9 | 11 |
| 4.6 | 5261 | 5282 | 5304 | 5326 | 5347 | 5369 | 5390 | 5412 | 5433 | 5454 | 5476 | 2 | 4 | 6 | 9 | 11 |
| 4.7 | 5476 | 5497 | 5518 | 5539 | 5560 | 5581 | 5602 | 5623 | 5644 | 5665 | 5686 | 2 | 4 | 6 | 8 | 11 |
| 4.8 | 5686 | 5707 | 5728 | 5748 | 5769 | 5790 | 5810 | 5831 | 5851 | 5872 | 5892 | 2 | 4 | 6 | 8 | 10 |
| 4.9 | 5892 | 5913 | 5933 | 5953 | 5974 | 5994 | 6014 | 6034 | 6054 | 6074 | 1.6094 | 2 | 4 | 6 | 8 | 10 |

# Table 11. Log$_e$ (Base e = 2.71828+)

| | 0 | 1 | 2 | 3 | 4 | 5 | 6 | 7 | 8 | 9 | 10 | Tenths of the Tabular Difference 1 | 2 | 3 | 4 | 5 |
|---|---|---|---|---|---|---|---|---|---|---|---|---|---|---|---|---|
| **5.0** | 1.6094 | 6114 | 6134 | 6154 | 6174 | 6194 | 6214 | 6233 | 6253 | 6273 | 6292 | 2 | 4 | 6 | 8 | 10 |
| 5.1 | 6292 | 6312 | 6332 | 6351 | 6371 | 6390 | 6409 | 6429 | 6448 | 6467 | 6487 | 2 | 4 | 6 | 8 | 10 |
| 5.2 | 6487 | 6506 | 6525 | 6544 | 6563 | 6582 | 6601 | 6620 | 6639 | 6658 | 6677 | 2 | 4 | 6 | 8 | 10 |
| 5.3 | 6677 | 6696 | 6715 | 6734 | 6752 | 6771 | 6790 | 6808 | 6827 | 6845 | 6864 | 2 | 4 | 6 | 7 | 9 |
| 5.4 | 6864 | 6882 | 6901 | 6919 | 6938 | 6956 | 6974 | 6993 | 7011 | 7029 | 7047 | 2 | 4 | 6 | 7 | 9 |
| 5.5 | 7047 | 7066 | 7084 | 7102 | 7120 | 7138 | 7156 | 7174 | 7192 | 7210 | 7228 | 2 | 4 | 5 | 7 | 9 |
| 5.6 | 7228 | 7246 | 7263 | 7281 | 7299 | 7317 | 7334 | 7352 | 7370 | 7387 | 7405 | 2 | 4 | 5 | 7 | 9 |
| 5.7 | 7405 | 7422 | 7440 | 7457 | 7475 | 7492 | 7509 | 7527 | 7544 | 7561 | 7579 | 2 | 3 | 5 | 7 | 9 |
| 5.8 | 7579 | 7596 | 7613 | 7630 | 7647 | 7664 | 7681 | 7699 | 7716 | 7733 | 7750 | 2 | 3 | 5 | 7 | 9 |
| 5.9 | 7750 | 7766 | 7783 | 7800 | 7817 | 7834 | 7851 | 7867 | 7884 | 7901 | 1.7918 | 2 | 3 | 5 | 7 | 8 |
| **6.0** | 1.7918 | 7934 | 7951 | 7967 | 7984 | 8001 | 8017 | 8034 | 8050 | 8066 | 8083 | 2 | 3 | 5 | 7 | 8 |
| 6.1 | 8083 | 8099 | 8116 | 8132 | 8148 | 8165 | 8181 | 8197 | 8213 | 8229 | 8245 | 2 | 3 | 5 | 7 | 8 |
| 6.2 | 8245 | 8262 | 8278 | 8294 | 8310 | 8326 | 8342 | 8358 | 8374 | 8390 | 8405 | 2 | 3 | 5 | 6 | 8 |
| 6.3 | 8405 | 8421 | 8437 | 8453 | 8469 | 8485 | 8500 | 8516 | 8532 | 8547 | 8563 | 2 | 3 | 5 | 6 | 8 |
| 6.4 | 8563 | 8579 | 8594 | 8610 | 8625 | 8641 | 8656 | 8672 | 8687 | 8703 | 8718 | 2 | 3 | 5 | 6 | 8 |
| 6.5 | 8718 | 8733 | 8749 | 8764 | 8779 | 8795 | 8810 | 8825 | 8840 | 8856 | 8871 | 2 | 3 | 5 | 6 | 8 |
| 6.6 | 8871 | 8886 | 8901 | 8916 | 8931 | 8946 | 8961 | 8976 | 8991 | 9006 | 9021 | 2 | 3 | 5 | 6 | 8 |
| 6.7 | 9021 | 9036 | 9051 | 9066 | 9081 | 9095 | 9110 | 9125 | 9140 | 9155 | 9169 | 1 | 3 | 4 | 6 | 7 |
| 6.8 | 9169 | 9184 | 9199 | 9213 | 9228 | 9242 | 9257 | 9272 | 9286 | 9301 | 9315 | 1 | 3 | 4 | 6 | 7 |
| 6.9 | 9315 | 9330 | 9344 | 9359 | 9373 | 9387 | 9402 | 9416 | 9430 | 9445 | 1.9459 | 1 | 3 | 4 | 6 | 7 |
| **7.0** | 1.9459 | 9473 | 9488 | 9502 | 9516 | 9530 | 9544 | 9559 | 9573 | 9587 | 9601 | 1 | 3 | 4 | 6 | 7 |
| 7.1 | 9601 | 9615 | 9629 | 9643 | 9657 | 9671 | 9685 | 9699 | 9713 | 9727 | 9741 | 1 | 3 | 4 | 6 | 7 |
| 7.2 | 9741 | 9755 | 9769 | 9782 | 9796 | 9810 | 9824 | 9838 | 9851 | 9865 | 1.9879 | 1 | 3 | 4 | 6 | 7 |
| 7.3 | 1.9879 | 9892 | 9906 | 9920 | 9933 | 9947 | 9961 | 9974 | 9988 | 0001 | 2.0015 | 1 | 3 | 4 | 5 | 7 |
| 7.4 | 2.0015 | 0028 | 0042 | 0055 | 0069 | 0082 | 0096 | 0109 | 0122 | 0136 | 0149 | 1 | 3 | 4 | 5 | 7 |
| 7.5 | 0149 | 0162 | 0176 | 0189 | 0202 | 0215 | 0229 | 0242 | 0255 | 0268 | 0281 | 1 | 3 | 4 | 5 | 7 |
| 7.6 | 0281 | 0295 | 0308 | 0321 | 0334 | 0347 | 0360 | 0373 | 0386 | 0399 | 0412 | 1 | 3 | 4 | 5 | 7 |
| 7.7 | 0412 | 0425 | 0438 | 0451 | 0464 | 0477 | 0490 | 0503 | 0516 | 0528 | 0541 | 1 | 3 | 4 | 5 | 6 |
| 7.8 | 0541 | 0554 | 0567 | 0580 | 0592 | 0605 | 0618 | 0631 | 0643 | 0656 | 0669 | 1 | 3 | 4 | 5 | 6 |
| 7.9 | 0669 | 0681 | 0694 | 0707 | 0719 | 0732 | 0744 | 0757 | 0769 | 0782 | 2.0794 | 1 | 3 | 4 | 5 | 6 |
| **8.0** | 2.0794 | 0807 | 0819 | 0832 | 0844 | 0857 | 0869 | 0882 | 0894 | 0906 | 0919 | 1 | 2 | 4 | 5 | 6 |
| 8.1 | 0919 | 0931 | 0943 | 0956 | 0968 | 0980 | 0992 | 1005 | 1017 | 1029 | 1041 | 1 | 2 | 4 | 5 | 6 |
| 8.2 | 1041 | 1054 | 1066 | 1078 | 1090 | 1102 | 1114 | 1126 | 1138 | 1150 | 1163 | 1 | 2 | 4 | 5 | 6 |
| 8.3 | 1163 | 1175 | 1187 | 1199 | 1211 | 1223 | 1235 | 1247 | 1258 | 1270 | 1282 | 1 | 2 | 4 | 5 | 6 |
| 8.4 | 1282 | 1294 | 1306 | 1318 | 1330 | 1342 | 1353 | 1365 | 1377 | 1389 | 1401 | 1 | 2 | 4 | 5 | 6 |
| 8.5 | 1401 | 1412 | 1424 | 1436 | 1448 | 1459 | 1471 | 1483 | 1494 | 1506 | 1518 | 1 | 2 | 4 | 5 | 6 |
| 8.6 | 1518 | 1529 | 1541 | 1552 | 1564 | 1576 | 1587 | 1599 | 1610 | 1622 | 1633 | 1 | 2 | 3 | 5 | 6 |
| 8.7 | 1633 | 1645 | 1656 | 1668 | 1679 | 1691 | 1702 | 1713 | 1725 | 1736 | 1748 | 1 | 2 | 3 | 5 | 6 |
| 8.8 | 1748 | 1759 | 1770 | 1782 | 1793 | 1804 | 1815 | 1827 | 1838 | 1849 | 1861 | 1 | 2 | 3 | 5 | 6 |
| 8.9 | 1861 | 1872 | 1883 | 1894 | 1905 | 1917 | 1928 | 1939 | 1950 | 1961 | 2.1972 | 1 | 2 | 3 | 4 | 6 |
| **9.0** | 2.1972 | 1983 | 1994 | 2006 | 2017 | 2028 | 2039 | 2050 | 2061 | 2072 | 2083 | 1 | 2 | 3 | 4 | 6 |
| 9.1 | 2083 | 2094 | 2105 | 2116 | 2127 | 2138 | 2148 | 2159 | 2170 | 2181 | 2192 | 1 | 2 | 3 | 4 | 5 |
| 9.2 | 2192 | 2203 | 2214 | 2225 | 2235 | 2246 | 2257 | 2268 | 2279 | 2289 | 2300 | 1 | 2 | 3 | 4 | 5 |
| 9.3 | 2300 | 2311 | 2322 | 2332 | 2343 | 2354 | 2364 | 2375 | 2386 | 2396 | 2407 | 1 | 2 | 3 | 4 | 5 |
| 9.4 | 2407 | 2418 | 2428 | 2439 | 2450 | 2460 | 2471 | 2481 | 2492 | 2502 | 2513 | 1 | 2 | 3 | 4 | 5 |
| 9.5 | 2513 | 2523 | 2534 | 2544 | 2555 | 2565 | 2576 | 2586 | 2597 | 2607 | 2618 | 1 | 2 | 3 | 4 | 5 |
| 9.6 | 2618 | 2628 | 2638 | 2649 | 2659 | 2670 | 2680 | 2690 | 2701 | 2711 | 2721 | 1 | 2 | 3 | 4 | 5 |
| 9.7 | 2721 | 2732 | 2742 | 2752 | 2762 | 2773 | 2783 | 2793 | 2803 | 2814 | 2824 | 1 | 2 | 3 | 4 | 5 |
| 9.8 | 2824 | 2834 | 2844 | 2854 | 2865 | 2875 | 2885 | 2895 | 2905 | 2915 | 2925 | 1 | 2 | 3 | 4 | 5 |
| 9.9 | 2925 | 2935 | 2946 | 2956 | 2966 | 2976 | 2986 | 2996 | 3006 | 3016 | 2.3026 | 1 | 2 | 3 | 4 | 5 |